The Checkbook Series

Mathematics 3 Checkbook

J O Bird
BSc (Hons), CEng, MIEE, FCollP, FIMA, MIEIE

A J C May
BA, CEng, MIMechE, FITE, MBIM

NEWNES

Newnes
An imprint of Butterworth-Heinemann Ltd
Linacre House, Jordan Hill, Oxford OX2 8DP

 PART OF REED INTERNATIONAL BOOKS

OXFORD LONDON BOSTON
MUNICH NEW DELHI SINGAPORE SYDNEY
TOKYO TORONTO WELLINGTON

First published 1981
Reprinted 1985, 1986, 1989, 1991, 1992

British Library Cataloguing in Publication Data
Bird, J. O.
Mathematics checkbook
1. Shop mathematics — problems, exercises, etc.
I. Title
510'246 J1165 80-41923

ISBN 0 7506 0260 0

Printed and bound in Great Britain by
Hartnolls Ltd, Bodmin, Cornwall

Contents

SECTION III. TRIGONOMETRY

SECTION IV. CALCULUS

Note to Reader

Checkbooks are designed for students seeking technician or equivalent qualification through the courses of the Business and Technician Education Council (BTEC), the Scottish Technical Education Council, Australian Technical and Further Education Departments, East and West African Examinations Council and other comparable examining authorities in technical subjects.

Checkbooks use problems and worked examples to establish and exemplify the theory contained in technical syllabuses. *Checkbook* readers gain real understanding through seeing problems solved and through solving problems themselves. *Checkbooks* do not supplant fuller textbooks, but rather supplement them with an alternative emphasis and an ample provision of worked and unworked problems, essential data, short answer and multi-choice questions (with answers where possible).

Preface

This textbook of worked problems provides coverage of the BTEC Level **N III** units in Mathematics. However, it can also be regarded as a basic textbook in Mathematics for a much wider range of courses. This practical book contains nearly 400 detailed worked problems, followed by some 650 further problems with answers.

Each topic considered in the text is presented in a way that assumes in the reader only the knowledge obtained in their particular BTEC Level **N II** Mathematics units. The aim of the book is to extend the concepts of Level **N II** Mathematics, to develop the concepts and use of mathematical modelling, and to broaden the mathematical knowledge as a basis for further study. Mathematics 3 provides a follow-up to the checkbooks written for Mathematics 1 and 2.

The authors would like to express their appreciation for the friendly co-operation and helpful advice given to them by the publishers. Thanks are also due to Mrs. Elaine Woolley for the excellent typing of the manuscript. Finally, the authors would like to add a word of thanks to their wives, Elizabeth and Juliet, for their continued patience, help and encouragement during the preparation of this book.

J.O. Bird
A.J.C. May
Highbury College of Technology,
Portsmouth

1 Number (1) – Exponential functions and Naperian logarithms

A. MAIN POINTS CONCERNED WITH EXPONENTIAL FUNCTIONS AND NAPERIAN LOGARITHMS

1 An **exponential function** is one which contains e^x, e being a constant called the **exponent** and having an approximate value of 2.7183. The exponent arises from the natural laws of growth and decay and is used as a base of natural or Naperian logarithms.

2 **The natural laws of growth and decay** are of the form $y = Ae^{kx}$, where A and k are constants. The natural laws occur frequently in engineering and science and examples of quantities related by a natural law include:

(i)	Linear expansion	$l = l_0 e^{\alpha\theta}$
(ii)	Change in electrical resistance with temperature	$R_\theta = R_0 e^{\alpha\theta}$
(iii)	Tension in belts	$T_1 = T_0 e^{\mu\alpha}$
(iv)	Newton's law of cooling	$\theta = \theta_0 e^{-kt}$
(v)	Biological growth	$y = y_0 e^{kt}$
(vi)	Discharge of a capacitor	$q = Qe^{-t/CR}$
(vii)	Atmospheric pressure	$p = p_0 e^{-h/c}$
(viii)	Radioactive decay	$N = N_0 e^{-\lambda t}$
(ix)	Decay of current in an inductive circuit	$i = Ie^{-Rt/L}$

3 The **value of e^x** may be determined by using

(i) a calculator which possesses an 'e^x' function,

(ii) the power series $e^x = 1+x+\frac{x^2}{2!}+\frac{x^3}{3!}+\ldots\ldots$ (where 3! is 'factorial 3' and means 3 × 2 × 1),

(iii) Naperian logarithms (see para. 7), or

(iv) 4 figure tables of exponential functions which enable values of e^x and e^{-x} to be read over a range of x from 0.02 to 6.0
For example, $e^{0.36} = 1.4333$, $e^{4.3} = 73.700$, $e^{-0.47} = 0.6250$ and $e^{-2.6} = 0.0743$.
Also, from tables and laws of indices,

$$e^{0.68} = e^{0.6+0.08} = (e^{0.6})(e^{0.08}) = (1.8221)(1.0833)$$
$$= 1.9739 \text{ correct to 4 decimal places.}$$

(See *Problems 1 and 2*.)

4 **Graphs of exponential functions**

(i) Values of e^x and e^{-x}, obtained from 4 figure tables, correct to 2 decimal places, over the range $x = -3$ to $x = 3$, are shown in the table below.

x	−3.0	−2.5	−2.0	−1.5	−1.0	−0.5	0	0.5	1.0	1.5	2.0	2.5	3.0
e^x	0.05	0.08	0.14	0.22	0.37	0.61	1.00	1.65	2.72	4.48	7.39	12.18	20.09
e^{-x}	20.09	12.18	7.39	4.48	2.72	1.65	1.00	0.61	0.37	0.22	0.14	0.08	0.05

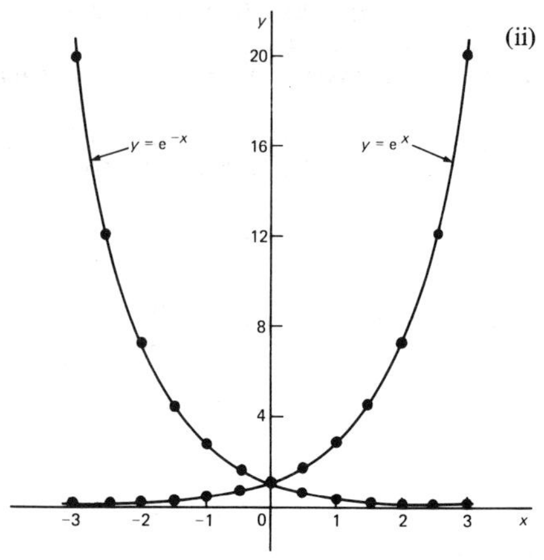

Fig 1

Fig 1 shows graphs of $y = e^x$ and $y = e^{-x}$.

(ii) A similar table may be drawn up for $y = 5e^{\frac{1}{2}x}$, a graph of which is shown in *Fig 2*. The gradient of the curve at any point, dy/dx, is obtained by drawing a tangent to the curve at that point and measuring the gradient of the tangent. For example:

when $x = 0$, $y = 5$ and $\dfrac{dy}{dx} = \dfrac{\text{BC}}{\text{AB}}$

$$= \frac{(6.2-3.7)}{1} = 2.5$$

and when $x = 2$, $y = 13.6$ and

$$\frac{dy}{dx} = \frac{\text{EF}}{\text{DE}} = \frac{(16.8-10)}{1} = 6.8$$

These two results each show that $\dfrac{dy}{dx} = \dfrac{1}{2}y$, and further determinations of the gradients of $y = 5e^{\frac{1}{2}x}$ would give the same result for each. In general, for all natural growth and decay laws of the form $y = Ae^{kx}$, where k is a positive constant for growth laws (as in *Fig 2*) and a negative constant for decay curves, $dy/dx = ky$, i.e., **the rate of change of the variable, y, is proportional to the variable itself.**

(iii) For any natural law of growth and decay of the form $dy/dx = ky$, the solution is always $y = Ae^{kx}$. (see *Problems 3 to 9*).

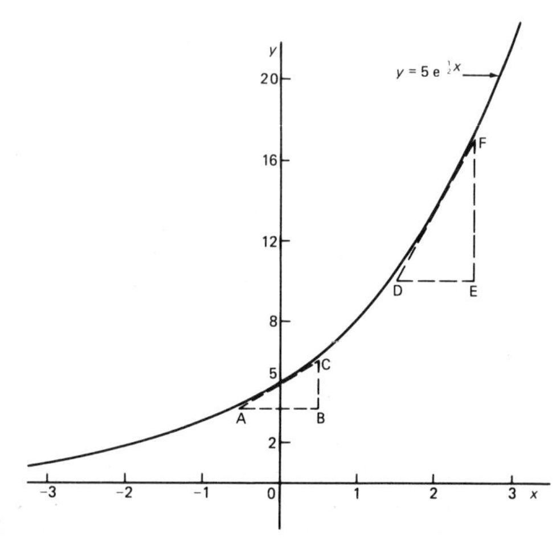

Fig 2

Hyperbolic or Naperian logarithms

5 (i) A **logarithm** of a number is the power to which a base has to be raised to be equal to the number. Thus, if $y = a^x$ then $x = \log_a y$.

(ii) Logarithms having a base of 10 are called **common logarithms** and the common logarithm of x is written as $\lg x$.

(iii) Logarithms having a base of e are called **hyperbolic, Naperian** or **natural logarithms** and the Naperian logarithm of x is written as $\log_e x$, or more commonly, $\ln x$.

6 The **value of a Naperian logarithm** may be obtained by using:

(i) a calculator possessing a '$\ln x$' function,

(ii) the change of base rule for logarithms, which states, $\log_a y = \dfrac{\log_b y}{\log_b a}$,

from which, $\ln y = \dfrac{\lg y}{\lg e} = \dfrac{\lg y}{0.4343} = 2.3026 \lg y$,

or (iii) 4 figure Naperian logarithm tables.

7 **Use of 4 figure Naperian logarithm tables**

(i) **For numbers from 1 to 10,** Naperian logarithm tables are used in a similar way to common logarithms. For example,

$\ln 3.4 = 1.2238$

$\ln 3.47 = 1.2442$

$\ln 3.478 = 1.2442+23$ (from the mean difference column equivalent to 8)

i.e. $\ln 3.478 = 1.2465$

Similarly, $\ln 5.731 = 1.7459$ and $\ln 9.159 = 2.2148$

(ii) **Numbers larger than 10** are initially expressed in standard form and the supplementary table of Naperian logarithms of 10^{+n} used. For example,

$\ln 64 = \ln(6.4 \times 10^1) = \ln 6.4+\ln 10^1$

$= 1.8563+2.3026 = 4.1589$

Similarly, $\ln 327.6 = \ln(3.276 \times 10^2) = \ln 3.276+\ln 10^2$

$= 1.1866+4.6052 = 5.7918.$

(iii) **Numbers smaller than 1** are initially expressed in standard form and the supplementary table of Naperian logarithms of 10^{-n} used.

For example, $\ln 0.064 = \ln(6.4\times10^{-2}) = \ln 6.4+\ln 10^{-2}$

$= 1.8563+\bar{5}.3948 = \bar{3}.2511$

$\bar{3}.2511$ is the same as $-3+0.2511 = -2.7489$ (which is the value given by a calculator).

Similarly, $\ln 0.003276 = \ln 3.276 \times 10^{-3}) = \ln 3.276+\ln 10^{-3}$

$= 1.1866+\bar{7}.0922 = \bar{6}.2788$ or -5.7212.

(iv) **The antilogarithm of a number between 0 and 2.3026** is obtained by finding the value of the logarithm within the table, the antilogarithm being obtained from the corresponding row and column. For example, if $\ln x = 1.4317$, the nearest number less than this is 1.4303 corresponding to 4.18. The difference between 1.4317 and 1.4303 is 14, which in the mean difference column corresponds to 6. Hence if $\ln x = 1.4317$ then $x = 4.186$.

(v) **To find the antilogarithm of a number greater than 2.3026,** for example, 5.3726, the nearest number less than 5.3726 is found in the supplementary table of Naperian logarithms of 10^{+n}; in this case it is 4.6052, corresponding to 10^2. The difference between 5.3726 and 4.6052 is 0.7674. The antilogarithm of 0.7674 is read directly from within the table as 2.154. Hence if $\ln x = 5.3726$ then $x = 2.154 \times 10^2 = 215.4$.

(vi) **To find the antilogarithm of a number less than 2.3026**, for example, –5.8314, the number is initially changed to $\overline{6}.1686$ and the nearest more negative number in the supplementary table of Naperian logarithms of 10^{-n} is found; in this case it is $\overline{7}.0922$, corresponding to 10^{-3}. The difference between $\overline{6}.1686$ and $\overline{7}.0922$ is 1.0764. The antilogarithm of 1.0764 is read directly from within the tables as 2.934. Hence if $\ln x = -5.8314$ then $x = 2.934 \times 10^{-3} = 0.002\,934$.

(vii) The procedure for determining antilogarithms is considerably quicker with a **calculator.** If $\ln x = y$ then $x = e^y$, from the definition of a logarithm. Thus, if $\ln x = 5.3726$, then $x = e^{5.3726} = 215.42224$, by calculator and if $\ln x = -5.8314$ then $x = e^{-5.8314} = 2.9339 \times 10^{-3} = 0.002\,933\,9$, by calculator. Alternatively, an 'invert ln x' function is available on many calculators.

(See *Problems 10 to 16.*)

B. WORKED PROBLEMS ON EXPONENTIAL FUNCTIONS AND NAPERIAN LOGARITHMS

Problem 1 Use exponential tables to determine the values of (a) $e^{0.18}$, (b) $e^{-1.4}$, (c) $3e^{-0.54}$, (d) $0.2e^8$, each correct to 4 significant figures.

From exponential tables:

(a) $e^{0.18} = 1.1972 = $ **1.197**, correct to 4 significant figures.

(b) $e^{-1.4} = $ **0.2466**

(c) $e^{-0.54} = (e^{-0.5})(e^{-0.04}) = (0.6065)(0.9608) = 0.582\,73$,
Hence $3e^{-0.54} = 3(0.582\,73) = $ **1.748**, correct to 4 significant figures.

(d) $e^8 = e^{4+4} = (e^4)(e^4) = (54.598)^2 = 2980.94$.
Hence $0.2e^8 = (0.2)(2980.94) = $ **596.2**, correct to 4 significant figures.

Problem 2 Determine the value of $5e^{0.5}$, correct to 5 significant figures by using the power series for e^x.

$$e^x = 1+x+\frac{x^2}{2!}+\frac{x^3}{3!}+ \ldots\ldots\ldots$$

$$\text{Hence } e^{0.5} = 1+0.5+\frac{(0.5)^2}{(2)(1)}+\frac{(0.5)^3}{(3)(2)(1)}+\frac{(0.5)^4}{(4)(3)(2)(1)}+\frac{(0.5)^5}{(5)(4)(3)(2)(1)}+\frac{(0.5)^6}{(6)(5)(4)(3)(2)(1)}$$

$$= 1+0.5+0.125+0.020\,833+0.002\,604\,2+0.000\,260\,4+0.000\,021\,7$$

i.e. $e^{0.5} = 1.648\,72$, correct to 6 significant figures.

Hence $5e^{0.5} = 5(1.64872) = $ **8.2436,** correct to 5 significant figures.

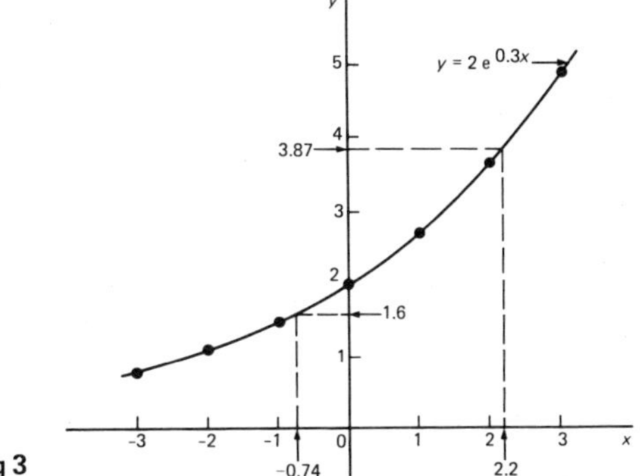

Fig 3

Problem 3 Plot a graph of $y = 2e^{0.3x}$ over a range of $x = -3$ to $x = 3$. Hence determine the value of y when $x = 2.2$ and the value of x when $y = 1.6$.

A table of values is drawn up as shown below.

x	-3	-2	-1	0	1	2	3
$0.3x$	-0.9	-0.6	-0.3	0	0.3	0.6	0.9
$e^{0.3x}$	0.407	0.549	0.741	1.00	1.350	1.822	2.460
$2e^{0.3x}$	0.81	1.10	1.48	2.00	2.70	3.64	4.92

A graph of $y = 2e^{0.3x}$ is shown plotted in *Fig 3*.
When $x = 2.2$, $y =$ **3.87** and when $y = 1.6$, $x =$ **−0.74**

Problem 4 Plot the graph of $y = \frac{1}{3}e^{-2x}$ over the range $x = -1.5$ to $x = 1.5$. Determine, from the graph, the value of y when $x = -1.2$ and the value of x when $y = 1.4$

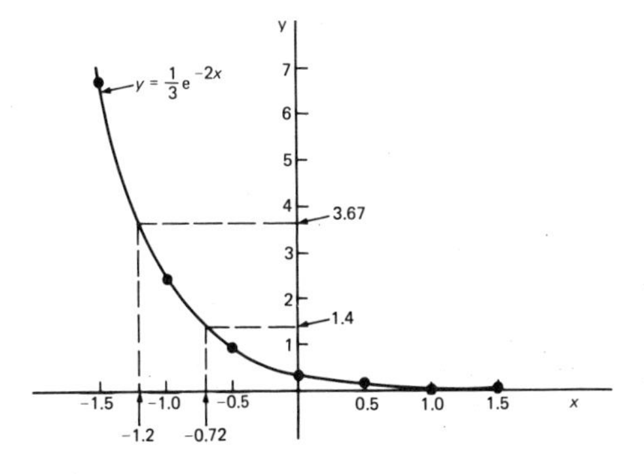

Fig 4

A table of values is drawn up as shown below.

x	-1.5	-1.0	-0.5	0	0.5	1.0	1.5
$-2x$	3	2	1	0	-1	-2	-3
e^{-2x}	20.086	7.389	2.718	1.00	0.368	0.135	0.050
$\frac{1}{3}e^{-2x}$	6.70	2.46	0.91	0.33	0.12	0.05	0.02

A graph of $y = \frac{1}{3}e^{-2x}$ is shown in *Fig 4*

When $x = -1.2$, $y =$ **3.67** and when $y = 1.4$, $x =$ **−0.72**

Problem 5 A natural law of growth is of the form $y = 4e^{0.2x}$. Plot a graph depicting this law for values of x from $x = -3$ to $x = 3$. From the graph determine (a) the value of y when x is 2.2, (b) the value of x when y is 3.4 and (c) the rate of change of y with respect to x (i.e. dy/dx) at $x = -2$

A table of values is drawn up as shown below.

x	-3	-2	-1	0	1	2	3
$0.2x$	-0.6	-0.4	-0.2	0	0.2	0.4	0.6
$e^{0.2x}$	0.549	0.670	0.819	1.00	1.221	1.492	1.822
$4e^{0.2x}$	2.20	2.68	3.28	4.00	4.88	5.97	7.29

A graph of $y = 4e^{0.2x}$ is shown in *Fig 5*. From the graph:

(a) $x = 2.2$, $y =$ **6.2**,

(b) when $y = 3.4$, $x =$ **−0.8**,

(c) at $x = -2$, gradient

$$(\text{i.e. } \frac{dy}{dx}) = \frac{BC}{AB} = \frac{1.08}{2} = \mathbf{0.54}$$

[From para. 4, when $y = Ae^{kx}$ then $dy/dx = ky$

In this case $A = 4$, $k = 0.2$ thus $y = 4e^{0.2x}$.

Hence, when $x = -2$, $dy/dx = (0.2)4e^{0.2(-2)} = 0.54$]

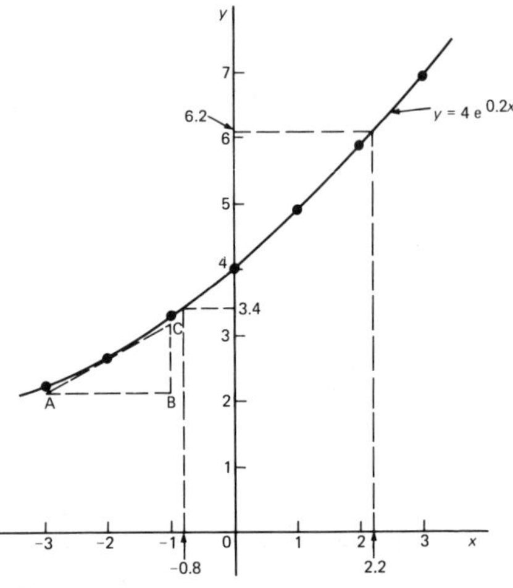

Fig 5

Problem 6 The decay of voltage, V volts, across a capacitor at time t seconds is given by $v = 250e^{-t/3}$. Draw a graph showing the natural decay curve over the first 6 s. From the graph find (a) the voltage after 3.4 s, and (b) the time when the voltage is 150 V. (c) Determine the rate of change of voltage after 2 s and after 4 s.

A table of values is drawn up as shown below.

t	0	1	2	3	4	5	6
$e^{-t/3}$	1.00	0.7165	0.5134	0.3679	0.2636	0.1889	0.1353
$v = 250e^{-t/3}$	250.0	179.1	128.4	91.97	65.90	47.22	33.83

The natural decay curve of $v = 250e^{-t/3}$ is shown in *Fig 6*.

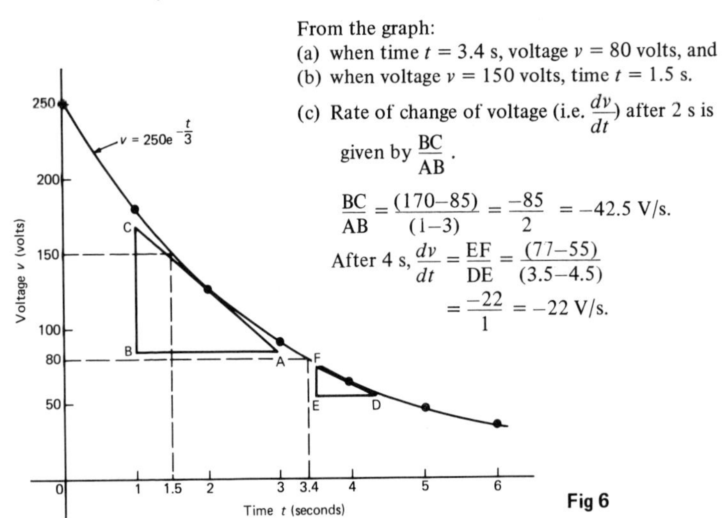

Fig 6

From the graph:
(a) when time $t = 3.4$ s, voltage $v = 80$ volts, and
(b) when voltage $v = 150$ volts, time $t = 1.5$ s.
(c) Rate of change of voltage (i.e. $\frac{dv}{dt}$) after 2 s is given by $\frac{BC}{AB}$.

$$\frac{BC}{AB} = \frac{(170-85)}{(1-3)} = \frac{-85}{2} = -42.5 \text{ V/s.}$$

After 4 s, $\frac{dv}{dt} = \frac{EF}{DE} = \frac{(77-55)}{(3.5-4.5)}$

$$= \frac{-22}{1} = -22 \text{ V/s.}$$

Problem 7 A law of natural growth is of the form $dy/dx = 8y$. Determine the solution of this equation.

An equation of the form $dy/dx = ky$ has the solution $y = Ae^{kx}$, where A and k are constants (see para. 4). With the equation $dy/dx = 8y$, $k = 8$ thus the solution is $y = Ae^{8x}$.

Problem 8 The rate of change of charge on a capacitor dQ/dt is proportional to the charge Q and is given by $R\frac{dQ}{dt} + \frac{Q}{C} = 0$, where R and C are constants. Solve the equation for Q.

Rearranging $R\frac{dQ}{dt} + \frac{Q}{C} = 0$ gives $R\frac{dQ}{dt} = -\frac{Q}{C}$

Thus $\frac{dQ}{dt} = -\frac{Q}{CR}$, i.e. $\frac{dQ}{dt} = \left(-\frac{1}{CR}\right) Q$ which is of the form $\frac{dy}{dx} = ky$,

where $k = -\frac{1}{CR}$.

The solution is $Q = Ae^{(-\frac{1}{CR})t} = Ae^{-\frac{t}{CR}}$.

Problem 9 The rate of decay of radioactive atoms, dN/dt, is proportional to the number of radioactive atoms N and is given by $dN/dt + 0.8 \times 10^4 N = 0$. Solve the equation for N.

Rearranging $\frac{dN}{dt} + 0.8 \times 10^4 N = 0$ gives $\frac{dN}{dt} = -0.8 \times 10^4 N$, which is of the form $\frac{dy}{dx} = ky$, where $k = -0.8 \times 10^4$.

Hence the solution is $N = Ae^{-0.8 \times 10^4 t}$

Problem 10 Use tables to determine (a) ln 7.483; (b) ln 748.3; (c) ln 0.007 483.

(a) ln 7.4 = 2.0015
ln 7.48 = 2.0122
ln 7.483 = 2.0122+4 (from the mean difference column equivalent to 3)
i.e. ln 7.483 = **2.0126**

(b) From para. 7(ii), ln 748.3 = ln(7.483×10^2) = ln 7.483+ln 10^{-2}
= 2.0126+4.6052 = **6.6178**

(c) From para. 7(iii), ln 0.007 483 = ln(7.483×10^{-3}) = ln 7.483+ln 10^{-3}
= $2.0126 + \bar{7}.0922$ = $\mathbf{\bar{5}.1048}$ **or −4.8952**

Problem 11 Determine from tables the value of x when ln x is (a) 1.7321; (b) 8.4628; (c) −8.4217.

(a) Since 1.7321 lies between 0 and 2.3026 the value of x can be read directly from within the tables of Naperian logarithms. The nearest number less than 1.7321 is 1.7317 corresponding to 5.65. The difference between 1.7321 and 1.7317 is 4 which corresponds to 2 in the mean difference column.
Hence if ln x = 1.7321, then x = **5.652**.
(Alternatively, if ln x = 1.7321, then $x = e^{1.7321}$ = 5.652 511 7 by calculator– or an 'invert ln x' is available on many calculators.)

(b) The nearest number less than 8.4628 in the supplementary table of Naperian logarithms of 10^{+n} is 6.9078 corresponding to 10^3. The difference between

8.4628 and 6.9078 is 1.5550. The antilogarithm of 1.5550 is found from within the tables and is 4.735.
Hence if $\ln x = 8.4628$, then $x = 4.735 \times 10^3 =$ **4735**.
(Alternatively, $x = e^{8.4628} = 4735.2983$ by calculator.)

(c) $-8.4217 = -9+0.5783 = \bar{9}.5783$
The nearest more negative number to $\bar{9}.5783$ in the supplementary table of Naperian logarithms of 10^{-n} is $\overline{10}.7897$, corresponding to 10^{-4}. The difference between $\bar{9}.5783$ and $\overline{10}.7897$ is 0.7886. The antilogarithm of 0.7886 is found from within the tables and is 2.200.
Hence if $\ln x = -8.4217$, then $x = \mathbf{2.200 \times 10^{-4}}$.

Problem 12 Use common logarithms to evaluate ln 36.25.

From the change of base rule, $\ln y = 2.3026 \lg y$ (from para. 6)

Hence $\ln 36.25 = (2.3026)(\lg 36.25)$

$= (2.3026)(1.5593) = \mathbf{3.5904}$, correct to 4 decimal places,

which may be checked by using Naperian logarithms or a calculator.

Problem 13 Use Naperian logarithms to evaluate (a) $e^{4.137}$ and (b) $-4.5e^{-2.4379}$, each correct to 3 significant figures.

(a) Let $y = e^{4.137}$. Taking Naperian logarithms of each side gives:
$\ln y = \ln e^{4.137} = 4.137 \ln e$ by the laws of logarithms.
Thus $\ln y = 4.137$, since $\ln e = 1$.
$\ln y = 2.3026+(4.137-2.3026)$
$= 2.3026+1.8344$.
Taking antilogarithms of both sides gives $y = 10^1 \times 6.261$.
i.e. $e^{4.137} = \mathbf{62.6}$, correct to 3 significant figures.

(b) Let $y = e^{-2.4379}$. Taking Naperian logarithms of each side gives:
$\ln y = -2.4379 = \bar{3}.5621 = \bar{5}.3948+(\bar{3}.5621-\bar{5}.3948)$
i.e. $\ln y = \bar{5}.3948+2.1673$
Taking antilogarithms of both sides gives $y = 10^{-2} \times 8.734$.
Thus $-4.5e^{-2.4379} = (-4.5)(8.734 \times 10^{-2}) = \mathbf{-0.393}$, correct to 3 significant figures.

Problem 14 The resistance R of an electrical conductor at temperature θ°C is given by $R = R_0 e^{\alpha\theta}$, where α is a constant and $R_0 = 5 \times 10^3$ ohms. Determine the value of α when $R = 6 \times 10^3$ ohms and $\theta = 1500$°C. Also, find the temperature when the resistance R is 5.4×10^3 ohms.

Transposing $R = R_0 e^{\alpha\theta}$ gives $\dfrac{R}{R_0} = e^{\alpha\theta}$

Taking Naperian logarithms of both sides gives: $\ln\left(\dfrac{R}{R_0}\right) = \ln e^{\alpha\theta} = \alpha\theta$

Hence $\alpha = \frac{1}{\theta} \ln \left(\frac{R}{R_0}\right) = \frac{1}{1500} \ln \left(\frac{6 \times 10^3}{5 \times 10^3}\right) = \frac{1}{1500} (0.1823)$

Hence $\boldsymbol{\alpha} = \mathbf{1.215 \times 10^{-4}}$ (by calculator or tables).

From above, $\ln \left(\frac{R}{R_0}\right) = \alpha\theta$, hence $\theta = \frac{1}{\alpha} \ln \left(\frac{R}{R_0}\right)$

When $R = 5.4 \times 10^3$, $\alpha = 1.215 \times 10^{-4}$ and $R_0 = 5 \times 10^3$,

$$\theta = \frac{1}{1.215 \times 10^{-4}} \ln \left(\frac{5.4 \times 10^3}{5 \times 10^3}\right) = \frac{10^4}{1.215} (7.696 \times 10^{-2}) = \mathbf{633.4°C.}$$

Problem 15 In an experiment involving Newton's law of cooling, the temperature θ (°C) is given by $\theta = \theta_0 e^{-kt}$. Find the value of constant k when $\theta_0 = 56.6$°C, $\theta = 16.5$°C and $t = 83.0$ seconds.

Transposing $\theta = \theta_0 e^{-kt}$ gives $\frac{\theta}{\theta_0} = e^{-kt}$ from which $\frac{\theta_0}{\theta} = \frac{1}{e^{-kt}} = e^{kt}$.

Taking Naperian logarithms of both sides gives: $\ln \left(\frac{\theta_0}{\theta}\right) = kt$,

from which, $k = \frac{1}{t} \ln \left(\frac{\theta_0}{\theta}\right) = \frac{1}{83.0} \ln \left(\frac{56.6}{16.5}\right) = \frac{1}{83.0} (1.2326)$

Hence $k = \mathbf{1.485 \times 10^{-2}}$.

Problem 16 The current i amperes flowing in a capacitor at time t seconds is given by $i = 8.0(1-e^{-t/CR})$, where the circuit resistance R is 25×10^3 ohms and capacitance C is 16×10^{-6} farads. Determine (a) the current i after 0.5 seconds and (b) the time for the current to reach 6.0 A.

(a) Current $i = 8.0(1-e^{-t/CR}) = 8.0 \left[1-e^{-\frac{0.5}{(16 \times 10^{-6})(25 \times 10^3)}}\right] = 8.0(1-e^{-1.25})$

$= 8.0(1-0.2865) = 8.0(0.7135) = $ **5.708 amperes.**

(b) Transposing $i = 8.0(1-e^{-t/CR})$ gives $\frac{i}{8.0} = 1-e^{-t/CR}$,

from which, $e^{-t/CR} = 1-\frac{i}{8.0} = \frac{8.0-i}{8.0}$

Taking the reciprocal of both sides gives: $e^{t/CR} = \frac{8.0}{8.0-i}$

Taking Naperian logarithms of both sides gives: $\frac{t}{CR} = \ln \left(\frac{8.0}{8.0-i}\right)$

Hence $t = CR \ln \left(\frac{8.0}{8.0-i}\right) = (16 \times 10^{-6})(25 \times 10^3) \ln \left(\frac{8.0}{8.0-6.0}\right)$

when $i = 6.0$ amperes,

i.e. $t = \frac{400}{10^3} \ln \left(\frac{8.0}{2.0}\right) = 0.4 \ln 4.0 = 0.4(1.3863) = \mathbf{0.5545\ s}$

C. FURTHER PROBLEMS ON EXPONENTIAL FUNCTIONS AND NAPERIAN LOGARITHMS

In *Problems 1 and 2* use exponential tables to evaluate the given functions correct to 4 significant figures. Check your results by using a calculator.

1 (a) $e^{4.4}$; (b) $e^{-0.25}$; (c) $e^{0.92}$

[(a) 81.45; (b) 0.7788; (c) 2.509]

2 (a) $e^{-1.8}$; (b) $e^{-0.78}$; (c) e^{10}

[(a) 0.1653; (b) 0.4584; (c) 22 030]

3 Use the power series for e^x to determine, correct to 4 significant figures, (a) e^2; (b) $e^{-0.3}$ and check your result by using exponential tables and/or a calculator.

[(a) 7.389; (b) 0.7408]

4 Evaluate, correct to 4 significant figures, (a) $3.5e^{2.8}$; (b) $-\frac{6}{5}e^{-1.5}$; (c) $2.16e^{5.7}$

[(a) 57.56; (b) –0.2678; (c) 645.6]

5 Plot a graph of $y = 3e^{0.2x}$ over the range $x = -3$ to $x = 3$. Hence determine the value of y when $x = 1.4$ and the value of x when $y = 4.5$.

[3.97; 2.03]

6 Plot a graph of $y = \frac{1}{2}e^{1.5x}$ over a range $x = -1.5$ to $x = 1.5$ and hence determine the value of y when $x = -0.8$ and the value of x when $y = 3.5$.

[1.66; –1.30]

7 Plot a graph of $y = 2.5e^{-0.15x}$ over a range $x = -8$ to $x = 8$. Determine from the graph the value of y when $x = -6.2$ and the value of x when $y = 5.4$.

[6.34; –5.13]

8 Draw a graph of $y = 2(2e^{-x} - 3e^{2x})$ over a range of $x = -3$ to $x = 3$. Determine the value of y when $x = -2.2$ and the value of x when $y = 17.4$.

[36.0; –1.49]

9 In a chemical reaction the amount of starting material C cm^3 left after t minutes is given by $C = 40e^{-0.006t}$. Plot a graph of C against t and determine (a) the concentration C after 1 hour; (b) the time taken for the concentration to decrease by half; (c) Determine the rate of change of C with t after 40 mins.

[(a) 27.9 cm^3; (b) 115.5 mins; (c) –0.189 cm^3/min]

10 The rate at which a body cools is given by $\theta = 250e^{-0.05t}$, where the excess of temperature of a body above its surroundings at time t minutes is θ°C. Plot a graph showing this natural decay curve for the first hour of cooling and hence determine the rate of cooling after (a) 15 minutes; (b) 45 minutes.

[(a) –5.90°C/min; (b) –1.32°C/min]

11 The tensions in two sides of a belt, T and T_0 newtons, passing round a pulley wheel and in contact with the pulley for an angle θ radians is given by $T = T_0e^{0.3\theta}$. Plot a graph depicting this relationship over a range $\theta = 0$ to $\theta = 2.0$ radians, given $T_0 = 50$ N. From the graph determine the value of $dT/d\theta$ when $\theta = 1.2$ radians.

[21.5 N/rad]

12 The voltage drop, v volts, across an inductor is related to time, t ms, by $v = 30 \times 10^3 e^{-t/10}$. Plot a graph of v against t from $t = 0$ to $t = 10$ ms. Use the graph to determine the rate of change of voltage with time (i.e. dv/dt) when $t = 5.5$ ms.

[–1731 V/ms]

13 Determine the solution of the following equations:

(a) $\frac{dT}{d\theta} = 0.4T$; (b) $\frac{dm}{di} - 2m = 0$; (c) $\frac{dv}{dx} + 9.81v = 0$; (d) $\frac{1}{5}\frac{dm}{d\theta} - \frac{1}{2}m = 0$

[(a) $T = Ae^{0.4\theta}$; (b) $m = Ae^{2i}$; (c) $v = Ae^{-9.81x}$; (d) $m = Ae^{\frac{5}{2}\theta}$]

14 The change of length l of a bar of metal with respect to temperature θ is directly proportional to its length and may be represented by $(dl/d\theta) - \alpha l = 0$, where α is a constant equal to 2.5×10^{-6}. Solve the differential equation for l.

[$l = Ae^{2.5 \times 10^{-6}\theta}$]

15 The rate of change of voltage across an electrical circuit, dV/dt, is directly proportional to the applied voltage V such that $7.5V - 5.0(dV/dt) = 0$. Solve the equation for V.

[$V = Ae^{1.5t}$]

In *Problems 16 to 18* use 4 figure tables to evaluate the given functions.

16 (a) ln 1.73; (b) ln 5.413; (c) ln 9.412

[(a) 0.5481; (b) 1.6887; (c) 2.2420]

17 (a) ln 17.3; (b) ln 541.3; (c) 9412

[(a) 2.8507; (b) 6.2939; (c) 9.1498]

18 (a) ln 0.173; (b) ln 0.005 413; (c) ln 0.094 12

[(a) $\bar{2}.2455$ or -1.7545; (b) $\bar{6}.7809$ or -5.2191; (c) $\bar{3}.6368$ or -2.3632]

In *Problems 19 to 21* use 4 figure tables to find the value of x when ln x has the value shown.

19 (a) 0.4317; (b) 1.2047; (c) 2.0491

[(a) 1.540; (b) 3.336; (c) 7.761]

20 (a) 3.1429; (b) 6.3312; (c) 10.3171

[(a) 23.17; (b) 561.8; (c) 30 250]

21 (a) $\bar{1}.3617$; (b) $\bar{6}.3173$; (c) -3.3117

[(a) 0.5282; (b) 0.003 404; (c) 0.036 45]

22 Use common logarithms to evaluate (a) ln 29.41 and (b) ln 110.5, each correct to 4 significant figures.

[(a) 3.381; (b) 4.705]

23 Use Naperian logarithms to evaluate, correct to 3 significant figures,

(a) $e^{3.912}$; (b) $2.7e^{-1.8463}$; (c) $-\frac{15}{8}e^{2.3131}$

[(a) 50.0; (b) 0.426; (c) -18.9]

24 Evaluate $0.26e^{-\frac{3x}{7}}$ when x has a value of (a) -3.68; (b) 2.417; (c) 14.

[(a) 1.2587; (b) 0.092 28; (c) 6.445×10^{-4}]

25 Two quantities x and y are related by the equation $y = ae^{-kx}$, where a and k are constants. (a) Determine the value of y when $a = 2.114$, $k = -3.20$ and $x = 1.429$. (b) Determine the value of x when $y = 115.4$, $a = 17.8$ and $k = 4.65$.

[(a) 204.7; (b) -0.4020]

26 The pressure p pascals at height h metres above ground level is given by $p = p_0 e^{-h/C}$, where p_0 is the pressure at ground level and C is a constant. When p_0 is 1.012×10^5 Pa and the pressure at a height of 1420 m is 9.921×10^4 Pa, determine the value of C.

[71 500]

27 The length l metres of a metal bar at temperature t°C is given by $l = l_0 e^{\alpha t}$, where l_0 and α are constants. Determine (a) the value of α when $l = 1.993$ m, $l_0 = 1.894$ m and $t = 250$°C, and (b) the value of l_0 when $l = 2.416$, $t = 310$°C and $\alpha = 1.682 \times 10^{-4}$.

[(a) 2.038×10^{-4}; (b) 2.293 m]

28 The temperature θ_2°C of an electrical conductor at time t seconds is given by $\theta_2 = \theta_1(1 - e^{-t/T})$, where θ_1 is the initial temperature and T seconds is a constant. Determine (a) θ_1 when $\theta_2 = 50$°C, $t = 30$ s and $T = 80$ s, and (b) the time t for θ_2 to fall to half the value of θ_1 if T remains at 80 s.

[(a) 159.9°C; (b) 55.45 s]

29 Quantities x and y are related by $y = 8.317(1 - e^{cx/t})$, where c and t are constants. Determine (a) the value of y when $c = 2.9 \times 10^{-3}$, $x = 841.2$ and $t = 4.379$, and (b) the value of t when $y = -83.68$, $x = 841.2$ and $c = 2.9 \times 10^{-2}$.

[(a) −6.201; (b) 10.15]

30 The voltage drop, v volts, across an inductor L henrys at time t seconds is given by $v = 200e^{-Rt/L}$, where $R = 150\ \Omega$ and $L = 12.5 \times 10^{-3}$ H. Determine (a) the voltage when $t = 160 \times 10^{-6}$ s and (b) the time for the voltage to reach 85 V.

[(a) 29.32 volts; (b) 71.31×10^{-6} s]

2 Number (2) – Complex numbers

A. MAIN POINTS CONCERNED WITH COMPLEX NUMBERS

1 If the quadratic equation $x^2+2x+5=0$ is solved using the quadratic formula then

$$x=\frac{-2\pm\sqrt{[(2)^2-(4)(1)(5)]}}{2(1)}=\frac{-2\pm\sqrt{-16}}{2}=\frac{-2\pm\sqrt{[(16)(-1)]}}{2}$$
$$=\frac{-2\pm\sqrt{16}\sqrt{-1}}{2}=\frac{-2\pm4\sqrt{-1}}{2}=-1\pm2\sqrt{-1}$$

It is not possible to evaluate $\sqrt{-1}$ in real terms. However, if an operator j is defined as $j=\sqrt{-1}$ then the solution may be expressed as $x=-1\pm j2$.

2 $-1+j2$ and $-1-j2$ are known as **complex numbers.** Both solutions are of the form $a+jb$, 'a' being termed the **real part** and jb the **imaginary part**. A complex number of the form $a+jb$ is called a **cartesian complex number**.

3 In pure mathematics the symbol i is used to indicate $\sqrt{-1}$ (i being the first letter of the word imaginary). However i is the symbol of electric current, and to avoid possible confusion the next letter in the alphabet, j, is used to represent $\sqrt{-1}$.

4 A complex number may be represented pictorially on rectangular or cartesian axes. The horizontal (or x) axis is used to represent the real axis and the vertical (or y) axis is used to represent the imaginary axis. Such a diagram is called an **Argand diagram**. In *Fig 1*, the point A represents the complex number $(3+j2)$ and is obtained by plotting the co-ordinates $(3, j2)$ as in graphical work. *Fig 1* also shows the Argand points B, C and D representing the complex numbers $(-2+j4)$, $(-3-j5)$ and $(1-j3)$ respectively.

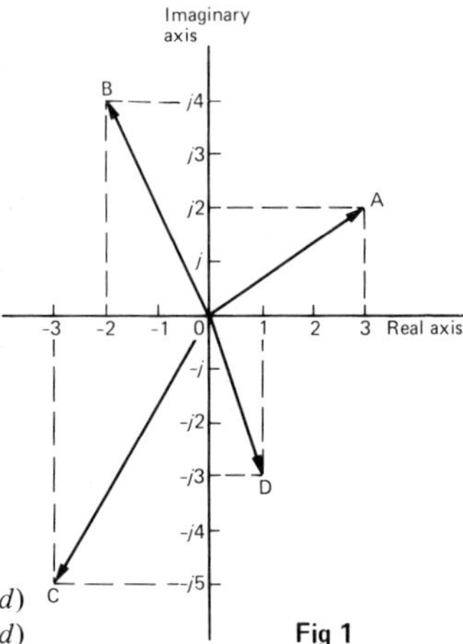

Fig 1

Addition and subtraction of complex numbers

5 Two complex numbers are added/subtracted by adding/subtracting separately the two real parts and the two imaginary parts.
For example, if $Z_1=a+jb$ and $Z_2=c+jd$,
then $Z_1+Z_2=(a+jb)+(c+jd)=(a+c)+j(b+d)$
and $Z_1-Z_2=(a+jb)-(c+jd)=(a-c)+j(b-d)$

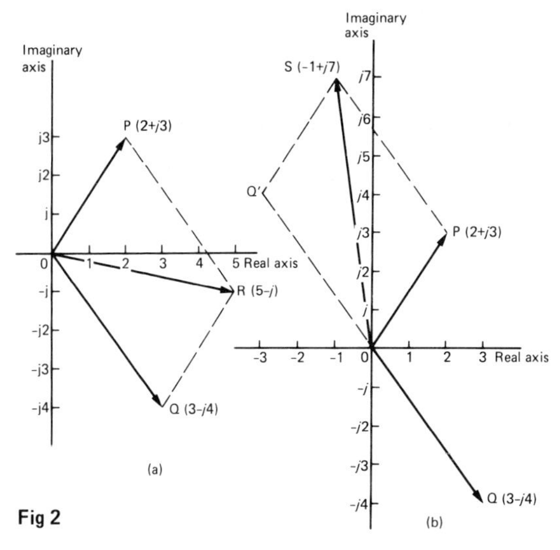

Fig 2

Thus, for example, $(2+j3)+(3-j4) = 2+j3+3-j4 = 5-j1$
and $(2+j3)-(3-j4) = 2+j3-3+j4 = -1+j7$

6 The addition and subtraction of complex numbers may be achieved graphically as shown in the Argand diagram of *Fig 2*. $(2+j3)$ is represented by vector OP and $(3-j4)$ by vector OQ. In *Fig 2(a)* by vector addition (i.e. the diagonal of the parallelogram **OP+OQ = OR**. R is the point $(5, -j1)$. Hence $(2+j3)+(3-j4) = 5-j1$, as in para. 5.

In *Fig 2(b)*, vector OQ is reversed (shown as OQ′) since it is being subtracted. (Note $OQ = 3-j4$ and $OQ' = -OQ = -(3-j4) = -3+j4$).
OP−OQ = OP+OQ′ = OS. S is found to be the Argand point $(-1, j7)$.
Hence $(2+j3)-(3-j4) = -1+j7$, as in para. 5.

7 **Multiplication of complex numbers** is achieved by assuming all quantities involved are real and then using $j^2 = -1$ to simplify.
Hence $(a+jb)(c+jd) = ac+a(jd)+(jb)c+(jb)(jd)$
$= ac+jad+jbc+j^2bd = (ac-bd)+j(ad+bc)$, since $j^2 = -1$.
Thus $(3+j2)(4-j5) = 12-j15+j8-j^2 10 = (12--10)+j(-15+8) = 22-j7$.

8 The **complex conjugate** of a complex number is obtained by changing the sign of the imaginary part. Hence the complex conjugate of $a+jb$ is $a-jb$. The product of a complex number and its complex conjugate is always a real number. For example, $(3+j4)(3-j4) = 9-j12+j12-j^2 16 = 9+16 = 25$.
($(a+jb)(a-jb)$ may be evaluated 'on sight' as a^2+b^2).

9 **Division of complex numbers** is achieved by multiplying both numerator and denominator by the complex conjugate of the denominator.

For example, $\frac{2-j5}{3+j4} = \frac{2-j5}{3+j4} \times \frac{(3-j4)}{(3-j4)} = \frac{6-j8-j15+j^2 20}{3^2+4^2} = \frac{-14-j23}{25}$

$$= \frac{-14}{25} - j\frac{23}{25} \text{ or } -0.56-j0.92$$

(See *Problems 1 to 6*)

10 **Complex equations**

If two complex numbers are equal, then their real parts are equal and their imaginary parts are equal. Hence if $a+jb = c+jd$, then $a = c$ and $b = d$. (See *Problems 7 and 8*.)

11 **The polar form of a complex number**

(i) Let a complex number Z be $x+jy$ as shown in the Argand diagram of *Fig 3*. Let distance OZ be r and the angle OZ makes with the positive real axis be θ. From trigonometry, $x = r\cos\theta$ and $y = r\sin\theta$.
Hence $Z = x+jy = r\cos\theta + jr\sin\theta = r(\cos\theta + j\sin\theta)$.
$Z = r(\cos\theta + j\sin\theta)$ is usually abbreviated to $Z = r\angle\theta$ which is known as the **polar form** of a complex number.

(ii) r is called the **modulus** (or magnitude) of Z and is written as mod Z or $|Z|$.
r is determined using Pythagoras' theorem on triangle OAZ in *Fig 3*, i.e. $r = \sqrt{(x^2+y^2)}$.

(iii) θ is called the **argument** (or amplitude) of Z and is written as arg. Z. By trigonometry on triangle OAZ, arg $Z = \theta = \textbf{arctan}\, y/x$.

(iv) Whenever changing from cartesian form to polar form, or vice-versa, a sketch is invaluable for determining the quadrant in which the complex number occurs.

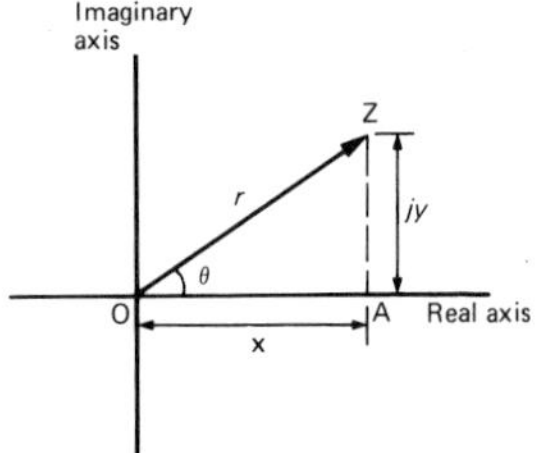

Fig 3

12 **Multiplication and division in polar form**

If $Z_1 = r_1\angle\theta_1$ and $Z_2 = r_2\angle\theta_2$ then:

(i) $Z_1 Z_2 = r_1 r_2 \angle(\theta_1 + \theta_2)$, and (ii) $\dfrac{Z_1}{Z_2} = \dfrac{r_1}{r_2}\angle(\theta_1 - \theta_2)$.

(See *Problems 9 to 14*)

13 **De Moivre's theorem** states: $[r\angle\theta]^n = r^n\angle n\theta$, which is true for all positive negative or fractional values of n, and is thus useful for determining powers and roots of complex numbers. For example,
$[3\angle 20°]^4 = 3^4\angle(4 \times 20°) = 81\angle 80°$.
The **square root of a complex number** is determined by letting $n = ½$ in

De Moivre's theorem, i.e., $\sqrt{[r\angle\theta]} = \sqrt{[r\angle\theta]^{\frac{1}{2}}} = r^{\frac{1}{2}}\angle\frac{1}{2}\theta = \sqrt{r}\angle\frac{\theta}{2}$.

There are two square roots of a real number, equal in size but opposite in sign. (See *Problems 15 and 16*)

14 There are several applications of complex numbers in science and engineering, in particular in electrical alternating current theory and in mechanical vector analysis.

The effect of multiplying a phasor by j is to rotate it in a positive direction (i.e. anticlockwise) on an Argand diagram through 90° without altering its length. Similarly, multiplying a phasor by $-j$ rotates the phasor through $-90°$. These facts are used in a.c. theory since certain quantities in the phasor diagrams lie at 90° to each other. For example, in the R–L series circuit shown in *Fig 4(a)*, V_L leads I by 90° (i.e. I lags V_L by 90°) and may be written as jV_L, the vertical axis being regarded as the imaginary axis of an Argand diagram. Thus $V_R + jV_L = V$ and since $V_R = IR$, $V = IX_L$ (where X_L is the inductive reactance ($2\pi fL$) ohms) and $V = IZ$ (where Z is the impedance) then $R+jX_L = Z$.

Similarly, for the R–C circuit shown in *Fig 4(b)*, V_C lags I by 90° (i.e. I leads

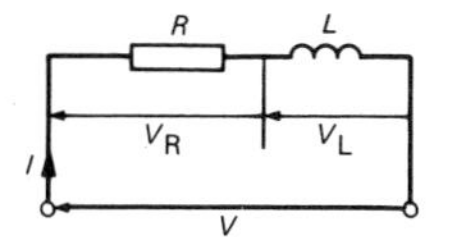

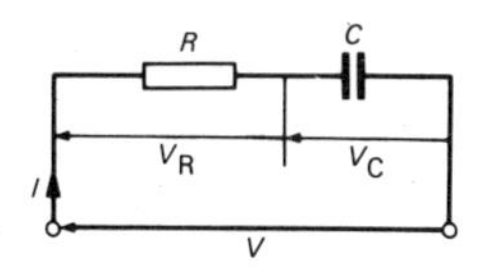

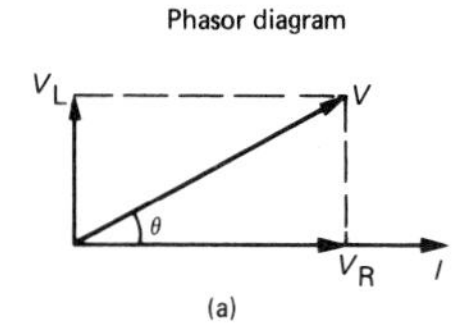

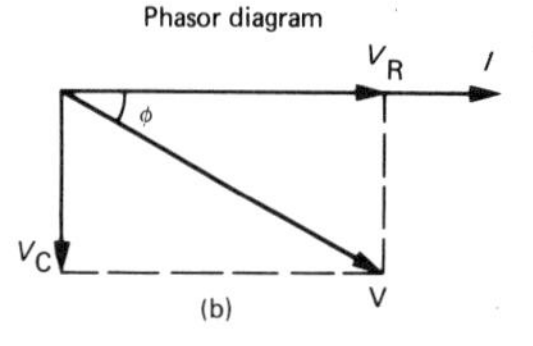

Fig 4

V_C by 90°) and $V_R - jV_C = V$, from which $R - jX_C = Z$ (where X_C is the capacitive reactance $\frac{1}{2\pi fC}$ Ω). (See *Problems 17 to 22*)

B. WORKED PROBLEMS ON COMPLEX NUMBERS

Problem 1 Solve the quadratic equation $x^2+4 = 0$.

Since $x^2+4 = 0$ then $x^2 = -4$ and $x = \sqrt{-4}$
i.e., $x = \sqrt{[(-1)(4)]} = \sqrt{(-1)}\sqrt{4} = j(\pm 2) = \pm j2$, (since $j = \sqrt{-1}$).
(Note that $\pm j2$ may also be written as $\pm 2j$)

Problem 2 Solve the quadratic equation $2x^2+3x+5 = 0$.

Using the quadratic formula, $x = \frac{-3 \pm \sqrt{[(3)^2-4(2)(5)]}}{2(2)} = \frac{-3 \pm \sqrt{-31}}{4}$

$= \frac{-3 \pm \sqrt{(-1)}\sqrt{31}}{4} = \frac{-3 \pm j\sqrt{31}}{4}$

Hence $\boldsymbol{x = -\frac{3}{4} \pm j\frac{\sqrt{31}}{4}}$ or $\boldsymbol{-0.750 \pm j1.392}$, correct to 3 decimal places.

(Note, a graph of $y = 2x^2+3x+5$ does not cross the x-axis and hence $2x^2+3x+5 = 0$ has no real roots.)

Problem 3 Evaluate (a) j^3 (b) j^4 (c) j^{23} (d) $\frac{-4}{j^9}$

(a) $j^3 = j^2 \times j = (-1) \times j = \boldsymbol{-j}$, since $j^2 = -1$
(b) $j^4 = j^2 \times j^2 = (-1) \times (-1) = \boldsymbol{1}$
(c) $j^{23} = (j^4)^5 \times (j^3) = (1)^5 \times (-j) = \boldsymbol{-j}$
(d) $j^9 = (j^4)^2(j) = (1)^2(j) = j$

Hence $\frac{-4}{j^9} = \frac{-4}{j} = \frac{-4}{j} \times \frac{-j}{-j} = \frac{4j}{-j^2} = \frac{4j}{-(-1)} = \boldsymbol{4j}$

Problem 4 Given $Z_1 = 2+j4$ and $Z_2 = 3-j$ determine (a) Z_1+Z_2, (b) Z_1-Z_2, (c) Z_2-Z_1 and show the results on an Argand diagram.

(a) $Z_1+Z_2 = (2+j4)+(3-j) = (2+3)+j(4-1) = \mathbf{5+j3}$
(b) $Z_1-Z_2 = (2+j4)-(3-j) = (2-3)+j(4--1) = \mathbf{-1+j5}$
(c) $Z_2-Z_1 = (3-j)-(2+j4) = (3-2)+j(-1-4) = \mathbf{1-j5}$
Each result is shown in the Argand diagram of *Fig 5.*

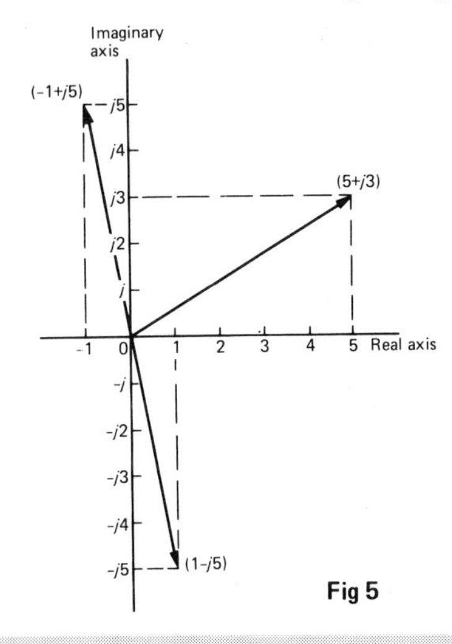

Fig 5

Problem 5 If $Z_1 = 1-j3$, $Z_2 = -2+j5$ and $Z_3 = -3-j4$, determine in $a+jb$ form:
(a) Z_1Z_2 (b) $\dfrac{Z_1}{Z_3}$ (c) $\dfrac{Z_1Z_2}{Z_1+Z_2}$ (d) $Z_1Z_2Z_3$.

(a) $Z_1Z_2 = (1-j3)(-2+j5) = -2+j5+j6-j^2 15 = (-2+15)+j(5+6)$, since $j^2 = -1$,
$= \mathbf{13+j11}$.

(b) $$\frac{Z_1}{Z_3} = \frac{1-j3}{-3-j4} = \frac{1-j3}{-3-j4} \times \frac{-3+j4}{-3+j4} = \frac{-3+j4+j9-j^2 12}{3^2+4^2}$$
$$= \frac{9+j13}{25} = \frac{9}{25} + j\frac{13}{25} \text{ or } \mathbf{0.36+j0.52}$$

(c) $$\frac{Z_1Z_2}{Z_1+Z_2} = \frac{(1-j3)(-2+j5)}{(1-j3)+(-2+j5)} = \frac{13+j11}{-1+j2}, \text{ from part (a),}$$
$$= \frac{13+j11}{-1+j2} \times \frac{-1-j2}{-1-j2} = \frac{-13-j26-j11-j^2 22}{1^2+2^2}$$
$$= \frac{9-j37}{5} = \frac{9}{5} - j\frac{37}{5} \text{ or } \mathbf{1.8-j7.4}$$

(d) $Z_1Z_2Z_3 = (13+j11)(-3-j4)$, since $Z_1Z_2 = 13+j11$, from part (a),
$= -39-j52-j33-j^2 44 = (-39+44)-j(52+33) = \mathbf{5-j85}$

Problem 6 Evaluate: (a) $\dfrac{2}{(1+j)^4}$ (b) $j\left(\dfrac{1+j3}{1-j2}\right)^2$

(a) $(1+j)^2 = (1+j)(1+j) = 1+j+j+j^2 = 1+j+j-1 = j2$

$(1+j)^4 = \left[(1+j)^2\right]^2 = (j2)^2 = j^2 4 = -4$

Hence $\dfrac{2}{(1+j)^4} = \dfrac{2}{-4} = -\dfrac{1}{2}$

(b) $\dfrac{1+j3}{1-j2} = \dfrac{1+j3}{1-j2} \times \dfrac{1+j2}{1+j2} = \dfrac{1+j2+j3+j^2 6}{1^2+2^2} = \dfrac{-5+j5}{5} = -1+j1 = -1+j$

$\left(\dfrac{1+j3}{1-j2}\right)^2 = (-1+j)^2 = (-1+j)(-1+j) = 1-j-j+j^2 = -j2$

Hence $j\left(\dfrac{1+j3}{1-j2}\right)^2 = j(-j2) = -j^2 2 = \mathbf{2}$, since $j^2 = -1$

Problem 7 Solve the complex equations: (a) $2(x+jy) = 6-j3$
(b) $(1+j2)(-2-j3) = a+jb$

(a) $2(x+jy) = 6-j3$. Hence $2x+j2y = 6-j3$
Equating the real parts gives: $2x = 6$, i.e., $\boldsymbol{x = 3}$
Equating the imaginary parts gives: $2y = -3$, i.e., $y = -\dfrac{3}{2}$

(b) $(1+j2)(-2-j3) = a+jb$
$-2-j3-j4-j^2 6 = a+jb$
Hence $4-j7 = a+jb$
Equating real and imaginary terms gives: $\boldsymbol{a = 4}$ and $\boldsymbol{b = -7}$

Problem 8 Solve the equations:
(a) $(2-j3) = \sqrt{(a+jb)}$; (b) $(x-j2y)+(y-j3x) = 2+j3$

(a) $(2-j3) = \sqrt{(a+jb)}$
Hence $(2-j3)^2 = a+jb$ i.e. $(2-j3)(2-j3) = a+jb$
Hence $4-j6-j6+j^2 9 = a+jb$
and $-5-j12 = a+jb$
Thus $\boldsymbol{a = -5}$ and $\boldsymbol{b = -12}$

(b) $(x-j2y)+(y-j3x) = 2+j3$
Hence $(x+y)+j(-2y-3x) = 2+j3$
Equating real and imaginary parts gives: $x+y = 2$ (1)
and $-3x-2y = 3$ (2)
i.e., two simultaneous equations to solve.
Multiplying equation (1) by 2 gives: $2x+2y = 4$ (3)
Adding equations (2) and (3) gives: $-x = 7$, i.e., $\boldsymbol{x = -7}$
From equation (1), $\boldsymbol{y = 9}$, which may be checked in equation (2).

Problem 9 Determine the modulus and argument of the complex number $Z = 2+j3$, and express Z in polar form.

$Z = 2+j3$ lies in the first quadrant as shown in *Fig 6.*
Modulus $|Z| = r = \sqrt{(2^2+3^2)} = \mathbf{\sqrt{13}}$ **or 3.606**, correct to 3 decimal places.

Argument, $\arg Z = \theta = \arctan\frac{3}{2} = \mathbf{56° \; 19'}$.

In polar form, $2+j3$ is written as $\mathbf{3.606 \angle 56° \; 19'}$.

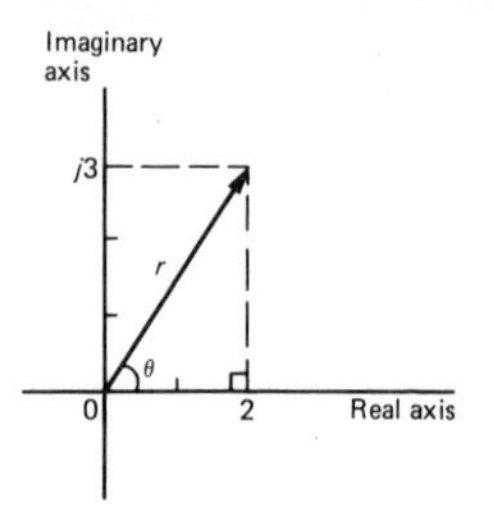

Fig 6

Problem 10 Express the following complex numbers in polar form:
(a) $3+j4$; (b) $-3+j4$; (c) $-3-j4$; (d) $3-j4$.

(a) $3+j4$ is shown in *Fig 7* and lies in the first quadrant. Modulus, $r = \sqrt{(3^2+4^2)} = 5$. Argument. $\theta = \arctan\frac{4}{3} = 53° \; 8'$.
Hence $\mathbf{3+j4 = 5\angle 53° \; 8'}$.

(b) $-3+j4$ is shown in *Fig 7* and lies in the second quadrant. Modulus, $r = 5$ and angle $\alpha = 53° \; 8'$, from part (a). Argument $= 180° - 53° \; 8'$ $= 126° \; 52'$ (i.e. the argument must be measured from the positive real axis).
Hence $\mathbf{-3+j4 = 5\angle 126° \; 52'}$.

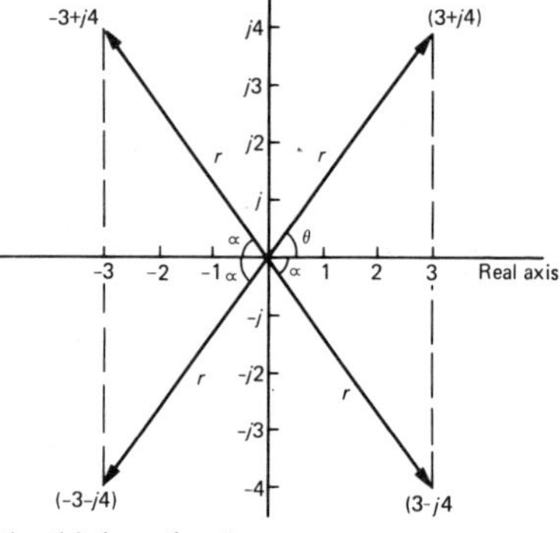

Fig 7

(c) $-3-j4$ is shown in *Fig 7* and lies in the third quadrant.
Modulus, $r = 5$ and $\alpha = 53° \; 8'$, as above.
Hence the argument $= 180° + 53° \; 8' = 233° \; 8'$ which is the same as $-126° \; 52'$
Hence $\mathbf{(-3-j4) = 5\angle 233° \; 8'}$ **or** $\mathbf{5\angle -126° \; 52'}$.
(By convention the principal value is normally used, i.e., the numerically least value, such that $-\pi \leqslant \theta \leqslant \pi$).

(d) $3-j4$ is shown in *Fig 7* and lies in the fourth quadrant.
Modulus, $r = 5$ and angle $\alpha = 53° \; 8'$, as above.
Hence $\mathbf{(3-j4) = 5\angle -53° \; 8'}$.

Problem 11 Convert (a) $4\angle 30°$; (b) $7\angle -145°$ into $a+jb$ form, correct to 4 significant figures.

(a) $4\angle 30°$ is shown in *Fig 8(a)* and lies in the first quadrant.
Using trigonometric ratios, $x = 4 \cos 30° = 3.464$ and $y = 4 \sin 30° = 2.000$
Hence $\mathbf{4\angle 30° = 3.464+j2.000}$

(b) $7\angle -145°$ is shown in *Fig 8(b)* and lies in the third quadrant.
Angle $\alpha = 180° - 145° = 35°$
Hence $x = 7 \cos 35° = 5.734$ and $y = 7 \sin 35° = 4.015$
Hence $\mathbf{7\angle -145° = -5.734-j4.015}$

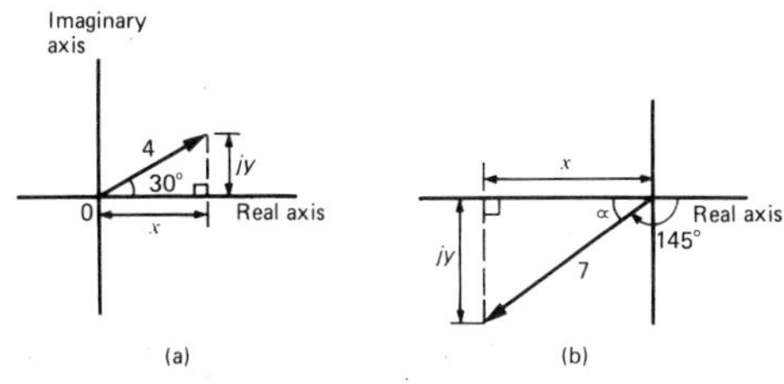

Fig 8

Problem 12 Determine, in polar form:
(a) $8\angle 25° \times 4\angle 60°$; (b) $3\angle 16° \times 5\angle -44° \times 2\angle 80°$.

(a) $8\angle 25° \times 4\angle 60° = (8 \times 4)\angle(25°+60°) = \mathbf{32\angle 85°}$.
(b) $3\angle 16° \times 5\angle -44° \times 2\angle 80° = (3\times 5\times 2)\angle[16°+(-44°)+80°] = \mathbf{30\angle 52°}$.

Problem 13 Evaluate in polar form (a) $\dfrac{16\angle 75°}{2\angle 15°}$; (b) $\dfrac{10\angle\frac{\pi}{4} \times 12\angle\frac{\pi}{2}}{6\angle -\frac{\pi}{3}}$

(a) $\dfrac{16\angle 75°}{2\angle 15°} = \dfrac{16}{2}\angle(75°-15°) = \mathbf{8\angle 60°}$

(b) $\dfrac{10\angle\frac{\pi}{4} \times 12\angle\frac{\pi}{2}}{6\angle -\frac{\pi}{3}} = \dfrac{10 \times 12}{6}\angle\left[\frac{\pi}{4}+\frac{\pi}{2}-(-\frac{\pi}{3})\right] = \mathbf{20\angle\frac{13\pi}{12}}$ **or** $\mathbf{20\angle 195°}$

or $\mathbf{20\angle -165°}$.

Problem 14 Evaluate, in polar form $2\angle 30°+5\angle -45°-4\angle 120°$.

Addition and subtraction in polar form is not possible directly. Each complex number has to be converted into cartesian form first.

$2\angle 30° = 2(\cos 30°+j\sin 30°) = 2\cos 30°+j2\sin 30° = 1.732+j1.000$
$5\angle -45° = 5(\cos(-45°)+j\sin(-45°))=5\cos 45°-j5\sin 45°=3.536-j3.536$
$4\angle 120° = 4(\cos 120°+j\sin 120°)=4\cos 120°+j4\sin 120°=-4\cos 60°+j4\sin 60°$
$=-2.000+j3.464$

Hence $2\angle 30°+5\angle -45°-4\angle 120° = (1.732+j1.000)+(3.536-j3.536)-(-2.000+j3.464)$
$= 7.268-j6.000$, which lies in the fourth quadrant
$=\sqrt{[(7.268)^2+6.000)^2]}\angle \arctan\left(\dfrac{-6.000}{7.268}\right)$
$= \mathbf{9.425\angle -39°\ 32'}$.

Problem 15 Determine (a) $[2\angle 35°]^5$; (b) $(-2+j3)^6$, in polar form.

(a) $[2\angle 35°]^5 = 2^5\angle(5 \times 35°)$, from De Moivre's theorem
$= \mathbf{32\angle 175°}$

(b) $(-2+j3) = \sqrt{[(-2)^2+(3)^2]}\angle\arctan\dfrac{3}{-2} = \sqrt{13}\angle 123°\ 41'$, since $-2+j3$ lies in the second quadrant.

$(-2+j3)^6 = [\sqrt{13}\angle 123° 41']^6 = (\sqrt{13})^6 \angle(6 \times 123° 41')$, by De Moivre's theorem

$= 2197\angle 742° 6'$

$= 2197\angle 382° 6'$ (since $742° 6' \equiv 742° 6' - 360° = 382° 6'$)

$= \mathbf{2197\angle 22° 6'}$ (since $382° 6' \equiv 382° 6' - 360° = 22° 6'$).

Problem 16 Determine the two square roots of the complex number $(5+j12)$ in polar and cartesian forms and show the roots on an Argand diagram.

$(5+j12) = \sqrt{[5^2+12^2]} \angle \arctan \dfrac{12}{5} = 13\angle 67° 23'$

When determining square roots two solutions result. To obtain the second solution one way is to express $13\angle 67° 23'$ also as $13\angle(67° 23'+360°)$, i.e., $13\angle 427° 23'$. When the angle is divided by 2 an angle less than 360° is obtained.

$$\text{Hence } \sqrt{(5+j12)} = \sqrt{[13\angle 67° 23']} \text{ and } \sqrt{[13\angle 427° 23']}$$

$$= [13\angle 67° 23']^{1/2} \text{ and } [13\angle 427° 23']^{1/2}$$

$$= 13^{1/2} \angle\left(\frac{1}{2} \times 67° 23'\right) \text{ and } 13^{1/2} \angle\left(\frac{1}{2} \times 427° 23'\right)$$

$$= \sqrt{13}\angle 33° 42' \text{ and } \sqrt{13}\angle 213° 42'$$

$$= 3.61\angle 33° 42' \text{ and } 3.61\angle 213° 42'$$

Thus, in polar form, the two roots are $3.61\angle 33° 42'$ and $3.61\angle -146° 18'$.

$3.61\angle 33° 42' = 3.61(\cos 33° 42' + j \sin 33° 42') = 3.0+j2.0$

$3.61\angle -146° 18' = 3.61(\cos(-146° 18') + j \sin(-146° 18')) = -3.0-j2.0$

Thus, in cartesian form the two roots are $\pm(3.0+j2.0)$

From the Argand diagram shown in *Fig 9* the two roots are seen to be 180° apart, which is always true when finding square roots of complex numbers.

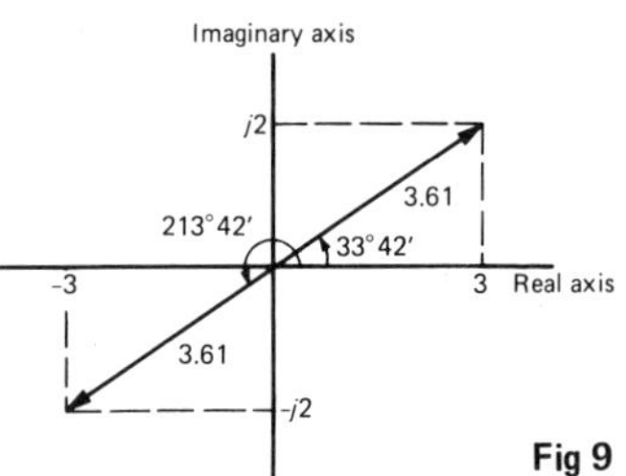

Fig 9

Problem 17 Determine the resistance and series inductance (or capacitance) for each of the following impedances, assuming a frequency of 50 Hz:
(a) $(4.0+j7.0)\ \Omega$; (b) $-j20\ \Omega$; (c) $10\angle 30°\ \Omega$; (d) $15\angle -60°\ \Omega$.

(a) Impedance, $Z = 4.0+j7.0\ \Omega$. Hence **resistance = 4.0 Ω** and reactance = 7.00 Ω. Since the imaginary part is positive, the reactance is inductive, i.e., $X_L = 7.0\ \Omega$.

Since $X_L = 2\pi f L$ then inductance, $L = \dfrac{X_L}{2\pi f} = \dfrac{7.0}{2\pi(50)} = \mathbf{0.0223\ H\ or\ 22.3\ mH.}$

(b) Impedance, $Z = -j20$, i.e. $Z = 0-j20$. Hence **resistance = 0** and reactance = 20 Ω. Since the imaginary part is negative, the reactance is capacitive, i.e., $X_C = 20\ \Omega$.

Since $X_C = \dfrac{1}{2\pi f C}$ then

capacitance, $C = \dfrac{1}{2\pi f X_C} = \dfrac{1}{2\pi(50)(20)}\ \text{F} = \dfrac{10^6}{2\pi(50)(20)}\ \mu\text{F} = \mathbf{159.2\ \mu F.}$

(c) Impedance, $Z = 10\angle 30° = 10(\cos 30° + j \sin 30°) = 8.66+j5.0\ \Omega$. Hence **resistance = 8.66 ohms** and inductive reactance, $X_L = 5.0\ \Omega$.

Since $X_L = 2\pi fL$ then inductance $L = \frac{X_L}{2\pi f} = \frac{5.0}{2\pi(50)}$ = **0.0159 H or 15.9 mH.**

(d) Impedance, $Z = 15\angle{-60°} = 15[\cos(-60°)+j\sin(-60°)] = 7.50-j12.99\ \Omega$.
Hence **resistance = 7.50 ohms** and capacitive reactance, $X_C = 12.99\ \Omega$.

Since $X_C = \frac{1}{2\pi fC}$ then capacitance, $C = \frac{1}{2\pi fX_C} = \frac{10^6}{2\pi(50)(12.99)}\ \mu F$ = **245 μF.**

Problem 18 An alternating voltage of 240 V, 50 Hz is connected across an impedance of $(60-j100)\ \Omega$. Determine (a) the resistance; (b) the capacitance; (c) the magnitude of the impedance and its phase angle and (d) the current flowing.

(a) Impedance $Z = (60-j100)\ \Omega$. Hence **resistance = 60 Ω.**
(b) Capacitive reactance $X_C = 100\ \Omega$.

Since $X_C = \frac{1}{2\pi fC}$ then

capacitance, $C = \frac{1}{2\pi fX_C} = \frac{1}{2\pi(50)(100)}\ F = \frac{10^6}{2\pi(50)(100)}\ \mu F$ = **31.83 μF.**

(c) Magnitude of impedance $|Z| = \sqrt{[(60)^2+(-100)^2]}$ = **116.6 Ω.**

Phase angle, $\arg Z = \arctan\left(\frac{-100}{60}\right)$ = **−59° 2′.**

(d) Current flowing, $I = \frac{V}{Z} = \frac{240\angle 0°}{116.6\angle{-59° 2'}}$ = **2.058∠59° 2′ A.**

The circuit and phasor diagrams are as shown in *Fig 4(b)*, page 17.

Problem 19 A resistance of 50 Ω is connected in series with an inductance of 0.20 H. If the terminal voltage is 220 V, 50 Hz determine (a) the inductive reactance; (b) the impedance; (c) the current flowing and its phase angle relative to the terminal voltage; (d) the voltage across the resistance and (e) the voltage across the inductance.

The circuit and phasor diagrams are as shown in *Fig 4(a)*.
(a) Inductive reactance, $X_L = 2\pi fL = 2\pi(50)(0.20)$ = **62.83 Ω.**

(b) Impedance $Z = R+jX_L = 50+j62.83 = \sqrt{[(50)^2+(62.83)^2]}\angle\arctan\frac{62.83}{50}$

$= 80.30\angle 51°\ 29'\ \Omega$.

(c) Current $I = \frac{V}{Z}$. Taking the terminal voltage V as the reference quantity,
i.e. $220\angle 0°$ V, then current $I = \frac{220\angle 0°}{80.30\angle 51°\ 29'} = 2.74\angle{-51°\ 29'}$ A,

i.e. **the current is 2.74 A lagging the terminal voltate by 51° 29′.**
(d) Voltage across resistance $V_R = IR$
$= (2.74\angle{-51°\ 29'})(50\angle 0°)$ = **137.0∠−51° 29′ V.**
(e) Voltage across inductance $V_L = IX_L$
$= (2.74\angle{-51°\ 29'})(62.83\angle 90°)$ = **172.2∠38° 31′ V.**
The phasor sum of V_R and V_L is the terminal voltage V as shown in *Fig 10*.

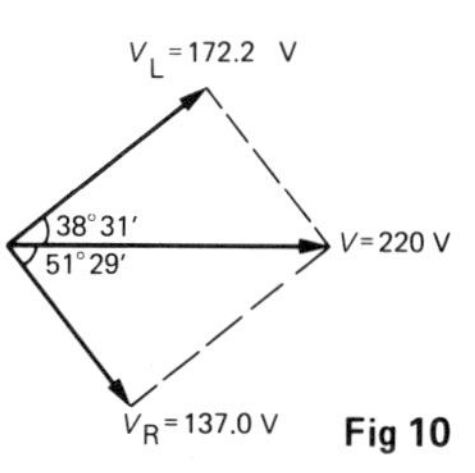

Fig 10

Problem 20 A coil of resistance 75 Ω and inductance 150 mH in series with an 8 μF capacitor is connected to a 500 V, 200 Hz supply. Calculate (a) the current flowing and its phase angle; and (b) the power factor of the circuit.

The circuit diagram is shown in *Fig 11*.

(a) Inductive reactance, $X_L = 2\pi fL = 2\pi(200)(150 \times 10^{-3}) = 188.5\ \Omega$.

Capacitive reactance, $X_C = \dfrac{1}{2\pi fC} = \dfrac{1}{2\pi(200)8 \times 10^{-6}} = 99.47\ \Omega$.

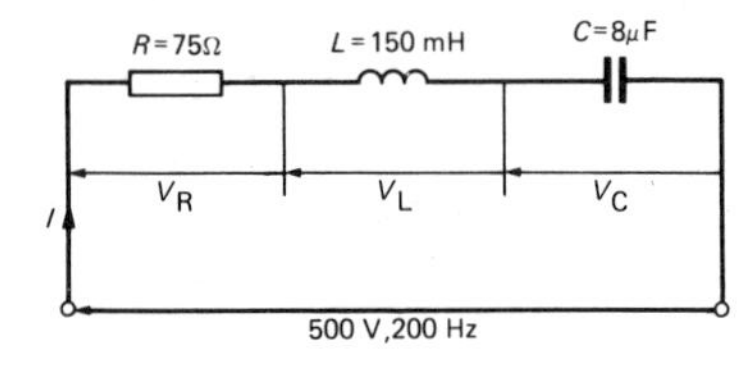

Fig 11

Impedance $Z = R+j(X_L - X_C) = 75+j(188.5-99.47) = (75+j89.03)\ \Omega$

$$= \sqrt{[(75)^2+(89.03)^2]} \angle \arctan\left(\frac{89.03}{75}\right)$$

$$= 116.4\angle 49^\circ\ 53'\ \Omega.$$

Current $I = \dfrac{V}{Z} = \dfrac{500\angle 0^\circ}{116.4\angle 49^\circ\ 53'} = 4.296\angle -49^\circ\ 53'$ A

i.e. **the current flowing is 4.296 A, lagging the voltage by 49° 53′.**
The phasor diagram is shown in *Fig 12.*

(b) Power factor $= \dfrac{R}{Z} = \dfrac{75}{116.4} =$ **0.644 lagging**

(or power factor = $\cos\phi = \cos 49^\circ\ 53'$
= **0.644 lagging**).

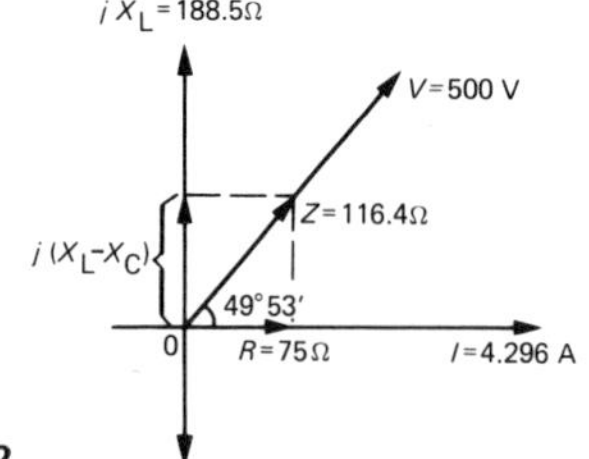

Fig 12

Problem 21 For the parallel circuit shown in *Fig 13* determine the value of current I and its phase relative to the 240 V supply, using complex numbers.

Current $I = \dfrac{V}{Z}$. Impedance Z for the three-branch parallel circuit is given by:

$\dfrac{1}{Z} = \dfrac{1}{Z_1} + \dfrac{1}{Z_2} + \dfrac{1}{Z_3}$, where $Z_1 = 4+j3$, $Z_2 = 10$ and $Z_3 = 12-j5$.

Admittance, $Y_1 = \dfrac{1}{Z_1} = \dfrac{1}{4+j3} = \dfrac{1}{4+j3} \times \dfrac{4-j3}{4-j3} = \dfrac{4-j3}{4^2+3^2} = 0.160-j0.120$ siemens.

Admittance, $Y_2 = \dfrac{1}{Z_2} = \dfrac{1}{10} = 0.10$ siemens.

Admittance, $Y_3 = \dfrac{1}{Z_3} = \dfrac{1}{12-j5} = \dfrac{1}{12-j5} \times \dfrac{12+j5}{12+j5} = \dfrac{12+j5}{12^2+5^2} = 0.0710+j0.0296$ siemens

Total admittance, $Y = Y_1 + Y_2 + Y_3$
$= (0.160 - j0.120) + (0.10) + (0.0710 + j0.0296)$
$= 0.331 - j0.0904 = 0.343\angle -15°\ 17'$ siemens.

Current $I = \dfrac{V}{Z} = VY = (240\angle 0°)$
$(0.343\angle -15°\ 17') =$ $\mathbf{82.32\angle -15°\ 17'\ A}$

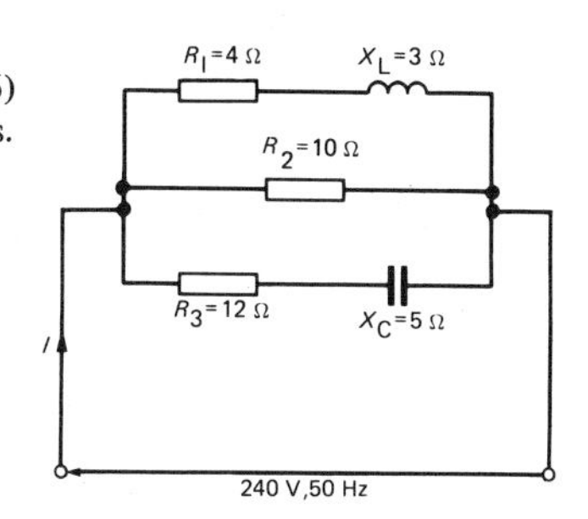

Fig 13

Problem 22 Determine the magnitude and direction of the resultant of the three coplanar forces given below, when they act at a point.
Force A, 10 N acting at 45° from the positive horizontal axis.
Force B, 8 N acting at 120° from the positive horizontal axis.
Force C, 15 N acting at 210° from the positive horizontal axis.

The space diagram is shown in *Fig 14*. The forces may be written as complex numbers.
Thus force A, $f_A = 10\angle 45°$, force B, $f_B = 8\angle 120°$
and force C, $f_C = 15\angle 210° = 15\angle -150°$
The resultant force

$$= f_A + f_B + f_C$$
$$= 10\angle 45° + 8\angle 120° + 15\angle -150°$$
$$= 10(\cos 45° + j \sin 45°) + 8(\cos 120° + j \sin 120°) + 15[\cos(-150°) + j \sin(-150°)]$$
$$= (7.071 + j7.071) + (-4.00 + j6.928) + (-12.99 - j7.50)$$
$$= -9.919 + j6.499$$

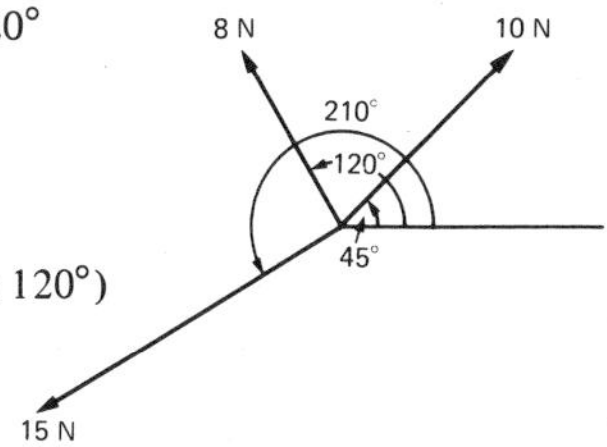

Fig 14

Magnitude of resultant force $= \sqrt{[(-9.919)^2 + (6.498)^2]} =$ **11.86 N**

Direction of resultant force $= \arctan\left(\dfrac{6.499}{-9.919}\right) =$ **146° 46′**,

(since $-9.919 + j6.499$ lies in the second quadrant).

C. FURTHER PROBLEMS ON COMPLEX NUMBERS

In *Problems 1 to 3*, solve the quadratic equations.

1 $x^2 + 25 = 0$ [$\pm j5$]

2 $2x^2 + 3x + 4 = 0$ [$-\dfrac{3}{4} \pm j\dfrac{\sqrt{23}}{4}$ or $-0.750 \pm j1.199$]

3 $4t^2 - 5t + 7 = 0$ [$\dfrac{5}{8} \mp j\dfrac{\sqrt{87}}{8}$ or $0.625 \pm j1.166$]

4 Evaluate (a) j^8; (b) $-\dfrac{1}{j^7}$; (c) $\dfrac{4}{2j^{13}}$; [(a) 1; (b) $-j$; (c) $-j2$]

5 Evaluate (a) $(3+j2)+(5-j)$ and (b) $(-2+j6)-(3-j2)$ and show the results on an Argand diagram. [(a) $8+j$; (b) $-5+j8$]

6 Write down the complex conjugates of (a) $3+j4$; (b) $2-j$. [(a) $3-j4$; (b) $2+j$]

In *Problems 7 to 11* evaluate in $a+jb$ form given $Z_1 = 1+j2$, $Z_2 = 4-j3$, $Z_3 = -2+j3$ and $Z_4 = -5-j$.

7 (a) $Z_1+Z_2-Z_3$; (b) $Z_2-Z_1+Z_4$ [(a) $7-j4$; (b) $-2-j6$]

8 (a) Z_1Z_2; (b) Z_3Z_4 [(a) $10+j5$; (b) $13-j13$]

9 (a) $Z_1Z_3+Z_4$; (b) $Z_1Z_2Z_3$ [(a) $-13-j2$; (b) $-35+j20$]

10 (a) $\dfrac{Z_1}{Z_2}$; (b) $\dfrac{Z_1+Z_3}{Z_2-Z_4}$ $\left[\text{(a) } \dfrac{-2}{25}+j\dfrac{11}{25}; \text{ (b) } \dfrac{-19}{85}+j\dfrac{43}{85}\right]$

11 (a) $\dfrac{Z_1Z_3}{Z_1+Z_3}$; (b) $Z_2+\dfrac{Z_1}{Z_4}+Z_3$ $\left[\text{(a) } \dfrac{3}{26}+j\dfrac{41}{26}; \text{ (b) } \dfrac{45}{26}-j\dfrac{9}{26}\right]$

12 Evaluate (a) $\dfrac{1-j}{1+j}$; (b) $\dfrac{1}{1+j}$ $\left[\text{(a) } -j; \text{ (b) } \dfrac{1}{2}-j\dfrac{1}{2}\right]$

13 Show that $\dfrac{-25}{2}\left(\dfrac{1+j2}{3+j4}-\dfrac{2-j5}{-j}\right) = 57+j24$

In *Problems 14 to 18* solve the complex equations.

14 $5(x+jy) = 10-j15$ [$x = 2; y = -3$]

15 $(2+j)(3-j2) = a+jb$ [$a = 8; b = -1$]

16 $\dfrac{2+j}{1-j} = j(x+jy)$ $\left[x = \dfrac{3}{2}; y = -\dfrac{1}{2}\right]$

17 $(2-j3) = \sqrt{(a+jb)}$ [$a = -5; b = -12$]

18 $(x-j2y)-(y-jx) = 2+j$ [$x = 3; y = 1$]

19 If $Z = R+j\omega L+1/j\omega C$, express Z in $(a+jb)$ form when $R = 10$, $L = 5$, $C = 0.04$ and $\omega = 4$. [$Z = 10+j13.75$]

20 Determine the modulus and argument of (a) $2+j4$; (b) $-5-j2$; (c) $j(2-j)$.
[(a) 4.472, 63° 26′; (b) 5.385, −158° 12′; (c) 2.236, 63° 26′]

In *Problems 21 and 22* express the given cartesian complex numbers in polar form, leaving answers in surd form.

21 (a) $2+j3$; (b) -4; (c) $-6+j$
[(a) $\sqrt{13}\angle 56°\ 19'$; (b) $4\angle 180°$; (c) $\sqrt{37}\angle 170°\ 32'$]

22 (a) $-j3$; (b) $(-2+j)^3$; (c) $j^3(1-j)$
[(a) $3\angle -90°$; (b) $\sqrt{125}\angle 100°\ 18'$; (c) $\sqrt{2}\angle -135°$]

In *Problems 23 and 24* convert the given polar complex numbers into $(a+jb)$ form giving answers correct to 4 significant figures.

23 (a) $5\angle 30°$; (b) $3\angle 60°$; (c) $7\angle 45°$
[(a) $4.330+j2.500$; (b) $1.500+j2.598$; (c) $4.950+j4.950$]

24 (a) $6\angle 125°$; (b) $4\angle \pi$; (c) $3.5\angle -120°$
[(a) $-3.441+j4.915$; (b) $-4.000+j0$; (c) $-1.750-j3.031$]

In *Problems 25 to 27*, evaluate in polar form.

25 (a) $3\angle 20° \times 15\angle 45°$; (b) $2.4\angle 65° \times 4.4\angle -21°$ [(a) $45\angle 65°$; (b) $10.56\angle 44°$]

26 (a) $6.4\angle 27° \div 2\angle -15°$; (b) $5\angle 30° \times 4\angle 80° \div 10\angle -40°$
[(a) $3.2\angle 42°$; (b) $2\angle 150°$]

27 (a) $4\angle\dfrac{\pi}{6}+3\angle\dfrac{\pi}{8}$; (b) $2\angle 120°+5.2\angle 58°-1.6\angle -40°$
[(a) $6.986\angle 26°\ 47'$; (b) $7.190\angle 85°\ 46'$]

28 Determine in polar form (a) $[1.5\angle 15°]^5$; (b) $(1+j2)^6$
[(a) $7.594\angle 75°$; (b) $125\angle 20°\ 37'$]

29 Determine in polar and cartesian forms (a) $[3\angle 41°]^4$; (b) $(-2-j)^5$
[(a) $81\angle 164°$, $-77.86+j22.33$; (b) $55.90\angle -47°\ 10'$, $38.00-j40.99$]

30 Convert $(3-j)$ into polar form and hence evalute $(3-j)^7$, giving the answer in polar form. [$\sqrt{10}\angle -18°\ 26'$, $3162\angle -129°\ 2'$]

In *Problems 31 to 33* determine the two square roots of the given complex numbers in cartesian form and show the results on an Argand diagram.

31 (a) $1+j$; (b) j [(a) $\pm$ (1.099+j0.455); (b) $\pm$ (0.707+j0.707)]

32 (a) $3-j4$; (b) $-1-j2$ [(a) $\pm$ (−2+j); (b) $\pm$ (−0.786+j1.272)]

33 (a) $7\angle 60°$; (b) $12\angle\frac{3\pi}{2}$ [(a) $\pm$ (2.291+j1.323); (b) $\pm$ (−2.449+j2.449)]

34 Determine the resistance R and series inductance L (or capacitance C) for each of the following impedances assuming the frequency to be 50 Hz.
(a) $(3+j8)\ \Omega$; (b) $(2-j3)\ \Omega$; (c) $j14\ \Omega$; (d) $-j150\ \Omega$; (e) $12\angle\frac{\pi}{6}\ \Omega$; (f) $8\angle -60°\ \Omega$.
[(a) $R = 3\Omega$, $L = 25.5$ mH; (b) $R = 2\ \Omega$, $C = 1061\ \mu$F
(c) $R = 0$, $L = 44.56$ mH; (d) $R = 0$, $C = 21.22\ \mu$F
(e) $R = 10.39\ \Omega$, $L = 19.10$ mH; (f) $R = 4\ \Omega$, $C = 459.4\ \mu$F]

35 Two impedances, $Z_1 = (3+j6)\ \Omega$ and $Z_2 = (4-j3)\ \Omega$ are connected in series to a supply voltage of 120 V. Determine the magnitude of the current and its phase angle relative to the voltage. [15.76 A, 23° 12′ lagging]

36 If the two impedances in *Problem 35* are connected in parallel determine the current flowing and its phase relative to the 120 V supply voltage. [27.25 A, 3° 22′ lagging]

37 An alternating voltage of 200 V, 50 Hz is applied across an impedance of $(30-j40)\ \Omega$. Calculate (a) the resistance; (b) the capacitance; (c) the current and (d) the phase angle between voltage and current.
[(a) 30 Ω; (b) 79.58 μF; (c) 4 A; (d) 53° 8′ leading]

38 A series circuit consists of a 12 Ω resistor, a coil of inductance 0.10 H and a capacitance of 160 μF. Calculate the current flowing and its phase relative to the supply voltage of 240 V, 50 Hz. Determine also the power factor of the circuit.
[14.42 A; 43° 51′ lagging; 0.721]

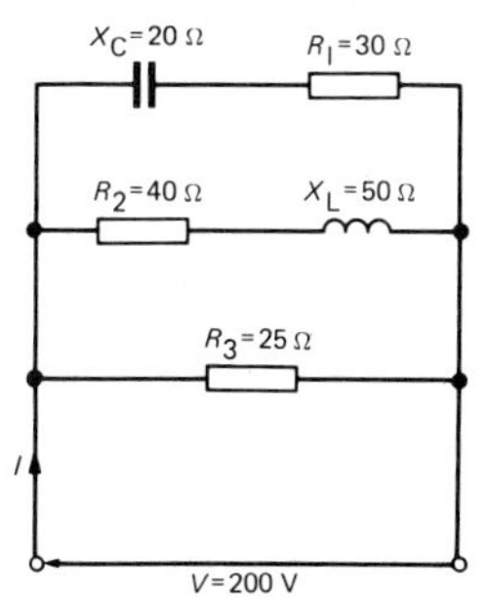

Fig 15

39 For the circuit shown in *Fig 15*, determine the current I flowing and its phase relative to the applied voltage. [14.6 A; 2° 30′ leading]

40 Determine, using complex numbers, the magnitude and direction of the resultant of the coplanar forces given below, which are acting at a point. Force A, 5 N acting horizontally; Force B, 9 N acting at an angle of 135° to force A; Force C, 12 N acting at an angle of 240° to force A. [8.393 N; 208° 40′ from force A]

3 Number (3) – Octal and binary numbering systems and coding

A. MAIN POINTS CONCERNED WITH OCTAL AND BINARY NUMBERING SYSTEMS AND CODING

1. The **denary** or decimal systems of numbers is the one in normal use in arithmetic and has a base of 10, that is, uses the ten digits 0, 1, 2, 3, 4, 5, 6, 7, 8 and 9.
   The **octal** system of numbers uses eight digits only, these being 0, 1, 2, 3, 4, 5, 6 and 7 and is said to have a base of 8. The relationship between an octal and a denary number is obtained by multiplying each digit by 8 raised to some power, the value of the power depending on the position of the digit. Thus,

$$37.42_8 = (3 \times 8^1)+(7 \times 8^0)+(4 \times 8^{-1})+(2 \times 8^{-2})$$

$$= 24+7+\frac{4}{8}+\frac{2}{64} = 31.53125_{10},$$

   The subscripts 8 and 10 indicating the bases. Octal to denary conversion of numbers is shown in *Problem 1*.

2 (i) An integer denary number is converted to an octal number by repeatedly dividing by 8, noting the remainder at each stage, the octal number being the remainder column read from the bottom upwards.
  (ii) A decimal fraction is converted to an octal number by multiplying by 8, listing the integer part resulting from multiplying and carrying forward the decimal fraction part for multiplication by 8 again. The fractional part of the octal number is obtained by reading down the column of integers obtained from multiplication, (see *Problem 2*).

3 The binary system of numbers uses two digits only, these being 0 and 1. The relationship between a binary and a denary number is obtained by multiplying each digit of the binary number by 2 raised to some power, the value of the power depending on the position of the digit. Thus,

$$101.01_2 = (1 \times 2^2)+(0 \times 2^1)+(1 \times 2^0)+(0 \times 2^{-1})+(1 \times 2^{-2})$$

$$= 4+0+1+0+\frac{1}{4} = 5.25_{10}.$$

4 Integer octal numbers may be converted to binary numbers by repeatedly dividing by 2, noting the remainder and reading the remainder column from the bottom upwards. However, the binary numbers corresponding to the octal numbers 0 to 7 soon become familiar with use and are as follows:

Octal Numbers	*Corresponding Binary Number*
0	0 0 0
1	0 0 1
2	0 1 0
3	0 1 1
4	1 0 0
5	1 0 1
6	1 1 0
7	1 1 1

An octal number may be written as a binary number by writing the corresponding three digit binary number for each octal digit. Thus

$435.7_8 = 100\ 011\ 101.111_2$

(See *Problem 3.*)

5 The procedure given in para. 4 is reversed when converting a binary number to an octal number. The binary number is split into groups of three bits (binary digits), starting at the binary point, and each group is converted to octal using the data given in para. 4. Thus:

$$1101110001.1111_2 = 001\ 101\ 110\ 001.\ 111\ 100$$
$$= 1\ 561.74_8$$

(See *Problem 4.*)

6 (i) Binary addition of two bits is according to the following rules, the order of the bits on the left-hand side being immaterial:

		Sum	*Carry*
0 + 0	=	0	0
0 + 1	=	1	0
1 + 1	=	0	1

(ii) Binary addition of three bits is according to the following rules, the order of the bits on the left-hand side being immaterial:

	Sum	*Carry*
0 + 0 + 0	0	0
0 + 0 + 1	1	0
0 + 1 + 1	0	1
1 + 1 + 1	1	1

An example of binary addition is given in *Problem 7.* When adding binary numbers A and B to give a result of C, i.e., $A+B = C$, A is called the **augend**, B is called the **addend** and C is called the **sum**. Thus for binary addition:

Augend + Addend = Sum

7 For binary subtraction, when evaluating 0–1, a 1 is borrowed from the next column on the left containing a 1, this becoming 2 for each move to the right. (see *Problem 8*). When subtracting a binary number B from binary number A, to give a result of C, i.e., $A-B = C$, A is called the **minuend**, B is called the **subtrahend** and C is called the **difference**. Thus for binary subtraction:

Minuend – Subtrahend = Difference

8 For octal addition, since the base is 8, 1 is carried to the next column on the left when the addition of two octal digits is 8 or more (see *Problem 9*).

9 In octal subtraction, when evaluating, say, 3–7, a 1 is borrowed from the next

column on the left containing 1 or more, this becoming 8 for each move to the right (see *Problem 10*).

10 In practice 'adder circuits' may be used in calculators and computers to perform both addition and subtraction of binary numbers. When the process is subtraction using adder circuits, a method of complements to subtract one binary number from another is used. Two 'complement methods' are in common use as shown in paras. 11 and 12.

11 **The one's complement method** (see *Problems 11 and 12*).

(a) Express both minuend and subtrahend so that they each have the same number of bits.

(b) Determine the one's complement of the subtrahend. This is done by writing 1 for 0 and 0 for 1 for each bit in the subtrahend.

(c) Add the minuend and one's complement of the subtrahend, to give a sum.

(d) If the sum in (c) contains an extra bit when compared with the number of bits in the minuend or subtrahend in (a), then the difference between the minuend and subtrahend is positive and is given by **'end-around carry'**, that is, by taking the 1 on the far left of the sum and adding it to the bit on the far right of the sum. If the sum in (c) does **not** contain an extra bit when compared with the number of bits in the minuend and subtrahend of (a), the difference between the minuend and subtrahend is negative and is given by the one's complement of the sum.

12 **The two's complement method** (see *Problems 13 and 14*).

(a) Express the minuend and subtrahend so that they each have the same number of bits.

(b) Add a sign bit, (0) for a positive number and (1) for a negative number, on the far left of both minuend and subtrahend.

(c) Determine the two's complement of the subtrahend. This is done by writing down the one's complement and then adding 1.

(d) Add the minuend and two's complement of the subtrahend to give a sum.

(e) When the sign bit is (0) the 1 on the left of the sign bit is disregarded and the difference between the minuend and subtrahend is positive and its value is given by the bits to the right of the sign bit. When the sign bit is (1), the difference is negative and the value is the two's complement of the sum.

13 The natural binary system of numbers requires a large number of bits to depict a denary number having several significant figures. For example, the denary number 3 418 is the binary number 110101011010 and this binary number gives no real idea of the size of the number. Also, registers for storing bits in calculators, computers and microprocessors are usually of uniform size, containing room to store 4 or 8 or 16 bits. For these reasons, various **codes** are used in practice, rather than the natural binary system of numbers.

14 It is usual, when expressing a denary number as a binary number, to use a **decade system**. Each decade contains four bits and is used to represent a denary number from 0–9. The units decade is used to represent denary numbers from 0–9, the ten's decade is used for denary numbers from 10–99, and so on. This system of coding is called the **binary-coded-decimal system**, usually abbreviated to **BCD system**. (See *Problem 15.*)

15 Another code in common use is the 'excess-three code', written as the **'X's 3 code'**, and when used as a BCD is called the **'X's 3 BCD'**. This code is formed by adding the binary number 0011 to the natural binary code. Thus the X's 3 binary numbers equivalent to the denary numbers 0, 1, 2, ... are 0011, 0100, 0101,

.... The advantage of this code is that there is no binary number 0000, which may occur in the event of electrical failure to a system. (See *Problem 17.*)

16 During the transmission of binary codes, errors can occur due to human errors and due to the signal being distorted in the transmitting medium. When this occurs, a 0 in the signal is received as a 1 and vice-versa. Error detecting codes may be used, giving an indication of an error having been made. The simplest of these error detecting codes uses five bits instead of four, four bits containing the data and the fifth bit being used for checking purposes, and called a **parity bit.** Two such codes are the odd parity system and the even parity system.

17 **The odd parity system.** The sum of the 1's in any group of five bits must add up to an odd decimal number, the fifth parity bit being selected to meet this requirement. Thus, 0100, 0110, 0111, 0000, 1111 is transmitted as:

Information	*Odd Parity Bit*	*Decimal Sum*
0100	0	1
0110	1	3
0111	0	3
0000	1	1
1111	1	5

i.e. 01000, 01101, 01110, 00001, 11111. If the five bits received do not contain an odd number of 1's then the recipient knows an error has been made. (See *Problem 18.*)

18 **The even parity system.** This is similar to the odd parity system except that the decimal sum of the five bits must be an even number. Thus 0100, 0110, 0111, 0000, 1111 is transmitted as 01001, 01100, 01111, 00000, 11110. If the five bits are received and the sum of the 1's is not an even decimal number, then the recipient knows an error has been made. (See *Problem 19.*)

19 Another error detecting code is called the **two-out-of-five code** in which five bits are used and the denary numbers 0 to 9 are each represented by two 1's and three 0's as shown below:

Two-out-of-five code	*Corresponding Denary Number*
11000	0
00011	1
00101	2
00110	3
01001	4
01010	5
01100	6
10001	7
10010	8
10100	9

If any set of five bits received contains more or less than two 1's then an error exists. (See *Problem 20.*)

B. WORKED PROBLEMS ON OCTAL AND BINARY NUMBERING SYSTEMS AND CODING

Problem 1 Convert (a) $4\ 375_8$ and (b) 21.64_8 to denary numbers.

(a) From para. 1, the denary equivalent of

$$4\,375_8 \text{ is } (4 \times 8^3)+(3 \times 8^2)+(7 \times 8^1)+(5 \times 8^0)$$

i.e. $4\,375_8 = (4 \times 512)+(3 \times 64)+(7 \times 8)+(5 \times 1)$

$= 2048+192+56+5 = \mathbf{2301_{10}}$

(b) From para. 1, the denary equivalent of

$$21.64_8 \text{ is } (2 \times 8^1)+(1 \times 8^0)+(6 \times 8^{-1})+(4 \times 8^{-2})$$

i.e. $21.64_8 = 16+1+0.125+0.0625 = \mathbf{17.1875_{10}}$.

Problem 2 Convert (a) 387_{10} and (b) 79.20215_{10} to octal numbers, correct to 3 octal places.

(a) From para. 2, 387_{10} is repeatedly divided by 8, the remainder being noted at each stage.

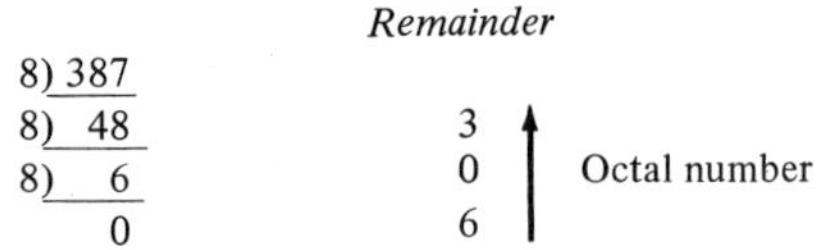

The corresponding octal number is given by the remainder column, read from the bottom upwards, i.e., $\mathbf{387_{10} = 603_8}$.

(b) The integer part of the number 79.20215_{10} is converted to an octal number as shown in part (a), i.e.

Remainder

	Remainder	
8) 79		
8) 9	7 ↑	
8) 1	1	Integer part of octal number
0	1	

Hence the integer part of the octal number is 117_8.

The fractional part is multiplied by 8, the fractional part resulting from each multiplication being carried forward, i.e.,

		Integer Part		*Fractional Part*
0.20215 × 8	=	1	.	6172
0.6172 × 8	=	4	.	9376
0.9376 × 8	=	7	.	5008
0.5008 × 8	=	4 ↓	.	0064

When expressing a denary number correct to 3 decimal places, the number in the third decimal place is increased by one if the number in the fourth decimal place belongs to the set of numbers 5, 6, 7, 8 or 9. In the octal system, a number is rounded up for the set of numbers 4, 5, 6 or 7. Thus, in this case 0.1474 becomes 0.150 due to rounding up.

Thus: $\mathbf{79.20215_{10} = 117.150_8}$, correct to 3 octal places.

Problem 3 Convert (a) 471_8 and (b) 56.42_8 to binary numbers.

(a) Using the table given in para. 4, each octal digit is converted to the corresponding three bit binary number.

Thus:	Octal number:	4	7	1
	Corresponding binary number:	100	111	001

i.e., $\mathbf{471_8 = 100111001_2}$

(b) Using the same procedure as above,

Octal number:	5	6	.	4	2
Corresponding binary number:	101	110	.	100	010

i.e., $\mathbf{56.42_8 = 101110.10001_2}$

Problem 4 Convert (a) 1101010110_2 and (b) 10110.01101_2 to octal numbers.

(a) For binary numbers having no fractional part, the binary numbers are split into groups of three bits, starting from the right, and the groups are converted to the corresponding octal numbers using the table in para. 4. Thus:

Binary number	001	101	010	110
Corresponding octal number	1	5	2	6

i.e. $\mathbf{1101010110_2 = 1526_8}$

(b) The binary number is split into groups of three bits, starting from the binary point. Thus:

Binary number:	010	110	.	011	010	
Corresponding octal number:	2	6	.	3	2	(from para. 4)

i.e., $\mathbf{10110.01101_2 = 26.32_8}$

Problem 5 Convert the denary numbers (a) 346_{10} and (b) 87.24_{10} to binary numbers, correct to 6 binary places, via octal numbers.

(a) The first step is to convert from denary to octal, (see *Problem 2*),

	Remainder
8) 346	
8) 43	2 ↑
8) 5	3 ↑
0	5 ↑

i.e., $346_{10} = 532_8$

The second step is to convert the octal number to a binary number, (see *Problem 3*),

$532_8 = 101\ 011\ 010_2$ i.e. $\mathbf{346_{10} = 101011010_2}$

(b) The denary number is converted to an octal number.

		Remainder			
Integer	8) 87		Fractional	$0.24 \times 8 =$	1.92 ↓
part:	8) 10	7 ↑	part:	$0.92 \times 8 =$	7.36 ↓
	8) 1	2 ↑		$0.36 \times 8 =$	2.88 ↓
	0	1 ↑		$0.88 \times 8 =$	7.04 ↓

i.e., $87.24 = 127.173_8$, correct to 3 octal places.

The octal number is converted to a binary number.

$127.173_8 = 001\ 010\ 111\ .\ 001\ 111\ 011_2$

i.e. $\mathbf{87.24_{10} = 1010111.001111_2}$, correct to 6 binary places.

Problem 6 Convert the binary numbers (a) 1011010111_2 and (b) 1010111.10111_2 to denary numbers, via octal.

(a) Grouping the binary numbers in threes from the right gives 001 011 010 111. Thus the corresponding octal number is $1\ 3\ 2\ 7_8$. The denary number corresponding to this octal number is $(1 \times 8^3)+(3 \times 8^2)+(2 \times 8^1)+(7 \times 8^0)$, (see *Problem 1*), i.e., 512+192+16+7 = 727. Thus $\mathbf{1011010111_2 = 727_{10}}$.

(b) Grouping the binary number in threes from the binary point gives 001 010 111 . 101 110. The corresponding octal number is 127.56_8.
The denary number corresponding to this octal number is:
$(1 \times 8^2)+(2 \times 8^1)+(7 \times 8^0)+(5 \times 8^{-1})+(6 \times 8^{-2})$
i.e. 64+16+7+0.625+0.09375.
Thus $\mathbf{1010111.10111_2 = 87.71875_{10}}$

Problem 7 Add the binary numbers 101.011_2 and 111.101_2.

The basic rules for binary addition and terms used are given in para. 6.

Column	7	6	5	4	3	2	1
Augend	0	1	0	1.	0	1	1
Addend	0	1	1	1.	1	0	1
Sum	1	1	0	1.	0	0	0
Carry	1	1	1	1	1	1	

The procedure is to apply two-bit addition to the augend and addend of column 1, i.e., 1+1 = sum 0, carry 1. 0 is written in the sum row and the carry 1 is written in the carry row under column 2. For the remaining columns three-bit addition is applied to the augend, addend and carry rows, giving the result: $\mathbf{101.011_2+111.101_2 = 1101_2}$.

Problem 8 Subtract 1101.101_2 from 10010.11_2.

Column	8	7	6	5	4	3	2	1
		~~2~~ 1	2	0	2		0	2
Minuend	~~1~~	0	0	~~1~~	0.	1	~~1~~	0
Subtrahend	0	1	1	0	1.	1	0	1
Difference	0	0	1	0	1.	0	0	1

With reference to para. 7, starting with column 1, minuend minus subtrahend is 0–1; borrow 1 from column 2, leaving 0 for the minuend in column 2, and this borrowed 1 becomes 2 in column 1; 2–1 gives a difference of 1 in column 1. The remaining columns where 0–1 arises are dealt with in a similar way, giving the result: $\mathbf{10010.11_2-1101.101_2 = 101.001_2}$.

Problem 9 Add 4674_8 and 5377_8.

Column	5	4	3	2	1
Augend	0	4	6	7	4
Addend	0	5	3	7	7
Sum	1	2	2	7	3
Carry	1	1	1	1	

With reference to para. 8, for column 1, 4+7 = 11. Since the sum exceeds 8, 1 is carried to column 2 and 11–8, i.e., 3, is the sum written in column 1. For columns 2 to 5 the augend, addend and carry rows are added.
Hence $\mathbf{4674_8+5377_8 = 12273_8}$.

Problem 10 Take 3725_8 from 5024_8.

Column	4	3	2	1
Minuend	~~5~~ 4	~~0~~ ~~8~~ 7	~~2~~ 1+8	~~4~~ 4+8
Subtrahend	3	7	2	5
Difference	1	0	7	7

With reference to para. 9, for column 1, 1 is borrowed from column 2, making the minuend 4+8, i.e., 12. 12–5 is 7 and this is the difference for column 1. For columns 2 to 4 a similar procedure is carried out when the minuend is less than the subtrahend.

Hence $5024_8 - 3725_8 = 1077_8$

Problem 11 Determine the value of $1011_2 - 110_2$ by the one's complement method.

The procedure given in para. 11 is followed, for a minuend of 1011 and a subtrahend of 110.

(a) 1011–110 = 1011–0110; (b) The one's complement of the subtrahend is 1001.

(c) Adding the minuend and the one's complement of the subtrahend gives:

```
     1011
     1001
Sum 10100
```

(d) An extra bit is generated in the sum and the value of the difference is positive and is obtained by 'end-around carry', i.e.,

```
     ~~1~~0110
      └──→1
Sum     101
```

Thus, $1011_2 - 110_2 = 101_2$.

Problem 12 Use the one's complement method to find the value of $10011_2 - 11010_2$.

The procedure given in para. 11 is followed for a minuend of 10011 and a subtrahend of 11010.

(a) This step is not necessary since both minuend and subtrahend have the same number of bits.

(b) The one's complement of the subtrahend is 00101.

(c) Adding the minuend and one's complement of the subtrahend gives:

```
     10011
     00101
Sum  11000
```

(d) An extra bit is **not** generated in the sum, hence the difference is negative and its value is the one's complement of the sum, i.e.

$10011_2 - 11010_2 = -111_2$

Problem 13 Use the two's complement method to find the value of $1101011_2 - 110100_2$.

The procedure given in para. 12 is followed for a minuend of 1101011 and a subtrahend of 110100.

(a) 1101011–110100 = 1101011–0110100
(b) Adding the sign bits gives (0)1101011–(0)0110100, the sign bit being (0) since both numbers are positive, i.e., $(+1101011_2)$–$(+110100_2)$.
(c) The two's complement of the subtrahend is obtained from the one's complement plus one, i.e. (1)1001011+1, that is, (1)1001100.
(d) Adding the minuend and the two's complement of the subtrahend gives:

	(0)1101011
	(1)1001100
Sum	1(0)0110111

(e) The sign bit is (0), hence the 1 on the left of the sign bit is disregarded, the value is positive and is given by the bits on the right of the sign bit, i.e. $\mathbf{1101011_2 - 110100_2 = 110111_2}$.

Problem 14 Determine the value of $101101_2 - 110110_2$ by using the two's complement method.

The procedure given in para. 12 is followed.
(a) This step is not necessary since both minuend and subtrahend have the same number of bits.
(b) Adding the sign bits gives (0)101101–(0)110110.
(c) The two's complement of the subtrahend is (1)001001+1, i.e., (1)001010.
(d) Adding the minuend and the two's complement of the subtrahend gives:

	(0)101101
	(1)001010
Sum	(1)110111

(e) The sign bit is (1), hence the value is negative and is the two's complement of the sum, i.e.,
$\mathbf{101101_2 - 110110_2 = -((0)001000+1) = -1001_2}$.

Problem 15 Encode 2719_{10} as a binary-coded-decimal number.

A BCD code uses decades for the denary units, tens, hundreds and thousands values and a four-bit natural binary code for the values. Thus:
2719_{10} = 0010 0111 0001 1001, when written as a BCD.

Problem 16 Express the BCD number 1000 0110 0101 0011 as a denary number.

The first group of four bits is the thousands decade of the corresponding denary number, the second group is the hundreds decade, and so on. Thus:
1000 0110 0101 0011 = 8653_{10}.

Problem 17 Encode 5208_{10} as an excess-three, binary-coded-decimal.

The X's 3 code is obtained by adding 0011 to the natural binary code, hence:

Denary Number	*Natural Binary Number*	*X's 3 Number*
5	0101	0101+0011 = 1000
2	0010	0010+0011 = 0101
0	0000	0000+0011 = 0011
8	1000	1000+0011 = 1011

i.e. **5208_{10} = 1000 0101 0011 1011, when expressed as an X's 3 BCD.**

Problem 18 Encode 2976_{10} into a binary-coded-decimal with odd parity.

For odd parity an extra bit is added to the BCD code, to make the sum of the 1's in any group of five bits an odd number. Thus:

Denary Number	*BCD*	*Odd Parity Bit*
2	0010	0
9	1001	1
7	0111	0
6	0110	1

i.e. 2976_{10} = **00100 10011 01110 01101 when expressed as a BCD with odd parity.**

Problem 19 Encode 4725_{10} into an excess-three, binary-coded-decimal with even parity.

For even parity, an extra bit is added to the X's 3, BCD code to make the sum of the 1's in any group of five bits an even number. Thus:

Denary Number	*BCD*	*X's 3 BCD*	*Even Parity Bit*
4	0100	0111	1
7	0111	1010	0
2	0010	0101	0
5	0101	1000	1

i.e., 4725_{10} = **01111 10100 01010 10001 when expressed as an X's 3, BCD with even parity.**

Problem 20 Encode 5728_{10} into a two-out-of-five code.

The two-out-of-five code is given in para. 19.

Denary Number	*Two-out-of-five code*
5	01010
7	10001
2	00101
8	10010

Thus 5728_{10} = 01010 10001 00101 10010 when expressed as a two-out-of-five code.

C. FURTHER PROBLEMS ON OCTAL AND BINARY NUMBERING SYSTEMS AND CODING

In *Problems 1 to 3*, convert the octal numbers given to denary numbers.

1 (a) 6_8; (b) 35_8; (c) 417_8 [(a) 6_{10}; (b) 29_{10}; (c) 271_{10}]

2 (a) 0.1_8; (b) 0.34_8; (c) 0.574_8, correct to 4 decimal places.
[(a) 0.1250_{10}; (b) 0.4375_{10}; (c) 0.7422_{10}]

3 (a) 17.6_8; (b) 63.42_8; (c) 275.35_8
[(a) 15.75_{10}; (b) 51.53125_{10}; (c) 189.453125_{10}]

In *Problems 4 to 6*, convert the denary numbers given to octal numbers, giving the answers to the fractional parts correct to 3 octal places.

4 (a) 7_{10}; (b) 49_{10}; (c) 572_{10} [(a) 7_8; (b) 61_8; (c) 1074_8]

5 (a) 0.31_{10}; (b) 0.872_{10}; (c) 0.529_{10} [(a) 0.237_8; (b) 0.676_8; (c) 0.417_8]

6 (a) 19.7_{10}; (b) 84.72_{10}; (c) 521.27_{10}
[(a) 23.546_8; (b) 124.561_8; (c) 1011.212_8]

7 Convert the octal numbers (a) 3746_8 and (b) 251.07_8 to binary numbers.
[(a) 11111100110_2; (b) 10101001.000111_2]

8 Convert the binary numbers (a) 1011011_2 and (b) 10011.1101_2 to octal numbers. [(a) 133_8; (b) 23.64_8]

In *Problems 9 and 10*, convert the denary numbers given to binary numbers, via octal, expressing the answers to fractional parts correct to 6 bits.

9 (a) 38_{10}; (b) 343_{10}; (c) 1265_{10}
[(a) 100110_2; (b) 101010111_2; (c) 10011110001_2]

10 (a) 0.36_{10}; (b) 7.43_{10}; (c) 41.248_{10}
[(a) 0.010111_2; (b) 111.011100_2; (c) 101001.010000_2]

In *Problems 11 and 12*, convert the binary numbers given to denary numbers, via octal.

11 (a) 101110_2; (b) 10101101_2; (c) 1110011011010_2
[(a) 46_{10}; (b) 173_{10}; (c) 7386_{10}]

12 (a) 10.0101_2; (b) 1101.1101_2; (c) 10001.0011_2
[(a) 2.3125_{10}; (b) 13.8125_{10}; (c) 17.1875_{10}]

13 Add 110110_2 and 101010_2. [1100000_2]

14 Determine the value of $1011101_2+1101101_2$. [11001010_2]

15 Evaluate $10110001.101_2+10110110.0111_2$ [101101000.0001_2]

16 Take 1101101_2 from 101100011_2 [11110110_2]

17 Determine the value of $110010010_2-11101111_2$ [10100011_2]

18 Evaluate $101.11001_2-100.0011_2$ [1.10011_2]

19 Add 4713_8 and 5626_8. [12541_8]

20 Determine the value of 746_8+3652_8. [4620_8]

21 Evaluate $163.27_8+721.66_8$ [1105.15_8]

22 Take 3274_8 from 4152_8 [656_8]

23 Determine the value of $31417_8-17375_8$ [12022_8]

24 Evaluate $273.46_8-175.67_8$ [75.57_8]

In *Problems 25 to 27*, use the one's complement method to find the value required.

25 1101_2-110_2 [111_2]

26 $1101001_2-1011100_2$ [1101_2]

27 $101011101_2-110110011_2$ [-1010110_2]

In *Problems 28 to 30*, use the two's complement method to find the values required.

28 11010_2-1011_2 [1111_2]

29 $10110101_2-1001101_2$ [1101000_2]

30 $10111010_2-11010110_2$ [-11100_2]

31 Encode the denary numbers given as binary-coded-decimals.
(a) 271; (b) 4320; (c) 89.17; (d) 471.56
[(a) 0010 0111 0001; (b) 0100 0011 0010 0000
(c) 1000 1001.0001 0111; (d) 0100 0111 0001.0101 0110]

32 Express the BCD numbers given as denary numbers.
(a) 0101 0110 0111; (b) 0100 1001 1000 0000;
(c) 1000 0111 0011.0010; (d) 1001 0001.0000 0101
[(a) 567_{10}; (b) 4980_{10}; (c) 873.2_{10}; (d) 91.05_{10}]

33 Encode the denary numbers given as an excess-three-binary-coded-decimal.
(a) 274; (b) 4971; (c) 58.63
[(a) 0101 1010 0111; (b) 0111 1100 1010 0100; (c) 1000 1011.1001 0110]

34 Encode the denary numbers given into a binary-coded-decimal with odd parity.
(a) 284; (b) 5937; (c) 601.53
[(a) 00100 10000 01000; (b) 01011 10011 00111 01110;
(c) 01101 00001 00010.01011 00111]

35 Encode the denary numbers given into an excess-three-binary-coded-decimal with even parity.
(a) 461; (b) 5197; (c) 68.03
[(a) 01111 10010 01001; (b) 10001 01001 11000 10100;
(c) 10010 10111.00110 01100]

36 Encode the denary numbers given into a two-out-of-five code.
(a) 237; (b) 4591; (c) 60.84
[(a) 00101 00110 10001; (b) 01001 01010 10100 00011;
(c) 01100 11000.10010 01001]

4 Algebra (1) – The binomial theorem

A. MAIN POINTS CONCERNED WITH THE BINOMIAL THEOREM

1 A **binomial expression** is one which contains two terms connected by a plus or minus sign. Thus $(p+q)$, $(a+x)^2$, $(2x+y)^3$ are examples of binomial expressions.

2 Expanding $(a+x)^n$ for integer values of n from 0 to 6 gives the following results:

$$
\begin{aligned}
(a+x)^0 &= 1\\
(a+x)^1 &= a+x\\
(a+x)^2 &= (a+x)(a+x) = a^2+2ax+x^2\\
(a+x)^3 &= (a+x)^2(a+x) = a^3+3a^2x+3ax^2+x^3\\
(a+x)^4 &= (a+x)^3(a+x) = a^4+4a^3x+6a^2x^2+4ax^3+x^4\\
(a+x)^5 &= (a+x)^4(a+x) = a^5+5a^4x+10a^3x^2+10a^2x^3+5ax^4+x^5\\
(a+x)^6 &= (a+x)^5(a+x) = a^6+6a^5x+15a^4x^2+20a^3x^3+15a^2x^4+6ax^5+x^6
\end{aligned}
$$

3 From the results of para. 2 the following patterns emerge:

(i) 'a' decreases in power moving from left to right.

(ii) 'x' increases in power moving from left to right.

(iii) The coefficients of each term of the expansions are symmetrical about the middle coefficient when n is even and symmetrical about the two middle coefficients when n is odd.

(iv) The coefficients are shown separately in *Table 1* and this arrangement is known as **Pascal's triangle.** A coefficient of a term may be obtained by adding the two adjacent coefficients immediately above in the previous row. This is shown by the triangles in *Table 1*, where, for example, $1+3 = 4$, $10+5 = 15$, and so on.

(v) Pascal's triangle method is used for expansion of the form $(a+x)^n$ for integer values of n less than about 8. The way in which Pascal's triangle can be used to obtain a binomial expansion is shown in *Problems 1 and 2.*

4 The **binomial theorem** is a formula for raising a binomial expression to any power without lengthy multiplication. The general binomial expansion of $(a+x)^n$ is given by:

$$(a+x)^n = a^n + na^{n-1}x + \frac{n(n-1)}{2!}a^{n-2}x^2 + \frac{n(n-1)(n-2)}{3!}a^{n-3}x^3 + \ldots\ldots + x^n$$

where 3! denotes $3 \times 2 \times 1$ and is termed 'factorial 3'.

With the binomial theorem n may be a fraction, a decimal fraction or a positive or negative integer. (See *Problems 3 and 4.*)

TABLE 1

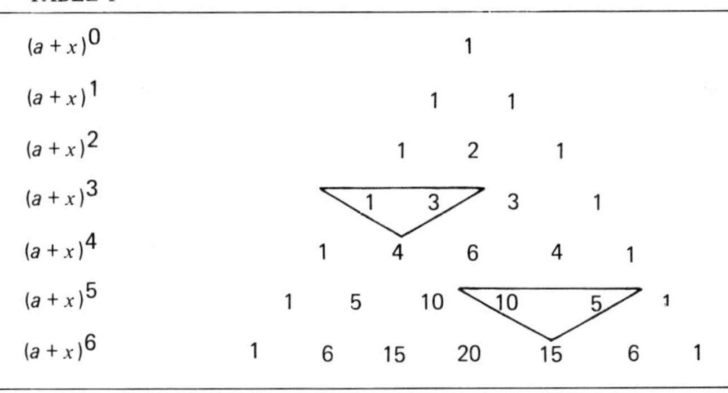

$(a+x)^0$	1
$(a+x)^1$	1 1
$(a+x)^2$	1 2 1
$(a+x)^3$	1 3 3 1
$(a+x)^4$	1 4 6 4 1
$(a+x)^5$	1 5 10 10 5 1
$(a+x)^6$	1 6 15 20 15 6 1

5 The r'th term of the expansion $(a+x)^n$ is:

$$\frac{n(n-1)(n-2)\ldots\ldots \text{ to } (r-1) \text{ terms}}{(r-1)!}\, a^{n-(r-1)}x^{r-1}$$

(see *Problems 5 and 6*).

6 If $a = 1$ in the binomial expansion of $(a+x)^n$ then:

$$(1+x)^n = 1+nx+\frac{n(n-1)}{2!}x^2+\frac{n(n-1)(n-2)}{3!}x^3+\ldots\ldots\ldots,$$

which is valid for $-1<x<1$.

When x is small compared with 1 then: $(1+x)^n \simeq 1+nx$.

(See *Problems 7 to 9.*)

7 Binomial expansions may be used for numerical approximations, for calculations with small variations and in probability theory (see chapter 24). (See *Problems 10 and 11.*)

B WORKED PROBLEMS ON THE BINOMIAL THEOREM

Problem 1 Use the Pascal's triangle method to determine the expansion of $(a+x)^7$.

From *Table 1*, the row of Pascal's triangle corresponding to $(a+x)^6$ is as shown in (1) below. Adding adjacent coefficients gives the coefficients of $(a+x)^7$ as shown in (2) below.

	1	6	15	20	15	6	1	(1)
1	7	21	35	35	21	7	1	(2)

The first and last terms of the expansion of $(a+x)^7$ are a^7 and x^7 respectively. The powers of 'a' decrease and the powers of 'x' increase moving from left to right.

Hence $\mathbf{(a+x)^7 = a^7+7a^6x+21a^5x^2+35a^4x^3+35a^3x^4+21a^2x^5+7ax^6+x^7}$.

Problem 2 Determine, using Pascal's triangle method, the expansion of $(2p-3q)^5$.

Comparing $(2p-3q)^5$ with $(a+x)^5$ shows that $a = 2p$ and $x = -3q$.
Using Pascal's triangle method:

$$(a+x)^5 = a^5 + 5a^4x + 10a^3x^2 + 10a^2x^3$$

Hence $(2p-3q)^5 = (2p)^5 + 5(2p)^4(-3q) + 10(2p)^3(-3q)^2 + 10(2p)^2(-3q)^3$
$+5(2p)(-3q)^4 + (-3q)^5$

i.e. $\mathbf{(2p-3q)^5 = 32p^5 - 240p^4q + 720p^3q^2 - 1080p^2q^3 + 810pq^4 - 243q^5}$

Problem 3 Use the binomial theorem to determine the expansion of $(2+x)^7$.

The binomial expansion of $(a+x)^n = a^n + na^{n-1}x + \frac{n(n-1)}{2!}a^{n-2}x^2$
$+ \frac{n(n-1)(n-2)}{3!}a^{n-3}x^3 + \ldots$

When $a = 2$ and $n = 7$:

$$(2+x)^7 = 2^7 + 7(2)^6x + \frac{(7)(6)}{(2)(1)}(2)^5x^2 + \frac{(7)(6)(5)}{(3)(2)(1)}(2)^4x^3 + \frac{(7)(6)(5)(4)}{(4)(3)(2)(1)}(2)^3x^4$$
$$+ \frac{(7)(6)(5)(4)(3)}{(5)(4)(3)(2)(1)}(2)^2x^5 + \frac{(7)(6)(5)(4)(3)(2)}{(6)(5)(4)(3)(2)(1)}(2)x^6$$
$$+ \frac{(7)(6)(5)(4)(3)(2)(1)}{(7)(6)(5)(4)(3)(2)(1)}x^7$$

i.e. $\mathbf{(2+x)^7 = 128 + 448x + 672x^2 + 560x^3 + 280x^4 + 84x^5 + 14x^6 + x^7}$.

Problem 4 Expand $\left(c - \frac{1}{c}\right)^5$ using the binomial theorem.

Using the binomial expansion of $(a+x)^n$, where $a = c$, $x = -\frac{1}{c}$ and $n = 5$ gives:

$$\left(c - \frac{1}{c}\right)^5 = c^5 + 5c^4\left(-\frac{1}{c}\right) + \frac{(5)(4)}{(2)(1)}c^3\left(-\frac{1}{c}\right)^2 + \frac{(5)(4)(3)}{(3)(2)(1)}c^2\left(-\frac{1}{c}\right)^3$$
$$+ \frac{(5)(4)(3)(2)}{(4)(3)(2)(1)}c\left(-\frac{1}{c}\right)^4 + \frac{(5)(4)(3)(2)(1)}{(5)(4)(3)(2)(1)}\left(-\frac{1}{c}\right)^5$$

i.e. $\mathbf{\left(c - \frac{1}{c}\right)^5 = c^5 - 5c^3 + 10c - \frac{10}{c} + \frac{5}{c^3} - \frac{1}{c^5}}$

Problem 5 Without fully expanding $(3+x)^7$, determine the fifth term.

The r'th term of the expansion $(a+x)^n$ is given by:

$$\frac{n(n-1)(n-2)\ldots\ldots \text{to } (r-1) \text{ terms}}{(r-1)!}a^{n-(r-1)}x^{r-1}.$$

Substituting $n = 7$, $a = 3$ and $r-1 = 5-1 = 4$ gives: $\frac{(7)(6)(5)(4)}{(4)(3)(2)(1)}(3)^{7-4}x^4$

i.e. the fifth term of $(3+x)^7 = 35(3)^3x^4 = \mathbf{945x^4}$

Problem 6 Find the middle term of $\left(2p - \frac{1}{2q}\right)^{10}$.

In the expansion of $(a+x)^{10}$ there are 10+1, i.e. 11 terms. Hence the middle term is the sixth. Using the general expression for the r'th term where $a = 2p$, $x = -\frac{1}{2q}$, $n = 10$ and $r-1 = 5$ gives:

$$\frac{(10)(9)(8)(7)(6)}{(5)(4)(3)(2)(1)}(2p)^{10-5}\left(-\frac{1}{2q}\right)^5 = 252(32p^5)\left(-\frac{1}{32q^5}\right)$$

Hence the middle term of $\left(2p - \frac{1}{2q}\right)^{10}$ is $\mathbf{-252\frac{p^5}{q^5}}$

Problem 7 Evaluate $(1.002)^9$ using the binomial theorem correct to (a) 3 decimal places and (b) 7 significant figures.

$$(1+x)^n = 1+nx + \frac{n(n-1)}{2!}x^2 + \frac{n(n-1)(n-2)}{3!}x^3 + \ldots\ldots\ldots\ldots$$

$(1.002)^9 = (1+0.002)^9$. Substituting $x = 0.002$ and $n = 9$ in the general expansion for $(1+x)^n$ gives:

$$\begin{aligned}(1+0.002)^9 &= 1+9(0.002) + \frac{(9)(8)}{(2)(1)}(0.002)^2 + \frac{(9)(8)(7)}{(3)(2)(1)}(0.002)^3 + \ldots\\ &= 1+0.018+0.000144+0.000000672+\ldots\ldots\\ &= 1.018144672\ldots\ldots\end{aligned}$$

Hence $(1.002)^9$ = **1.018, correct to 3 decimal places**

= **1.018145, correct to 7 significant figures.**

Problem 8 Evaluate $(0.97)^6$ correct to 4 significant figures using the binomial expansion.

$(0.97)^6$ is written as $(1-0.03)^6$. Using the expansion of $(1+x)^n$ where $n = 6$ and $x = -0.03$ gives:

$$\begin{aligned}(1-0.03)^6 &= 1+6(-0.03)+\frac{(6)(5)}{(2)(1)}(-0.03)^2 + \frac{(6)(5)(4)}{(3)(2)(1)}(-0.03)^3\\ &\quad + \frac{(6)(5)(4)(3)}{(4)(3)(2)(1)}(-0.03)^4 + \ldots\\ &= 1-0.18+0.0135-0.00054+0.00001215-\ldots\ldots\\ &\simeq 0.83297215\end{aligned}$$

i.e. $\mathbf{(0.97)^6}$ = **0.8330, correct to 4 significant figures.**

Problem 9 Determine the value of $(3.039)^4$, correct to 6 significant figures using the binomial theorem.

$(3.039)^4$ may be written in the form $(1+x)^n$ as:

$$(3.039)^4 = (3+0.039)^4 = \left[3\left(1+\frac{0.039}{3}\right)\right]^4 = 3^4(1+0.013)^4$$

$$\begin{aligned}(1+0.013)^4 &= 1+4(0.013)+\frac{(4)(3)}{(2)(1)}(0.013)^2 + \frac{(4)(3)(2)}{(3)(2)(1)}(0.013)^3 + \ldots\ldots\\ &= 1+0.052+0.001014+0.000008788+\ldots\ldots\\ &= 1.0530228 \text{ correct to 8 significant figures.}\end{aligned}$$

Hence $(3.039)^4 = 3^4(1.0530228)$ = **85.2948, correct to 6 significant figures.**

Problem 10 The radius of a cylinder is reduced by 4% and its height is increased by 2%. Determine the approximate percentage change in (a) its volume and (b) its curved surface area, (neglecting the products of small quantities).

Volume of cylinder $= \pi r^2 h$

Let r and h be the original values of radius and height.

The new values are $0.96r$ or $(1-0.04)r$ and $1.02h$ or $(1+0.02)h$

(a) New volume $= \pi[(1-0.04)r]^2 [(1+0.02)h]$

$= \pi r^2 h(1-0.04)^2 (1+0.02)$

Now $(1-0.04)^2 = 1-(2 \times 0.04)+(0.04)^2 = (1-0.08)$, neglecting powers of small terms.

Hence new volume $\simeq \pi r^2 h(1-0.08)(1+0.02)$

$\simeq \pi r^2 h(1-0.08+0.02)$, neglecting products of small terms

$\simeq \pi r^2 h(1-0.06)$ or $0.94\pi r^2 h$, i.e., 94% of the original volume.

Hence the volume is reduced by approximately 6%.

(b) Curved surface area of cylinder $= 2\pi rh$

New surface area $= 2\pi[(1-0.04)r] [(1+0.02)h]$

$= 2\pi rh(1-0.04)(1+0.02)$

$\simeq 2\pi rh(1-0.04+0.02)$, neglecting products of small terms

$\simeq 2\pi rh(1-0.02)$ or $0.98(2\pi rh)$, i.e., 98% of the original surface area.

Hence the curved surface area is reduced by approximately 2%.

Problem 11 The second moment of area of a rectangle through its centroid is given by $bl^3/12$. Determine the approximate change in the second moment of area if b is increased by 3.5% and l is reduced by 2.5%.

New values of b and l are $(1+0.035)b$ and $(1-0.025)l$ respectively.

New second moment of area $= \frac{1}{12}[(1+0.035)b] [(1-0.025)l]^3$

$= \frac{bl^3}{12}(1+0.035)(1-0.025)^3$

$\simeq \frac{bl}{12}(1+0.035)(1-0.075)$, neglecting powers of small terms

$\simeq \frac{bl^3}{12}(1+0.035-0.075)$, neglecting products of small terms

$\simeq \frac{bl^3}{12}(1-0.040)$ or $(0.96)\frac{bl^3}{12}$, i.e. 96% of the original second moment of area

Hence the second moment of area is reduced by approximately 4%.

C. FURTHER PROBLEMS ON THE BINOMIAL THEOREM

1 Use Pascal's triangle to expand $(x-y)^7$.

[$x^7-7x^6y+21x^5y^2-35x^4y^3+35x^3y^4-21x^2y^5+7xy^6-y^7$]

2 Expand $(2a+3b)^5$ using Pascal's triangle.

[$32a^5+240a^4b+720a^3b^2+1080a^2b^3+810ab^4+243b^5$]

3 Use the binomial theorem to expand (a) $(a+2x)^4$; (b) $(2-x)^6$.

[(a) $a^4+8a^3x+24a^2x^2+32ax^3+16x^4$;
(b) $64-192x+240x^2-160x^3+60x^4-12x^5+x^6$]

4 Expand (a) $(2x-3y)^4$; (b) $\left(2x+\frac{2}{x}\right)^5$.

$$\left[\begin{array}{l}\text{(a) } 16x^4-96x^3y+216x^2y^2-216xy^3+81y^4;\\ \text{(b) } 32x^5+160x^3+320x+\dfrac{320}{x}+\dfrac{160}{x^3}+\dfrac{32}{x^5}\end{array}\right]$$

5 Expand $(p+2q)^{11}$ as far as the fifth term.
[$p^{11}+22p^{10}q+220p^9q^2+1320p^8q^3+5280p^7q^4$]

6 Determine the sixth term of $(3p+\frac{q}{3})^{13}$ [$34749p^8q^5$]

7 Determine the middle term of $(2a-5b)^8$. [$700\,000a^4b^4$]

8 Use the binomial theorem to determine, correct to 4 decimal places:
(a) $(1.003)^8$; (b) $(1.042)^7$ [(a) 1.0243; (b) 1.3337]

9 Use the binomial theorem to determine, correct to 5 significant figures:
(a) $(0.98)^7$; (b) $(2.01)^9$ [(a) 0.86813; (b) 535.51]

10 Evaluate $(4.044)^6$ correct to 3 decimal places. [4373.880]

11 Pressure p and volume v are related by $pv^3 = c$, where c is a constant. Determine the approximate percentage change in c when p is increased by 3% and v decreased by 1.2%. [0.6% decrease]

12 Kinetic energy is given by $\frac{1}{2}mv^2$. Determine the approximate change in the kinetic energy when mass m is increased by 2.5% and the velocity v is reduced by 3%. [3.5% decrease]

13 An error of +1.5% was made when measuring the radius of a sphere. Ignoring the products of small quantities determine the approximate error in calculating (a) the volume, and (b) the surface area. [(a) 4.5% increase; (b) 3.0% increase]

14 The power developed by an engine is given by $I = k\,PLAN$, where k is a constant. Determine the approximate percentage change in the power when P and A are each increased by 2.5% and L and N are each decreased by 1.4%.
[2.2% increase]

15 The radius of a cone is increased by 2.7% and its height reduced by 0.9%. Determine the approximate percentage change in its volume, neglecting the products of small terms. [4.5% increase]

5 Algebra (2) – Partial fractions

A. MAIN POINTS CONCERNED WITH PARTIAL FRACTIONS

1. By algebraic addition, $\dfrac{1}{x-2}+\dfrac{3}{x+1}=\dfrac{(x+1)+3(x-2)}{(x-2)(x+1)}=\dfrac{4x-5}{x^2-x-2}$.

   The reverse process of moving from $\dfrac{4x-5}{x^2-x-2}$ to $\dfrac{1}{x-2}+\dfrac{3}{x+1}$ is called resolving into **partial fractions.**
2. In order to resolve an algebraic expression into partial fractions:
   (i) the denominator must factorise (in the above example, x^2-x-2 factorises as $(x-2)(x+1)$), and
   (ii) the numerator must be at least one degree less than the denominator (in the above example $(4x-5)$ is of degree 1 since the highest powered x term is x^1, and (x^2-x-2) is of degree 2).
3. When the degree of the numerator is equal to or higher than the degree of the denominator, the numerator must be divided by the denominator until the remainder is of less degree than the denominator (see *Problems 3 and 4*).
4. There are basically three types of partial fraction and the form of partial fraction used is summarised below, where $f(x)$ is assumed to be of less degree than the relevant denominator and A, B and C are constants to be determined.

Type	*Denominator containing*	*Expression*	*Form of partial fraction*
1	Linear factors (see *Problems 1 to 4*)	$\dfrac{f(x)}{(x+a)(x-b)(x+c)}$	$\dfrac{A}{(x+a)}+\dfrac{B}{(x-b)}+\dfrac{C}{(x+c)}$
2	Repeated linear factors (see *Problems 5 to 7*)	$\dfrac{f(x)}{(x+a)^3}$	$\dfrac{A}{(x+a)}+\dfrac{B}{(x+a)^2}+\dfrac{C}{(x+a)^3}$
3	Quadratic factors (see *Problems 8 and 9*)	$\dfrac{f(x)}{(ax^2+bx+c)(x+d)}$	$\dfrac{Ax+B}{(ax^2+bx+c)}+\dfrac{C}{(x+d)}$

(In the latter type, ax^2+bx+c is a quadratic expression which does not factorise without containing surds or imaginary terms.)

5. Resolving an algebraic expression into partial fractions is used as a preliminary to integrating certain functions.

B WORKED PROBLEMS ON PARTIAL FRACTIONS

Problem 1 Resolve $\dfrac{11-3x}{x^2+2x-3}$ into partial fractions.

The denominator factorises as $(x-1)(x+3)$ and the numerator is of less degree than the denominator. Thus $\dfrac{11-3x}{x^2+2x-3}$ may be resolved into partial fractions.

Let $\dfrac{11-3x}{(x-1)(x+3)} \equiv \dfrac{A}{(x-1)} + \dfrac{B}{(x+3)}$, where A and B are constants to be determined,

i.e. $\dfrac{11-3x}{(x-1)(x+3)} \equiv \dfrac{A(x+3)+B(x-1)}{(x-1)(x+3)}$, by algebraic addition.

Since the denominators are the same on each side of the identity then the numerators are equal to each other.
Thus, $11-3x \equiv A(x+3)+B(x-1)$.
To determine constants A and B, values of x are chosen to make the term in A or B equal to zero.
When $x = 1$, then $11-3(1) \equiv A(1+3)+B(0)$
i.e. $8 = 4A$
i.e. $A = 2$
When $x = -3$, then $11-3(-3) \equiv A(0)+B(-3-1)$
i.e. $20 = -4B$
i.e. $B = -5$

$$\text{Thus } \frac{11-3x}{x^2+2x-3} \equiv \frac{2}{(x-1)} + \frac{-5}{(x+3)} \equiv \mathbf{\frac{2}{(x-1)} - \frac{5}{(x+3)}}$$

$$\text{Check: } \frac{2}{x-1} - \frac{5}{x+3} = \frac{2(x+3)-5(x-1)}{(x-1)(x+3)} = \frac{11-3x}{x^2+2x-3}$$

Problem 2 Convert $\dfrac{2x^2-9x-35}{(x+1)(x-2)(x+3)}$ into the sum of three partial fractions.

$$\text{Let } \frac{2x^2-9x-35}{(x+1)(x-2)(x+3)} \equiv \frac{A}{(x+1)} + \frac{B}{(x-2)} + \frac{C}{(x+3)}$$

$$\equiv \frac{A(x-2)(x+3)+B(x+1)(x+3)+C(x+1)(x-2)}{(x+1)(x-2)(x+3)},$$

by algebraic addition.
Equating the numerators gives: $2x^2-9x-35 \equiv A(x-2)(x+3)+B(x+1)(x+3)$
$+C(x+1)(x-2)$

Let $x = -1$. Then $2(-1)^2-9(-1)-35 \equiv A(-3)(2)+B(0)(2)+C(0)(-3)$

i.e. $-24 = -6A$

i.e. $A = \dfrac{-24}{-6} = \mathbf{4}$

Let $x = 2$. Then $2(2)^2-9(2)-35 \equiv A(0)(5)+B(3)(5)+C(3)(0)$

i.e. $-45 = 15B$

i.e. $B = \dfrac{-45}{15} = \mathbf{-3}$

Let $x = -3$. Then $2(-3)^2 - 9(-3) - 35 \equiv A(-5)(0) + B(-2)(0) + C(-2)(-5)$

i.e. $\quad 10 = 10C$

i.e. $\quad C = 1$

Thus $\displaystyle \frac{2x^2-9x-35}{(x+1)(x-2)(x+3)} \equiv \frac{4}{(x+1)} - \frac{3}{(x-2)} + \frac{1}{(x+3)}$

Problem 3 Resolve $\dfrac{x^2+1}{x^2-3x+2}$ into partial fractions.

The denominator is of the same degree as the numerator. Thus dividing out gives:

$$
\begin{array}{r|l}
 & \quad 1 \\
\cline{2-2}
x^2-3x+2 & x^2 \qquad +1 \\
 & \underline{x^2-3x+2} \\
 & \qquad 3x-1
\end{array}
$$

Hence $\displaystyle \frac{x^2+1}{x^2-3x+2} \equiv 1 + \frac{3x-1}{x^2-3x+2} \equiv 1 + \frac{3x-1}{(x-1)(x-2)}$

Let $\displaystyle \frac{3x-1}{(x-1)(x-2)} \equiv \frac{A}{(x-1)} + \frac{B}{(x-2)} \equiv \frac{A(x-2)+B(x-1)}{(x-1)(x-2)}$

Equating numerators gives: $\quad 3x-1 \equiv A(x-2)+B(x-1)$

Let $x = 1$. Then $2 = -A$

i.e. $\quad A = -2$

Let $x = 2$. Then $5 = B$

Hence $\displaystyle \frac{3x-1}{(x-1)(x-2)} \equiv \frac{-2}{(x-1)} + \frac{5}{(x-2)}$

Thus $\displaystyle \frac{x^2+1}{x^2-3x+2} \equiv 1 - \frac{2}{(x-1)} + \frac{5}{(x-2)}$

Problem 4 Express $\dfrac{x^3-2x^2-4x-4}{x^2+x-2}$ in partial fractions.

The numerator is of higher degree than the denominator. Thus dividing out gives:

$$
\begin{array}{r|l}
 & x \;-\; 3 \\
\cline{2-2}
x^2+x-2 & x^3-2x^2-4x-4 \\
 & \underline{x^3+x^2\;\;-2x} \\
 & \quad -3x^2-2x-4 \\
 & \quad \underline{-3x^2-3x+6} \\
 & \qquad\qquad x-10
\end{array}
$$

Thus $\displaystyle \frac{x^3-2x^2-4x-4}{x^2+x-2} \equiv x-3+\frac{x-10}{x^2+x-2} \equiv x-3+\frac{x-10}{(x+2)(x-1)}$

Let $\displaystyle \frac{x-10}{(x+2)(x-1)} \equiv \frac{A}{(x+2)} + \frac{B}{(x-1)} \equiv \frac{A(x-1)+B(x+2)}{(x+2)(x-1)}$

Equating the numerators gives: $x-10 \equiv A(x-1)+B(x+2)$

Let $x = -2$. Then $-12 = -3A$

i.e. $\quad A = 4$

Let $x = 1$. Then $-9 = 3B$

i.e. $\quad B = -3$

Hence $\dfrac{x-10}{(x+2)(x-1)} \equiv \dfrac{4}{(x+2)} - \dfrac{3}{(x-1)}$

Thus $\mathbf{\dfrac{x^3-2x^2-4x-4}{x^2+x-2} \equiv x-3+\dfrac{4}{(x+2)} - \dfrac{3}{(x-1)}}$

Problem 5 Resolve $\dfrac{2x+3}{(x-2)^2}$ into partial fractions.

The denominator contains a repeated linear factor, $(x-2)^2$.

Let $\dfrac{2x+3}{(x-2)^2} \equiv \dfrac{A}{(x-2)} + \dfrac{B}{(x-2)^2} \equiv \dfrac{A(x-2)+B}{(x-2)^2}$

Equating the numerators gives: $2x+3 \equiv A(x-2)+B$
Let $x = 2$. Then $7 = A(0)+B$ i.e. $\mathbf{B = 7}$
$2x+3 \equiv A(x-2)+B \equiv Ax-2A+B$
Since an identity is true for all values of the unknown, the coefficients of similar terms may be equated.
Hence, equating the coefficients of x gives: $2 = \mathbf{A}$
[Also, as a check, equating the constant terms gives: $3 = -2A+B$
When $A = 2, B = 7$, R.H.S. $= -2(2)+7 = 3 =$ L.H.S.]

Hence $\mathbf{\dfrac{2x+3}{(x-2)^2} \equiv \dfrac{2}{(x-2)} + \dfrac{7}{(x-2)^2}}$

Problem 6 Express $\dfrac{5x^2-2x-19}{(x+3)(x-1)^2}$ as the sum of three partial fractions.

The denominator is a combination of a linear factor and a repeated factor.

Let $\dfrac{5x^2-2x-19}{(x+3)(x-1)^2} \equiv \dfrac{A}{(x+3)} + \dfrac{B}{(x-1)} + \dfrac{C}{(x-1)^2}$

$\equiv \dfrac{A(x-1)^2+B(x+3)(x-1)+C(x+3)}{(x+3)(x-1)^2}$, by algebraic addition.

Equating the numerators gives: $5x^2-2x-19 \equiv A(x-1)^2+B(x+3)(x-1)+C(x+3)$ (1)

Let $x = -3$. Then $5(-3)^2-2(-3)-19 \equiv A(-4)^2+B(0)(-4)+C(0)$
i.e. $32 = 16A$; i.e. $\mathbf{A = 2}$
Let $x = 1$. Then $5(1)^2-2(1)-19 \equiv A(0)^2+B(4)(0)+C(4)$
i.e. $-16 = 4C$; i.e. $\mathbf{C = -4}$
Without expanding the RHS of equation (1) it can be seen that equating the coefficients of x^2 gives: $5 = A+B$; Since $A = 2$, $\mathbf{B = 3}$.
[*Check*: Identity (1) may be expressed as:
$5x^2-2x-19 \equiv A(x^2-2x+1)+B(x^2+2x-3)+C(x+3)$
i.e. $5x^2-2x-19 \equiv Ax^2-2AX+A+Bx^2+2Bx-3B+Cx+3C$

Equating the x term coefficients gives: $-2 \equiv -2A+2B+C$
When $A = 2$, $B = 3$ and $C = -4$ then $-2A+2B+C = -2(2)+2(3)-4 = -2 =$ LHS
Equating the constant terms gives: $-19 \equiv A-3B+3C$
RHS $= 2-3(3)+3(-4) = 2-9-12 = -19 =$ LHS]

Hence $\mathbf{\dfrac{5x^2-2x-19}{(x+3)(x-1)^2} \equiv \dfrac{2}{(x+3)} + \dfrac{3}{(x-1)} - \dfrac{4}{(x-1)^2}}$

Problem 7 Resolve $\dfrac{3x^2+16x+15}{(x+3)^3}$ into partial fractions.

Let $\dfrac{3x^2+16x+15}{(x+3)^3} \equiv \dfrac{A}{(x+3)} + \dfrac{B}{(x+3)^2} + \dfrac{C}{(x+3)^3} \equiv \dfrac{A(x+3)^2+B(x+3)+C}{(x+3)^3}$

Equating the numerators gives: $3x^2+16x+15 \equiv A(x+3)^2+B(x+3)+C$ (1)

Let $x = -3$. Then $3(-3)^2+16(-3)+15 \equiv A(0)^2+B(0)+C$

i.e. $\mathbf{-6 = C}$

Identity (1) may be expanded as: $3x^2+16x+15 \equiv A(x^2+6x+9)+B(x+3)+C$

i.e. $3x^2+16x+15 \equiv Ax^2+6Ax+9A+Bx+3B+C$

Equating the coefficients of x^2 terms gives: $\mathbf{3 = A}$

Equating the coefficients of x terms gives: $16 = 6A+B$; Since $A = 3$, $\mathbf{B = -2}$.

[*Check*: equating the constant terms gives: $15 = 9A+3B+C$

When $A = 3$, $B = -2$ and $C = -6$, $9A+3B+C = 9(3)+3(-2)+(-6)$

$= 27-6-6 = 15 =$ LHS]

Thus $\mathbf{\dfrac{3x^2+16x+15}{(x+3)^3} \equiv \dfrac{3}{(x+3)} - \dfrac{2}{(x+3)^2} - \dfrac{6}{(x+3)^3}}$

Problem 8 Express $\dfrac{7x^2+5x+13}{(x^2+2)(x+1)}$ in partial fractions.

The denominator is a combination of a quadratic factor, (x^2+2), (which does not factorise without introducing imaginary surd terms), and a linear factor, $(x+1)$.

Let $\dfrac{7x^2+5x+13}{(x^2+2)(x+1)} \equiv \dfrac{Ax+B}{(x^2+2)} + \dfrac{C}{(x+1)} \equiv \dfrac{(Ax+B)(x+1)+C(x^2+2)}{(x^2+2)(x+1)}$

Equating numerators gives: $7x^2+5x+13 \equiv (Ax+B)(x+1)+C(x^2+2)$ (1)

Let $x = -1$. Then $7(-1)^2+5(-1)+13 \equiv (Ax+B)(0)+C(1+2)$

i.e. $15 = 3C$; i.e. $\mathbf{C = 5}$

Identity (1) may be expanded as: $7x^2+5x+13 \equiv Ax^2+Ax+Bx+B+Cx^2+2C$

Equating the coefficients of x^2 terms gives: $7 = A+C$; Since $C = 5$, $\mathbf{A = 2}$

Equating the coefficients of x terms gives: $5 = A+B$; Since $A = 2$, $\mathbf{B = 3}$

[*Check*: equating the constant terms gives: $13 = B+2C$

When $B = 3$ and $C = 5$, $B+2C = 3+10 = 13 =$ LHS]

Hence $\mathbf{\dfrac{7x^2+5x+13}{(x^2+2)(x+1)} \equiv \dfrac{2x+3}{(x^2+2)} + \dfrac{5}{(x+1)}}$

Problem 9 Resolve $\dfrac{3+6x+4x^2-2x^3}{x^2(x^2+3)}$ into three partial fractions.

Terms such as x^2 may be treated as $(x+0)^2$, i.e. they are repeated linear factors.

Let $\dfrac{3+6x+4x^2-2x^3}{x^2(x^2+3)} \equiv \dfrac{A}{x} + \dfrac{B}{x^2} + \dfrac{Cx+D}{(x^2+3)}$

$$\equiv \frac{Ax(x^2+3)+B(x^2+3)+(Cx+D)x^2}{x^2(x^2+3)}$$

Equating the numerators gives:

$$3+6x+4x^2-2x^3 \equiv Ax(x^2+3)+B(x^2+3)+(Cx+D)x^2$$
$$\equiv Ax^3+3Ax+Bx^2+3B+Cx^3+Dx^2$$

Let $x = 0$. Then $3 = 3B$

i.e. $\mathbf{B = 1}$

Equating the coefficients of x^3 terms gives: $-2 = A+C$ (1)

Equating the coefficients of x^2 terms gives: $4 = B+D$

Since $B = 1$, $\mathbf{D = 3}$

Equating the coefficients of x terms gives: $6 = 3A$

i.e. $\mathbf{A = 2}$

From equation (1), since $A = 2$, $\mathbf{C = -4}$

Hence $\dfrac{\mathbf{3+6x+4x^2-2x^3}}{\mathbf{x^2(x^2+3)}} \equiv \dfrac{2}{x}+\dfrac{1}{x^2}+\dfrac{-4x+3}{x^2+3} \equiv \dfrac{\mathbf{2}}{\mathbf{x}}+\dfrac{\mathbf{1}}{\mathbf{x^2}}+\dfrac{\mathbf{3-4x}}{\mathbf{x^2+3}}$

C FURTHER PROBLEMS ON PARTIAL FRACTIONS

Resolve the following into partial fractions.

1 $\dfrac{12}{x^2-9}$ $\left[\dfrac{2}{(x-3)}-\dfrac{2}{(x+3)}\right]$

2 $\dfrac{4(x-4)}{x^2-2x-3}$ $\left[\dfrac{5}{(x+1)}-\dfrac{1}{(x-3)}\right]$

3 $\dfrac{x^2-3x+6}{x(x-2)(x-1)}$ $\left[\dfrac{3}{x}+\dfrac{2}{(x-2)}-\dfrac{4}{(x-1)}\right]$

4 $\dfrac{3(2x^2-8x-1)}{(x+4)(x+1)(2x-1)}$ $\left[\dfrac{7}{(x+4)}-\dfrac{3}{(x+1)}-\dfrac{2}{(2x-1)}\right]$

5 $\dfrac{x^2+9x+8}{x^2+x-6}$ $\left[1+\dfrac{2}{(x+3)}+\dfrac{6}{(x-2)}\right]$

6 $\dfrac{x^2-x-14}{x^2-2x-3}$ $\left[1-\dfrac{2}{(x-3)}+\dfrac{3}{(x+1)}\right]$

7 $\dfrac{3x^3-2x^2-16x+20}{(x-2)(x+2)}$ $\left[3x-2+\dfrac{1}{(x-2)}-\dfrac{5}{(x+2)}\right]$

8 $\dfrac{4x-3}{(x+1)^2}$ $\left[\dfrac{4}{(x+1)}-\dfrac{7}{(x+1)^2}\right]$

9 $\dfrac{x^2+7x+3}{x^2(x+3)}$ $\left[\dfrac{1}{x^2}+\dfrac{2}{x}-\dfrac{1}{(x+3)}\right]$

10 $\dfrac{5x^2-30x+44}{(x-2)^3}$ $\left[\dfrac{5}{(x-2)}-\dfrac{10}{(x-2)^2}+\dfrac{4}{(x-2)^3}\right]$

11 $\dfrac{18+21x-x^2}{(x-5)(x+2)^2}$ $\left[\dfrac{2}{(x-5)}-\dfrac{3}{(x+2)}+\dfrac{4}{(x+2)^2}\right]$

12 $\dfrac{x^2-x-13}{(x^2+7)(x-2)}$ $\left[\dfrac{2x+3}{(x^2+7)}-\dfrac{1}{(x-2)}\right]$

13 $\dfrac{6x-5}{(x-4)(x^2+3)}$ $\left[\dfrac{1}{(x-4)}+\dfrac{2-x}{(x^2+3)}\right]$

14 $\dfrac{15+5x+5x^2-4x^3}{x^2(x^2+5)}$ $\left[\dfrac{1}{x}+\dfrac{3}{x^2}+\dfrac{2-5x}{(x^2+5)}\right]$

15 $\dfrac{x^3+4x^2+20x-7}{(x-1)^2(x^2+8)}$ $\left[\dfrac{3}{(x-1)}+\dfrac{2}{(x-1)^2}+\dfrac{1-2x}{(x^2+8)}\right]$

6 Algebra (3) – The theory of matrices and determinants

A MAIN POINTS CONCERNED WITH THE THEORY OF MATRICES AND DETERMINANTS

1 Matrices and determinants are mainly used at this level for the solution of linear simultaneous equations. The coefficients of the variables for linear simultaneous equations may be shown in matrix form. The coefficients of x and y in the simultaneous equations

$$\begin{aligned} x+2y &= 3 \\ 4x-5y &= 6 \end{aligned} \quad \text{become} \quad \begin{pmatrix} 1 & 2 \\ 4 & -5 \end{pmatrix} \quad \text{in matrix notation.}$$

Similarly, the coefficients of p, q and r in the equations

$$\begin{aligned} 1.3p-2.0q+r &= 7 \\ 3.7p+4.8q-7r &= 3 \\ -4.1p+3.8q+12r &= -6 \end{aligned} \quad \text{become} \quad \begin{pmatrix} 1.3 & -2.0 & 1 \\ 3.7 & 4.8 & -7 \\ -4.1 & 3.8 & 12 \end{pmatrix} \quad \text{in matrix form.}$$

2 The numbers within a matrix are called an **array** and the coefficients forming the array are called the **elements** of the matrix. The number of rows in a matrix is usually specified by m and the number of columns by n and a matrix referred to as an 'm by n' matrix. Thus, $\begin{pmatrix} 2 & 3 & 6 \\ 4 & 5 & 7 \end{pmatrix}$ is a '2 by 3' matrix.

3 Matrices cannot be expressed as a single numerical value, but they can often be simplified or combined, and unknown element values can be determined by comparison methods. Just as there are rules for addition, subtraction, multiplication and division of numbers in arithmetic, rules for these operations can be applied to matrices and the rules of matrices are such that they obey most of those governing the algebra of numbers.

4 (i) **Addition of matrices**

Corresponding elements in two matrices may be added to form a single matrix (see *Problem 1*).

(ii) **Subtraction of matrices**

If A is a matrix and B is another matrix, then $(A-B)$ is a single matrix formed by subtracting the elements of B from the corresponding elements of A (see *Problem 2*).

(iii) **Multiplication**

When a matrix is multiplied by a number, (called scalar multiplication), a single matrix results in which each element of the original matrix has been multiplied by the number, (see *Problem 3*). When a matrix A is multiplied by another matrix B, a single matrix results in which elements are obtained from the sum of the products of the corresponding rows of A and the corresponding columns of B.

Two matrices A and B may be multiplied together, provided the number of elements in the rows of matrix A are equal to the number of elements in the columns of matrix B. In general terms, when multiplying a matrix of dimensions (m by n) by a matrix of dimensions (n by r), the resulting matrix has dimensions (m by r). Thus a 2 by 3 matrix multiplied by a 3 by 1 matrix gives a matrix of dimensions 2 by 1. (See *Problems 4 to 7.*)

5 In algebra, the commutative law of multiplication states that $a \times b = b \times a$. For matrices, this law is only true in a few special cases, and in general $A \times B$ is **not** equal to $B \times A$. (See *Problem 8.*)

6 A **unit matrix,** I, is one which all elements of the leading diagonal (\) have a value of 1 and all other elements have a value of 0. Multiplication of a matrix by I is the equivalent of multiplying by 1 in arithmetic.

7 The **determinant** of a 2 by 2 matrix, $\begin{pmatrix} a & b \\ c & d \end{pmatrix}$ is defined as $(ad-bc)$.
The elements of the determinant of a matrix are written between vertical lines.
Thus, the determinant of $\begin{pmatrix} 3 & -4 \\ 1 & 6 \end{pmatrix}$ is written as $\begin{vmatrix} 3 & -4 \\ 1 & 6 \end{vmatrix}$
and is equal to $(3 \times 6)-(-4 \times 1)$, i.e. $18-(-4)$ or 22. Hence the determinant of a matrix can be expressed as a single numerical value,

i.e. $\begin{vmatrix} 3 & -4 \\ 1 & 6 \end{vmatrix} = 22.$

(See *Problem 9.*)

8 **The inverse or reciprocal of a 2 by 2 matrix**
The inverse of matrix A is A^{-1} such that $A \times A^{-1} = I$, the unit matrix.
Let matrix A be $\begin{pmatrix} 1 & 2 \\ 3 & 4 \end{pmatrix}$ and let the inverse matrix, A^{-1} be $\begin{pmatrix} a & b \\ c & d \end{pmatrix}$.
Then, since $A \times A^{-1} = I$,

$$\begin{pmatrix} 1 & 2 \\ 3 & 4 \end{pmatrix} \times \begin{pmatrix} a & b \\ c & d \end{pmatrix} = \begin{pmatrix} 1 & 0 \\ 0 & 1 \end{pmatrix}$$

Multiplying the matrices on the left side, (see *Problem 3*), gives

$$\begin{pmatrix} a+2c & b+2d \\ 3a+4c & 3b+4d \end{pmatrix} = \begin{pmatrix} 1 & 0 \\ 0 & 1 \end{pmatrix}$$

Equating corresponding elements gives:

$$b+2d = 0, \text{ i.e. } b = -2d$$

and $\quad 3a+4c = 0$, i.e. $a = -\frac{4}{3}c$.

Substituting for a and b gives:

$$\begin{pmatrix} -\frac{4}{3}c + 2c & -2d + 2d \\ 3(-\frac{4}{3}c) + 4c & 3(-2d) + 4d \end{pmatrix} = \begin{pmatrix} 1 & 0 \\ 0 & 1 \end{pmatrix},$$

i.e. $$\begin{pmatrix} \frac{2}{3}c & 0 \\ 0 & -2d \end{pmatrix} = \begin{pmatrix} 1 & 0 \\ 0 & 1 \end{pmatrix},$$

showing that $\frac{2}{3}c = 1$, i.e. $c = \frac{3}{2}$ and $-2d = 1$, i.e. $d = -\frac{1}{2}$.

Since $b = -2d$, $b = 1$ and since $a = -\frac{4}{3}c$, $a = -2$.

Thus the inverse of matrix $\begin{pmatrix} 1 & 2 \\ 3 & 4 \end{pmatrix}$ is $\begin{pmatrix} a & b \\ c & d \end{pmatrix}$

that is, $\begin{pmatrix} -2 & 1 \\ \frac{2}{3} & -\frac{1}{2} \end{pmatrix}$

There is, however, a quicker method of obtaining the inverse of a 2 by 2 matrix. For any matrix $\begin{pmatrix} p & q \\ r & s \end{pmatrix}$ the inverse may be obtained by:

(i) interchanging the positions of p and s;
(ii) changing the signs of q and r, and
(iii) multiplying this new matrix by the reciprocal of the determinant of $\begin{pmatrix} p & q \\ r & s \end{pmatrix}$.

Thus the inverse of matrix $\begin{pmatrix} 1 & 2 \\ 3 & 4 \end{pmatrix}$ is $\frac{1}{4-6}\begin{pmatrix} 4 & -2 \\ -3 & 1 \end{pmatrix} = \begin{pmatrix} -2 & 1 \\ \frac{3}{2} & -\frac{1}{2} \end{pmatrix}$
as obtained previously.
(See *Problem 10.*)

9 **The determinant of a 3 by 3 matrix**

(i) The **minor** of an element of a 3 by 3 matrix is the value of the 2 by 2 determinant obtained by covering up the row and column containing that element.

Thus for the matrix $\begin{pmatrix} 1 & 2 & 3 \\ 4 & 5 & 6 \\ 7 & 8 & 9 \end{pmatrix}$ the minor of element 4 is obtained by covering the row (4 5 6) and the column $\begin{pmatrix} 1 \\ 4 \\ 7 \end{pmatrix}$, leaving the 2 by 2 determinant $\begin{vmatrix} 2 & 3 \\ 8 & 9 \end{vmatrix}$, i.e., the minor of element 4 is $(2 \times 9)-(3 \times 8)$, i.e., -6.

(ii) The sign of a minor depends on its position within the matrix, the sign pattern being $\begin{pmatrix} + & - & + \\ - & + & - \\ + & - & + \end{pmatrix}$. Thus the signed-minor of element 4 in the matrix $\begin{pmatrix} 1 & 2 & 3 \\ 4 & 5 & 6 \\ 7 & 8 & 9 \end{pmatrix}$ is $-\begin{vmatrix} 2 & 3 \\ 8 & 9 \end{vmatrix} = -(-6) = 6$.

The signed-minor of an element is called the **cofactor** of the element.

(iii) The value of a 3 by 3 determinant is the sum of the products of the elements and their cofactors of **any row or any column** of the corresponding 3 by 3 matrix.

(See *Problem 11.*)

10 **The inverse or reciprocal of a 3 by 3 matrix**

The **adjoint** of a matrix A is obtained by:

(i) forming a matrix B of the cofactors of A, and
(ii) **transposing** matrix B to give B^T, where B^T is the matrix obtained by writing the rows of B as the columns of B^T. Then adj $A = B^T$. The inverse of matrix A, A^{-1} is given by $A^{-1} = \frac{\text{adj } A}{|A|}$,

where adj A is the adjoint of matrix A and $|A|$ is the determinant of matrix A.
(See *Problem 12.*)

B WORKED PROBLEMS ON THE THEORY OF MATRICES AND DETERMINANTS

Problem 1 Add the matrices (a) $\begin{pmatrix} 2 & -1 \\ -7 & 4 \end{pmatrix}$ and $\begin{pmatrix} -3 & 0 \\ 7 & -4 \end{pmatrix}$

and (b) $\begin{pmatrix} 3 & 1 & -4 \\ 4 & 3 & 1 \\ 1 & 4 & -3 \end{pmatrix}$ and $\begin{pmatrix} 2 & 7 & -5 \\ -2 & 1 & 0 \\ 6 & 3 & 4 \end{pmatrix}$

(a) Adding the corresponding elements gives:

$$\begin{pmatrix} 2 & -1 \\ -7 & 4 \end{pmatrix} + \begin{pmatrix} -3 & 0 \\ 7 & -4 \end{pmatrix} = \begin{pmatrix} 2+(-3) & -1+0 \\ -7+7 & 4+(-4) \end{pmatrix}$$

$$= \begin{pmatrix} -1 & -1 \\ 0 & 0 \end{pmatrix}$$

(b) Adding the corresponding elements gives:

$$\begin{pmatrix} 3 & 1 & -4 \\ 4 & 3 & 1 \\ 1 & 4 & -3 \end{pmatrix} + \begin{pmatrix} 2 & 7 & -5 \\ -2 & 1 & 0 \\ 6 & 3 & 4 \end{pmatrix} = \begin{pmatrix} 3+2 & 1+7 & -4+(-5) \\ 4+(-2) & 3+1 & 1+0 \\ 1+6 & 4+3 & -3+4 \end{pmatrix}$$

$$\begin{pmatrix} 5 & 8 & -9 \\ 2 & 4 & 1 \\ 7 & 7 & 1 \end{pmatrix}$$

Problem 2 Subtract (a) $\begin{pmatrix} -3 & 0 \\ 7 & -4 \end{pmatrix}$ from $\begin{pmatrix} 2 & -1 \\ -7 & 4 \end{pmatrix}$

and (b) $\begin{pmatrix} 2 & 7 & -5 \\ -2 & 1 & 0 \\ 6 & 3 & 4 \end{pmatrix}$ from $\begin{pmatrix} 3 & 1 & -4 \\ 4 & 3 & 1 \\ 1 & 4 & -3 \end{pmatrix}$

To find matrix A minus matrix B, the elements of B are taken from the corresponding elements of A. Thus:

$$\begin{pmatrix} 2 & -1 \\ -7 & 4 \end{pmatrix} - \begin{pmatrix} -3 & 0 \\ 7 & -4 \end{pmatrix} = \begin{pmatrix} 2-(-3) & -1-0 \\ -7-7 & 4-(-4) \end{pmatrix}$$

$$= \begin{pmatrix} 5 & -1 \\ -14 & 8 \end{pmatrix}$$

(b) $$\begin{pmatrix} 3 & 1 & -4 \\ 4 & 3 & 1 \\ 1 & 4 & -3 \end{pmatrix} - \begin{pmatrix} 2 & 7 & -5 \\ -2 & 1 & 0 \\ 6 & 3 & 4 \end{pmatrix} = \begin{pmatrix} 3-2 & 1-7 & -4-(-5) \\ 4-(-2) & 3-1 & 1-0 \\ 1-6 & 4-3 & -3-4 \end{pmatrix}$$

$$= \begin{pmatrix} 1 & -6 & 1 \\ 6 & 2 & 1 \\ -5 & 1 & -7 \end{pmatrix}$$

Problem 3 If $A=\begin{pmatrix} -3 & 0 \\ 7 & -4 \end{pmatrix}$, $B = \begin{pmatrix} 2 & -1 \\ -7 & 4 \end{pmatrix}$ and $C = \begin{pmatrix} 1 & 0 \\ -2 & -4 \end{pmatrix}$,
find (a) $A+B-C$ and (b) $2A-3B+4C$.

(a) $A+B = \begin{pmatrix} -1 & -1 \\ 0 & 0 \end{pmatrix}$, (from *Problem 1*)

Hence, $A+B-C = \begin{pmatrix}-1 & -1\\ 0 & 0\end{pmatrix} - \begin{pmatrix}1 & 0\\ -2 & -4\end{pmatrix}$

$$= \begin{pmatrix}-1-1 & -1-0\\ 0-(-2) & 0-(-4)\end{pmatrix} = \begin{pmatrix}-2 & -1\\ 2 & 4\end{pmatrix}$$

Alternatively $A+B-C = \begin{pmatrix}-3 & 0\\ 7 & -4\end{pmatrix} + \begin{pmatrix}2 & -1\\ -7 & 4\end{pmatrix} - \begin{pmatrix}1 & 0\\ -2 & -4\end{pmatrix}$

$$= \begin{pmatrix}-3+2-1 & 0+(-1)-0\\ 7+(-7)-(-2) & -4+4-(-4)\end{pmatrix}$$

$$= \begin{pmatrix}-2 & -1\\ 2 & 4\end{pmatrix}$$ as obtained previously.

(b) For scalar multiplication, each element is multiplied by the scalar quantity, hence

$$2A = 2\begin{pmatrix}-3 & 0\\ 7 & -4\end{pmatrix} = \begin{pmatrix}-6 & 0\\ 14 & -8\end{pmatrix}$$

$$3B = 3\begin{pmatrix}2 & -1\\ -7 & 4\end{pmatrix} = \begin{pmatrix}6 & -3\\ -21 & 12\end{pmatrix}$$

and $4C = 4\begin{pmatrix}1 & 0\\ -2 & -4\end{pmatrix} = \begin{pmatrix}4 & 0\\ -8 & -16\end{pmatrix}$

Hence $2A-3B+4C = \begin{pmatrix}-6 & 0\\ 14 & -8\end{pmatrix} - \begin{pmatrix}6 & -3\\ -21 & 12\end{pmatrix} + \begin{pmatrix}4 & 0\\ -8 & -16\end{pmatrix}$

$$= \begin{pmatrix}-6-6+4 & 0-(-3)+0\\ 14-(-21)+(-8) & -8-12+(-16)\end{pmatrix}$$

$$= \begin{pmatrix}-8 & 3\\ 27 & -36\end{pmatrix}$$

Problem 4 If $A = \begin{pmatrix}2 & 3\\ 1 & -4\end{pmatrix}$ and $B = \begin{pmatrix}-5 & 7\\ -3 & 4\end{pmatrix}$ find $A \times B$.

Let $A \times B = C$ where $C = \begin{pmatrix}C_{11} & C_{12}\\ C_{21} & C_{22}\end{pmatrix}$

C_{11} is the sum of the products of the first row of elements of A and the first column elements of B taken one at a time. Thus $C_{11}=(2 \times (-5))+(3 \times (-3))=-19$. C_{12} is the sum of the products of the first row elements of A and the second colum elements of B, taken one at a time, i.e., $C_{12} = (2 \times 7)+(3 \times 4) = 26$. C_{21} is the sum of the products of the second row elements of A and the first column elements of B, taken one at a time, i.e., $C_{21} = (1 \times (-5))+((-4) \times (-3)) = 7$. Finally, C_{22} is the sum of the products of the second row elements of A and the second column elements of B, taken one at a time, i.e. $C_{22} = (1 \times 7)+((-4) \times 4) = -9$.

Thus, $A \times B = \begin{pmatrix}-19 & 26\\ 7 & -9\end{pmatrix}$

Problem 5 Simplify $\begin{pmatrix}3 & 4 & 0\\ -2 & 6 & -3\\ 7 & -4 & 1\end{pmatrix} \times \begin{pmatrix}2\\ 5\\ -1\end{pmatrix}$

The sum of the products of the elements of each row of the first matrix and the elements of the second matrix, (called a column matrix), are taken one at a time. Thus:

$$\begin{pmatrix} 3 & 4 & 0 \\ -2 & 6 & -3 \\ 7 & -4 & 1 \end{pmatrix} \times \begin{pmatrix} 2 \\ 5 \\ -1 \end{pmatrix} = \begin{pmatrix} (3 \times 2) + (4 \times 5) + (0 \times (-1)) \\ (-2 \times 2) + (6 \times 5) + (-3 \times (-1)) \\ (7 \times 2) + (-4 \times 5) + (1 \times (-1)) \end{pmatrix}$$

$$= \begin{pmatrix} \mathbf{26} \\ \mathbf{29} \\ \mathbf{-7} \end{pmatrix}$$

Problem 6 If $A = \begin{pmatrix} 3 & 4 & 0 \\ -2 & 6 & -3 \\ 7 & -4 & 1 \end{pmatrix}$ and $B = \begin{pmatrix} 2 & -5 \\ 5 & -6 \\ -1 & -7 \end{pmatrix}$, find $A \times B$.

The sum of the products of the elements of each row of the first matrix and the elements of each column of the second matrix are taken one at a time. Thus:

$$\begin{pmatrix} 3 & 4 & 0 \\ -2 & 6 & -3 \\ 7 & -4 & 1 \end{pmatrix} \times \begin{pmatrix} 2 & -5 \\ 5 & -6 \\ -1 & -7 \end{pmatrix}$$

$$= \begin{pmatrix} [(3\times2)+(4\times5)+(0\times(-1))] & [(3\times(-5))+4\times(-6))+(0\times(-7))] \\ [(-2\times2)+(6\times5)+(-3\times(-1))] & [(-2\times(-5))+(6\times(-6))+(-3\times(-7))] \\ [(7\times2)+(-4\times5)+(1\times(-1))] & [(7\times(-5))+(-4\times(-6))+(1\times(-7))] \end{pmatrix}$$

$$= \begin{pmatrix} \mathbf{26} & \mathbf{-39} \\ \mathbf{29} & \mathbf{-5} \\ \mathbf{-7} & \mathbf{-18} \end{pmatrix}$$

Problem 7 Determine $\begin{pmatrix} 1 & 0 & 3 \\ 2 & 1 & 2 \\ 1 & 3 & 1 \end{pmatrix} \times \begin{pmatrix} 2 & 2 & 0 \\ 1 & 3 & 2 \\ 3 & 2 & 0 \end{pmatrix}$

The sum of the products of the elements of each row of the first matrix and the elements of each column of the second matrix are taken one at a time. Thus:

$$\begin{pmatrix} 1 & 0 & 3 \\ 2 & 1 & 2 \\ 1 & 3 & 1 \end{pmatrix} \times \begin{pmatrix} 2 & 2 & 0 \\ 1 & 3 & 2 \\ 3 & 2 & 0 \end{pmatrix} \quad \text{is equal to}$$

$$\begin{pmatrix} [(1\times2)+(0\times1)+(3\times3)] & [(1\times2)+(0\times3)+(3\times2)] & [(1\times0)+(0\times2)+(3\times0)] \\ [(2\times2)+(1\times1)+(2\times3)] & [(2\times2)+(1\times3)+(2\times2)] & [(2\times0)+(1\times2)+(2\times0)] \\ [(1\times2)+(3\times1)+(1\times3)] & [(1\times2)+(3\times3)+(1\times2)] & [(1\times0)+(3\times2)+(1\times0)] \end{pmatrix}$$

$$\text{i.e.,} \quad \begin{pmatrix} \mathbf{11} & \mathbf{8} & \mathbf{0} \\ \mathbf{11} & \mathbf{11} & \mathbf{2} \\ \mathbf{8} & \mathbf{13} & \mathbf{6} \end{pmatrix}$$

Problem 8 If $A = \begin{pmatrix} 2 & 3 \\ 1 & 0 \end{pmatrix}$ and $B = \begin{pmatrix} 2 & 3 \\ 0 & 1 \end{pmatrix}$ show that $A \times B \neq B \times A$.

$$A \times B = \begin{pmatrix} 2 & 3 \\ 1 & 0 \end{pmatrix} \times \begin{pmatrix} 2 & 3 \\ 0 & 1 \end{pmatrix} = \begin{pmatrix} [(2\times2)+(3\times0)] & [(2\times3)+(3\times1)] \\ [(1\times2)+(0\times0)] & [(1\times3)+(0\times1)] \end{pmatrix}$$

$$= \begin{pmatrix} 4 & 9 \\ 2 & 3 \end{pmatrix}$$

$$B \times A = \begin{pmatrix} 2 & 3 \\ 0 & 1 \end{pmatrix} \times \begin{pmatrix} 2 & 3 \\ 1 & 0 \end{pmatrix} = \begin{pmatrix} [(2\times2)+(3\times1)] & [(2\times3)+(3\times0)] \\ [(0\times2)+(1\times1)] & [(0\times3)+(1\times0)] \end{pmatrix}$$

$$= \begin{pmatrix} 7 & 6 \\ 1 & 0 \end{pmatrix}$$

Since $\begin{pmatrix} 4 & 9 \\ 2 & 3 \end{pmatrix} \neq \begin{pmatrix} 7 & 6 \\ 1 & 0 \end{pmatrix}$, then $A \times B \neq B \times A$.

Problem 9 Determine the value of $\begin{vmatrix} 3 & -2 \\ 7 & 4 \end{vmatrix}$

The vertical lines show that this is a determinant and the value of $\begin{vmatrix} p & q \\ r & s \end{vmatrix}$ is defined as $(p \times s)-(q \times r)$. (See para. 7.)

Thus $\begin{vmatrix} 3 & -2 \\ 7 & 4 \end{vmatrix} = (3 \times 4)-(-2 \times 7)$
$= 12-(-14) = \mathbf{26}$

Problem 10 Determine the inverse of $\begin{pmatrix} 3 & -2 \\ 7 & 4 \end{pmatrix}$

The inverse of matrix $\begin{pmatrix} p & q \\ r & s \end{pmatrix}$ is obtained by exchanging the positions of p and s, changing the signs of q and r and multiplying by the reciprocal of the determinant $\begin{vmatrix} p & q \\ r & s \end{vmatrix}$. Thus, the inverse of

$$\begin{pmatrix} 3 & -2 \\ 7 & 4 \end{pmatrix} = \frac{1}{(3\times4)-(-2\times7)} \begin{pmatrix} 4 & 2 \\ -7 & 3 \end{pmatrix} = \frac{1}{26} \begin{pmatrix} 4 & 2 \\ -7 & 3 \end{pmatrix}$$

$$= \begin{pmatrix} \frac{2}{13} & \frac{1}{13} \\ \frac{-7}{26} & \frac{3}{26} \end{pmatrix}$$

Problem 11 Find the value of $\begin{vmatrix} 3 & 4 & -1 \\ 2 & 0 & 7 \\ 1 & -3 & -2 \end{vmatrix}$

With reference to para. 9, the value of this determinant is the sum of the products of the elements and their cofactors, of any row or of any column. If the second row or second column is selected, the element 0 will make the product of the element and its cofactor zero and reduce the amount of arithmetic to be done to a minimum. Supposing a second row expansion is selected.

The minor of 2 is the value of the determinant remaining when the row and column containing the 2 (i.e. the second row and the first column), is covered up. Thus the cofactor of element 2 is $\begin{vmatrix} 4 & -1 \\ -3 & -2 \end{vmatrix}$ i.e., -11. The sign of element 2 is is minus, (see para. 9(ii)), hence the cofactor of element 2, (the signed-minor) is $+11$. Similarly the minor of element 7 is $\begin{vmatrix} 3 & 4 \\ 1 & -3 \end{vmatrix}$ i.e., -13, and its cofactor is $+13$. Hence the value of the sum of the products of the elements and their cofactors is $2 \times 11+7 \times 13$, i.e.

$$\begin{vmatrix} 3 & 4 & -1 \\ 2 & 0 & 7 \\ 1 & -3 & -2 \end{vmatrix} = 113$$

The same result will be obtained whichever row or column is selected. For example, the third column expansion is

$$+(-1)\begin{vmatrix} 2 & 0 \\ 1 & -3 \end{vmatrix} -7\begin{vmatrix} 3 & 4 \\ 1 & -3 \end{vmatrix} +(-2)\begin{vmatrix} 3 & 4 \\ 2 & 0 \end{vmatrix}$$

i.e., 6+91+16 = **113**, as obtained previously.

Problem 12 Determine the inverse of the matrix $\begin{pmatrix} 3 & 4 & -1 \\ 2 & 0 & 7 \\ 1 & -3 & -2 \end{pmatrix}$

The inverse of matrix A, $A^{-1} = \dfrac{\text{adj } A}{|A|}$

The adjoint of A is found by: (i) obtaining the matrix of the cofactors of the elements, and (ii) transposing this matrix.

The cofactor of element 3 is $+\begin{vmatrix} 0 & 7 \\ -3 & -2 \end{vmatrix}$, i.e. 21

The cofactor of element 4 is $-\begin{vmatrix} 2 & 7 \\ 1 & -2 \end{vmatrix}$, i.e. 11, and so on.

The matrix of cofactors is $\begin{pmatrix} 21 & 11 & -6 \\ 11 & -5 & 13 \\ 28 & -23 & -8 \end{pmatrix}$

The transpose of the matrix of cofactors, i.e. the adjoint of the matrix, is obtained by writing the rows as columns, and is $\begin{pmatrix} 21 & 11 & 28 \\ 11 & -5 & -23 \\ -6 & 13 & -8 \end{pmatrix}$

From *Problem 11*, the determinant of $\begin{vmatrix} 3 & 4 & -1 \\ 2 & 0 & 7 \\ 1 & -3 & -2 \end{vmatrix}$ is 113.

Hence the inverse of $\begin{pmatrix} 3 & 4 & -1 \\ 2 & 0 & 7 \\ 1 & -3 & -2 \end{pmatrix}$ is $\dfrac{\begin{pmatrix} 21 & 11 & 28 \\ 11 & -5 & -23 \\ -6 & 13 & -8 \end{pmatrix}}{113}$,

i.e. $\dfrac{1}{113}\begin{pmatrix} 21 & 11 & 28 \\ 11 & -5 & -23 \\ -6 & 13 & -8 \end{pmatrix}$

C FURTHER PROBLEMS ON THE THEORY OF MATRICES AND DETERMINANTS

In *Problems 1 to 40*, the matrices stated are:

$$A = \begin{pmatrix} 3 & -1 \\ -4 & 7 \end{pmatrix}, \quad B = \begin{pmatrix} \frac{1}{2} & \frac{2}{3} \\ -\frac{1}{3} & -\frac{3}{5} \end{pmatrix}, \quad C = \begin{pmatrix} -1.3 & 7.4 \\ 2.5 & -3.9 \end{pmatrix}$$

$$D = \begin{pmatrix} 4 & -7 & 6 \\ -2 & 4 & 0 \\ 5 & 7 & -4 \end{pmatrix}, \quad E = \begin{pmatrix} 3 & 6 & \frac{1}{2} \\ 5 & -\frac{2}{3} & 7 \\ -1 & 0 & \frac{3}{5} \end{pmatrix}, \quad F = \begin{pmatrix} 3.1 & 2.4 & 6.4 \\ -1.6 & 3.8 & -1.9 \\ 5.3 & 3.4 & -4.8 \end{pmatrix}$$

$$G = \begin{pmatrix} \frac{3}{4} \\ 1\frac{2}{5} \end{pmatrix}, \quad H = \begin{pmatrix} -2 \\ 5 \end{pmatrix}, \quad J = \begin{pmatrix} 4 \\ -11 \\ 7 \end{pmatrix}, \quad K = \begin{pmatrix} 1 & 0 \\ 0 & 1 \\ 1 & 0 \end{pmatrix}$$

In *Problems 1 to 17*, perform the matrix operation stated.

1 $A+B$ $\left[\begin{pmatrix} 3\frac{1}{2} & -\frac{1}{3} \\ -4\frac{1}{3} & 6\frac{2}{5} \end{pmatrix}\right]$

2 $D+E$ $\left[\begin{pmatrix} 7 & -1 & 6\frac{1}{2} \\ 3 & 3\frac{1}{3} & 7 \\ 4 & 7 & -3\frac{2}{5} \end{pmatrix}\right]$

3 $A-B$ $\left[\begin{pmatrix} 2\frac{1}{2} & -1\frac{2}{3} \\ -3\frac{2}{3} & 7\frac{3}{5} \end{pmatrix}\right]$

4 $D-E$ $\left[\begin{pmatrix} 1 & -13 & 5\frac{1}{2} \\ -7 & 4\frac{2}{3} & -7 \\ 6 & 7 & -4\frac{3}{5} \end{pmatrix}\right]$

5 $A+B-C$ $\left[\begin{pmatrix} 2.2 & 7.0\dot{6} \\ -6.8\dot{3} & 10.3 \end{pmatrix}\right]$

6 $D-E+F$ $\left[\begin{pmatrix} 4.1 & -10.6 & 11.9 \\ -8.6 & 8.4\dot{6} & -8.9 \\ 11.3 & 10.4 & -9.4 \end{pmatrix}\right]$

7 $5A+6B$ $\left[\begin{pmatrix} 18.0 & -1.0 \\ -22.0 & 31.4 \end{pmatrix}\right]$

8 $2D+3E-4F$ $\left[\begin{pmatrix} 4.6 & -5.6 & -12.1 \\ 17.4 & -9.2 & 28.6 \\ -14.2 & 0.4 & 13.0 \end{pmatrix}\right]$

9 $A \times H$ $\left[\begin{pmatrix} -11 \\ 43 \end{pmatrix}\right]$

10 $B \times G$ $\left[\begin{pmatrix} 1\frac{37}{120} \\ -1\frac{9}{100} \end{pmatrix}\right]$

11 $A \times B$ $\left[\begin{pmatrix} 1\frac{5}{6} & 2\frac{3}{5} \\ -4\frac{1}{3} & -6\frac{13}{15} \end{pmatrix}\right]$

12 $A \times C$ $\left[\begin{pmatrix} -6.4 & 26.1 \\ 22.7 & -56.9 \end{pmatrix}\right]$

13 $D \times J$ $\left[\begin{pmatrix} 135 \\ -52 \\ -85 \end{pmatrix}\right]$

14 $F \times J$ $\left[\begin{pmatrix} 30.8 \\ -61.5 \\ -49.8 \end{pmatrix}\right]$

15 $E \times K$ $\left[\begin{pmatrix} 3\frac{1}{2} & 6 \\ 12 & -\frac{2}{3} \\ -\frac{2}{5} & 0 \end{pmatrix}\right]$

16 $D \times E$ $\left[\begin{pmatrix} -29 & 28\frac{2}{3} & -43\frac{2}{5} \\ 14 & -14\frac{2}{3} & 27 \\ 54 & 25\frac{1}{3} & 49\frac{1}{10} \end{pmatrix}\right]$

17 $D \times F$ $\left[\begin{pmatrix} 55.4 & 3.4 & 10.1 \\ -12.6 & 10.4 & -20.4 \\ -16.9 & 25.0 & 37.9 \end{pmatrix}\right]$

18 Show that $A \times C \neq C \times A$.

$\left[A \times C = \begin{pmatrix} -6.4 & 26.1 \\ 22.7 & -56.9 \end{pmatrix}, \quad C \times A = \begin{pmatrix} -33.5 & -53.1 \\ 23.1 & -29.8 \end{pmatrix} \right.$

Hence they are not equal.]

19 Calculate the determinant of matrix A. [17]

20 Calculate the determinant of matrix B. $\left[-\frac{7}{90}\right]$

21 Calculate the determinant of matrix C. [−13.43]

22 Find the matrix of minors of matrix D. $\left[\begin{pmatrix} -16 & 8 & -34 \\ -14 & -46 & 63 \\ -24 & 12 & 2 \end{pmatrix}\right]$

23 Find the matrix of minors of matrix E. $\left[\begin{pmatrix} -\frac{2}{5} & 10 & -\frac{2}{3} \\ 3\frac{3}{5} & 2\frac{3}{10} & 6 \\ 42\frac{1}{3} & 18\frac{1}{2} & -32 \end{pmatrix}\right]$

24 Find the matrix of cofactors of matrix D. $\left[\begin{pmatrix} -16 & -8 & -34 \\ 14 & -46 & -63 \\ -24 & -12 & 2 \end{pmatrix}\right]$

25 Find the matrix of cofactors of matrix E. $\left[\begin{pmatrix} -\frac{2}{5} & -10 & -\frac{2}{3} \\ -3\frac{3}{5} & 2\frac{3}{10} & -6 \\ 42\frac{1}{3} & -18\frac{1}{2} & -32 \end{pmatrix}\right]$

26 Calculate the determinant of matrix D. $[-212]$

27 Calculate the determinant of matrix E. $\left[-61\frac{8}{15}\right]$

28 Calculate the determinant of matrix F. $[-242.83]$

29 Determine the inverse of matrix A. $\left[\begin{pmatrix} \frac{7}{17} & \frac{1}{17} \\ \frac{4}{17} & \frac{3}{17} \end{pmatrix}\right]$

30 Determine the inverse of matrix B. $\left[\begin{pmatrix} 7\frac{5}{7} & 8\frac{4}{7} \\ 4\frac{2}{7} & -6\frac{3}{7} \end{pmatrix}\right]$

31 Determine the inverse of matrix C. $\left[\begin{pmatrix} 0.290 & 0.551 \\ 0.186 & 0.097 \end{pmatrix} \text{ correct to 3 decimal places}\right]$

32 Write down the transpose of matrix D. $\left[\begin{pmatrix} 4 & -2 & 5 \\ -7 & 4 & 7 \\ 6 & 0 & -4 \end{pmatrix}\right]$

33 Write down the transpose of matrix E. $\left[\begin{pmatrix} 3 & 5 & -1 \\ 6 & -\frac{2}{3} & 0 \\ \frac{1}{2} & 7 & \frac{3}{5} \end{pmatrix}\right]$

34 Write down the transpose of matrix F. $\left[\begin{pmatrix} 3.1 & -1.6 & 5.3 \\ 2.4 & 3.8 & 3.4 \\ 6.4 & -1.9 & -4.8 \end{pmatrix}\right]$

35 Determine the adjoint of matrix D. $\left[\begin{pmatrix} -16 & 14 & -24 \\ -8 & -46 & -12 \\ -34 & -63 & 2 \end{pmatrix}\right]$

36 Determine the adjoint of matrix E. $\left[\begin{pmatrix} -\frac{2}{5} & -3\frac{3}{5} & 42\frac{1}{3} \\ -10 & 2\frac{3}{10} & -18\frac{1}{2} \\ -\frac{2}{3} & -6 & -32 \end{pmatrix}\right]$

37 Determine the adjoint of matrix F. $\left[\begin{pmatrix} -11.78 & 33.28 & -28.88 \\ -17.75 & -48.80 & -4.35 \\ -25.58 & 2.18 & 15.62 \end{pmatrix}\right]$

38 Find the inverse of matrix D. $\left[-\frac{1}{212}\begin{pmatrix} -16 & 14 & -24 \\ -8 & -46 & -12 \\ -34 & -63 & 2 \end{pmatrix}\right]$

39 Find the inverse of matrix E. $\left[-\frac{15}{923}\begin{pmatrix} -\frac{2}{5} & -3\frac{3}{5} & 43\frac{1}{3} \\ -10 & 2\frac{3}{10} & -18\frac{1}{2} \\ -\frac{2}{3} & -6 & -32 \end{pmatrix}\right]$

40 Find the inverse of matrix F. $\left[-\frac{1}{242.83}\begin{pmatrix} -11.78 & 33.28 & -28.88 \\ -17.75 & -48.80 & -4.35 \\ -25.58 & 2.18 & 15.62 \end{pmatrix}\right]$

7 Algebra (4) – The solution of linear equations by determinants

A MAIN POINTS CONCERNED WITH THE SOLUTION OF LINEAR EQUATIONS BY DETERMINANTS

1 There are certain **properties of determinants** which enable the value of a determinant to be found more simply. Some of these properties are given below.

(i) If all the elements in a row or column are interchanged with the corresponding elements in another row or column, the value of the determinant obtained is -1 times the value of the original determinant. Thus

$$\begin{vmatrix} a & b \\ c & d \end{vmatrix} \equiv (-1) \times \begin{vmatrix} b & a \\ d & c \end{vmatrix}$$

(ii) If two rows or two columns of a determinant are equal its value is equal to zero. Thus

$$\begin{vmatrix} a & a \\ c & c \end{vmatrix} \equiv 0$$

(iii) If all the elements in any row or any column of a determinant have a common factor, the elements in that row or column can be divided by the common factor and the factor becomes a factor of the determinant. Thus

$$\begin{vmatrix} a & b \\ b \times c & b \times d \end{vmatrix} \equiv b \times \begin{vmatrix} a & b \\ c & d \end{vmatrix}$$

(iv) The value of a determinant remains unaltered if a multiple of the elements in any row or any column are added to the corresponding elements of any other row or column. Thus

$$\begin{vmatrix} a & b \\ c & d \end{vmatrix} \equiv \begin{vmatrix} a & b+ka \\ c & d+kc \end{vmatrix}$$

2 The properties of determinants listed in para. 1 may be used to:

(i) reduce the size of elements within the determinant, and

(ii) introduce as many zero elements as is practical before evaluating the determinant (see *Problems 1 and 2*). Simplification is mainly achieved by using property (iv) in para. 1.

3 The procedure for solving linear simultaneous equations in two unknowns using matrices is:

(i) write the equations in the form

$$a_1x+b_1y = c_1$$
$$a_2x+b_2y = c_2$$

(ii) write the matrix equation corresponding to these equations,

i.e. $\begin{pmatrix} a_1 & b_1 \\ a_2 & b_2 \end{pmatrix} \times \begin{pmatrix} x \\ y \end{pmatrix} = \begin{pmatrix} c_1 \\ c_2 \end{pmatrix}$,

(iii) determine the inverse matrix of $\begin{pmatrix} a_1 & b_1 \\ a_2 & b_2 \end{pmatrix}$,

i.e. $\dfrac{1}{a_1 b_2 - b_1 a_2} \begin{pmatrix} b_2 & -b_1 \\ -a_2 & a_1 \end{pmatrix}$, (from chapter 6, para. 8),

(iv) multiply each side of (ii) by the inverse matrix, and

(v) solve for x and y by equating corresponding elements. (See *Problem 3.*)

4 The procedure for solving linear simultaneous equations in three unknowns by matrices is:

(i) write the equations in the form

$$a_1x+b_1y+c_1z = d_1$$
$$a_2x+b_2y+c_2z = d_2$$
$$a_3x+b_3y+c_3z = d_3$$

(ii) write the matrix equation corresponding to these equations, i.e.

$$\begin{pmatrix} a_1 & b_1 & c_1 \\ a_2 & b_2 & c_2 \\ a_3 & b_3 & c_3 \end{pmatrix} \times \begin{pmatrix} x \\ y \\ z \end{pmatrix} = \begin{pmatrix} d_1 \\ d_2 \\ d_3 \end{pmatrix},$$

(iii) determine the inverse matrix of $\begin{pmatrix} a_1 & b_1 & c_1 \\ a_2 & b_2 & c_2 \\ a_3 & b_3 & c_3 \end{pmatrix}$

(see chapter 6, para. 10)

(iv) multiply each side of (ii) by the inverse matrix, and

(v) solve for x, y and z by equating the corresponding elements. (See *Problem 4.*)

5 When solving linear simultaneous equations in two unknowns using determinants:

(i) write the equations in the form

$$a_1x+b_1y+c_1 = 0$$
$$a_2x+b_2y+c_2 = 0$$

and then

(ii) the solution is given by

$$\frac{x}{D_x} = \frac{-y}{D_y} = \frac{1}{D}$$

where $D_x = \begin{vmatrix} b_1 & c_1 \\ b_2 & c_2 \end{vmatrix}$ i.e. the determinant of the coefficients left when the $x-$ column is covered up.

$D_y = \begin{vmatrix} a_1 & c_1 \\ a_2 & c_2 \end{vmatrix}$ i.e. the determinant of the coefficients left when the $y-$ column is covered up, and

$D = \begin{vmatrix} a_1 & b_1 \\ a_2 & b_2 \end{vmatrix}$ i.e. the determinant of the coefficients left when the constants-column is covered up

(See *Problem 5.*)

6 When solving simultaneous equations in three unknowns using determinants:

(i) Write the equations in the form

$$\begin{aligned} a_1x+b_1y+c_1z+d_1 &= 0 \\ a_2x+b_2y+c_2z+d_2 &= 0 \\ a_3x+b_3y+c_3z+d_3 &= 0 \end{aligned}$$

and then

(ii) the solution is given by

$$\frac{x}{D_x} = \frac{-y}{D_y} = \frac{z}{D_z} = \frac{-1}{D}$$

where D_x is $\begin{vmatrix} b_1 & c_1 & d_1 \\ b_2 & c_2 & d_2 \\ b_3 & c_3 & d_3 \end{vmatrix}$ i.e. the determinant of the coefficients obtained by covering up the x-column,

D_y is $\begin{vmatrix} a_1 & c_1 & d_1 \\ a_2 & c_2 & d_2 \\ a_3 & c_3 & d_3 \end{vmatrix}$ i.e. the determinant of the coefficients obtained by covering up the y-column,

D_z is $\begin{vmatrix} a_1 & b_1 & d_1 \\ a_2 & b_2 & d_2 \\ a_3 & b_3 & d_3 \end{vmatrix}$ i.e. the determinant of the coefficients obtained by covering up the z-column,

and D is $\begin{vmatrix} a_1 & b_1 & c_1 \\ a_2 & b_2 & c_2 \\ a_3 & b_3 & c_3 \end{vmatrix}$ i.e. the determinant of the coefficients obtained by covering up the constants-column.

(See *Problem 6.*)

B WORKED PROBLEMS ON THE SOLUTION OF LINEAR EQUATIONS BY DETERMINANTS

Problem 1 Simplify and evaluate $\begin{vmatrix} 7 & 30 \\ 60 & 252 \end{vmatrix}$.

Using the properties listed in para. 1:

$$\begin{vmatrix} 7 & 30 \\ 60 & 252 \end{vmatrix} = 4 \times \begin{vmatrix} 7 & 30 \\ 15 & 63 \end{vmatrix}, \quad \text{(property iii).}$$

Taking 4 times column 1 from column 2, (property iv) gives:

$$4 \times \begin{vmatrix} 7 & (30-28) \\ 15 & (63-60) \end{vmatrix} = 4 \times \begin{vmatrix} 7 & 2 \\ 15 & 3 \end{vmatrix}$$

Taking 5 times column 2 from column 1, (property iv) gives:

$$4 \times \begin{vmatrix} (7-10) & 2 \\ (15-15) & 3 \end{vmatrix} = 4 \times \begin{vmatrix} -3 & 2 \\ 0 & 3 \end{vmatrix}$$

Hence, $\begin{vmatrix} 7 & 30 \\ 60 & 252 \end{vmatrix} = 4[(-3 \times 3)-0] = \mathbf{-36}.$

(If an electronic calculator is available, it is usually easier just to evaluate the original determinant, i.e., $(7 \times 252)-(30 \times 60) = \mathbf{-36}$).

Problem 2 Simplify and evaluate: $\begin{vmatrix} 2 & 7 & 26 \\ 1 & 2 & 6 \\ 4 & 11 & 40 \end{vmatrix}$

Taking 3 times column 2 from column 3 gives $\begin{vmatrix} 2 & 7 & (26-21) \\ 1 & 2 & (6-6) \\ 4 & 11 & (40-33) \end{vmatrix}$

i.e. $\begin{vmatrix} 2 & 7 & 5 \\ 1 & 2 & 0 \\ 4 & 11 & 7 \end{vmatrix}$

Taking twice column 1 from column 2 gives $\begin{vmatrix} 2 & (7-4) & 5 \\ 1 & (2-2) & 0 \\ 4 & (11-8) & 7 \end{vmatrix}$

i.e. $\begin{vmatrix} 2 & 3 & 5 \\ 1 & 0 & 0 \\ 4 & 3 & 7 \end{vmatrix}$. Since two of the elements in row 2 are zero, the value of this determinant is

$$-1 \times \begin{vmatrix} 3 & 5 \\ 3 & 7 \end{vmatrix} + 0 - 0 = -3 \times \begin{vmatrix} 1 & 5 \\ 1 & 7 \end{vmatrix}, \text{ (property (iii))}$$
$$= -3(7-5) = \mathbf{-6}$$

(As for second order determinants, it is often easier to just evaluate the original determinant when an electronic calculator is available, rather than simplifying it.)

Problem 3 Use matrices to solve the simultaneous equations:

$3x+5y-7 \quad = 0$ (1)

$4x-3y-19 \quad = 0$ (2)

The procedure given in para. 3 is followed.

(i) Writing the equations in the $a_1x+b_1y = c_1$ form gives:

$3x+5y = 7$

$4x-3y = 19$

(ii) The matrix equation is

$$\begin{pmatrix} 3 & 5 \\ 4 & -3 \end{pmatrix} \times \begin{pmatrix} x \\ y \end{pmatrix} = \begin{pmatrix} 7 \\ 19 \end{pmatrix}.$$

(iii) The inverse of matrix $\begin{pmatrix} 3 & 5 \\ 4 & -3 \end{pmatrix}$ is

$$\frac{1}{3 \times (-3)-5 \times 4} \begin{pmatrix} -3 & -5 \\ -4 & 3 \end{pmatrix} \text{ i.e. } \begin{pmatrix} \frac{3}{29} & \frac{5}{29} \\ \frac{4}{29} & \frac{-3}{29} \end{pmatrix}$$

(iv) Multiplying each side of (ii) by (iii) and remembering that $A \times A^{-1} = I$, the unit matrix gives:

$$\begin{pmatrix} 1 & 0 \\ 0 & 1 \end{pmatrix} \begin{pmatrix} x \\ y \end{pmatrix} = \begin{pmatrix} \frac{3}{29} & \frac{5}{29} \\ \frac{4}{29} & \frac{-3}{29} \end{pmatrix} \times \begin{pmatrix} 7 \\ 19 \end{pmatrix}$$

Thus $\quad \begin{pmatrix} x \\ y \end{pmatrix} = \begin{pmatrix} \frac{21}{29} + \frac{95}{29} \\ \frac{28}{29} - \frac{57}{29} \end{pmatrix}$

i.e. $\quad \begin{pmatrix} x \\ y \end{pmatrix} = \begin{pmatrix} 4 \\ -1 \end{pmatrix}$

(v) By comparing corresponding elements,

$x = 4$ and $y = -1$.

Checking: equation (1), $3 \times 4+5 \times (-1)-7 = 0 = \text{RHS}$
equation (2), $4 \times 4-3 \times (-1)-19 = 0 = \text{RHS}$

Problem 4 Use matrices to solve the simultaneous equations

$x+y+z-4 = 0$ (1)
$2x-3y+4z-33 = 0$ (2)
$3x-2y-2z-2 = 0$ (3)

The procedure given in para. 4 is followed.

(i) Writing the equations in the $a_1x+b_1y+c_1z = d_1$ form gives:

$x+y+z = 4$
$2x-3y+4z = 33$
$3x-2y-2z = 2$

(ii) The matrix equation is

$$\begin{pmatrix} 1 & 1 & 1 \\ 2 & -3 & 4 \\ 3 & -2 & -2 \end{pmatrix} \times \begin{pmatrix} x \\ y \\ z \end{pmatrix} = \begin{pmatrix} 4 \\ 33 \\ 2 \end{pmatrix}$$

(iii) The inverse matrix of $A = \begin{pmatrix} 1 & 1 & 1 \\ 2 & -3 & 4 \\ 3 & -2 & -2 \end{pmatrix}$ is given by $A^{-1} = \frac{\text{adj } A}{|A|}$

The adjoint of A is the transpose of the matrix of the cofactors of the elements, (see chapter 6, para 10). The matrix of cofactors is

$\begin{pmatrix} 14 & 16 & 5 \\ 0 & -5 & 5 \\ 7 & -2 & -5 \end{pmatrix}$ and the transpose of this matrix gives

$$\text{adj } A = \begin{pmatrix} 14 & 0 & 7 \\ 16 & -5 & -2 \\ 5 & 5 & -5 \end{pmatrix}.$$

The determinant of A, i.e. the sum of the products of elements and their cofactors, using a first row expansion is $1\begin{vmatrix} -3 & 4 \\ -2 & -2 \end{vmatrix} - 1\begin{vmatrix} 2 & 4 \\ 3 & -2 \end{vmatrix} + 1\begin{vmatrix} 2 & -3 \\ 3 & -2 \end{vmatrix}$,
i.e., $(1 \times 14)-(1 \times -16)+(1 \times 5)$, that is, 35.

Hence the inverse of A, $A^{-1} = \frac{1}{35}\begin{pmatrix} 14 & 0 & 7 \\ 16 & -5 & -2 \\ 5 & 5 & -5 \end{pmatrix}$

(iv) Multiplying each side of (ii) by (iii), remembering that $A \times A^{-1} = I$, the unit matrix gives

$$\begin{pmatrix} 1 & 0 & 0 \\ 0 & 1 & 0 \\ 0 & 0 & 1 \end{pmatrix} \times \begin{pmatrix} x \\ y \\ z \end{pmatrix} = \frac{1}{35} \begin{pmatrix} 14 & 0 & 7 \\ 16 & -5 & -2 \\ 5 & 5 & -5 \end{pmatrix} \times \begin{pmatrix} 4 \\ 33 \\ 2 \end{pmatrix}$$

$$\begin{pmatrix} x \\ y \\ z \end{pmatrix} = \frac{1}{35} \begin{pmatrix} (14\times4)+(0\times33)+(7\times2) \\ (16\times4)+(-5\times33)+((-2)\times2) \\ (5\times4)+(5\times33)+((-5)\times2) \end{pmatrix} = \frac{1}{35}\begin{pmatrix} 70 \\ -105 \\ 175 \end{pmatrix} = \begin{pmatrix} 2 \\ -3 \\ 5 \end{pmatrix}$$

(v) By comparing corresponding elements, $\boldsymbol{x = 2, y = -3, z = 5}$, which can be checked in the original equations.

Problem 5 The velocity of a car, accelerating at uniform acceleration a between two points, is given by $v = u+at$, where u is its velocity when passing the first point and t is the time taken to pass between the two points. If $v = 21$ m/s when $t = 3.5$ s and $v = 33$ m/s when $t = 6.1$ s, use determinants to find the values of u and a, each correct to 4 significant figures.

Substituting the given values in $v = u+at$ gives $u+3.5a = 21$ (1)

$u+6.1a = 33$ (2)

The procedure given in para. 5 is used.

(i) The equations are written in the form $a_1x+b_1y+c_1 = 0$,

i.e. $u+3.5a-21 = 0$; $u+6.1a-33 = 0$.

(ii) The solution is given by $\frac{u}{D_u} = \frac{-a}{D_a} = \frac{1}{D}$,

where D_u is the determinant of coefficients left when the u-column is covered up,

i.e. $D_u = \begin{vmatrix} 3.5 & -21 \\ 6.1 & -33 \end{vmatrix} = 12.6$

Similarly, $D_a = \begin{vmatrix} 1 & -21 \\ 1 & -33 \end{vmatrix} = -12$

and $D = \begin{vmatrix} 1 & 3.5 \\ 1 & 6.1 \end{vmatrix} = 2.6$

Thus $\frac{u}{12.6} = \frac{-a}{-12} = \frac{1}{2.6}$ i.e. $u = \frac{12.6}{2.6} = \mathbf{4.846\ m/s}$

and $a = \frac{12}{2.6} = \mathbf{4.615\ m/s^2}$, each correct to 4 significant figures.

Problem 6 A dc circuit comprises three closed loops. Applying Kirchhoff's laws to the closed loops gives the following equations for current flow in milliamperes:

$$2I_1+3I_2-4I_3 = 26$$
$$I_1-5I_2-3I_3 = -87$$
$$-7I_1+2I_2+6I_3 = 12.$$ Use determinants to solve for I_1, I_2 and I_3.

The procedure given in para. 6 is used.

(i) Writing the equations in the $a_1x+b_1y+c_1z+d_1 = 0$ form gives:

$$2I_1+3I_2-4I_3-26 = 0$$
$$I_1-5I_2-3I_3+87 = 0$$
$$-7I_1+2I_2+6I_3-12 = 0$$

(ii) The solution is given by

$$\frac{I_1}{D_{I_1}} = \frac{-I_2}{D_{I_2}} = \frac{I_3}{D_{I_3}} = \frac{-1}{D}$$, where D_{I_1} is the determinant of coefficients obtained by covering up the I_1 column, i.e.,

$$D_{I_1} = \begin{vmatrix} 3 & -4 & -26 \\ -5 & -3 & 87 \\ 2 & 6 & -12 \end{vmatrix}$$

The evaluation of this determinant may be simplified as follows:

Adding twice column 2 to column 3 gives $\begin{vmatrix} 3 & -4 & -34 \\ -5 & -3 & 81 \\ 2 & 6 & 0 \end{vmatrix}$

Taking 3 times column 1 from column 2 gives $\begin{vmatrix} 3 & -13 & -34 \\ -5 & 12 & 81 \\ 2 & 0 & 0 \end{vmatrix}$

Hence $D_{I_1} = 2[(-13 \times 81)-(12 \times (-34)] = -1290$

Also, $D_{I_2} = \begin{vmatrix} 2 & -4 & -26 \\ 1 & -3 & 87 \\ -7 & 6 & -12 \end{vmatrix} = 1806$

$D_{I_3} = \begin{vmatrix} 2 & 3 & -26 \\ 1 & -5 & 87 \\ -7 & 2 & -12 \end{vmatrix} = -1161$

and $D = \begin{vmatrix} 2 & 3 & -4 \\ 1 & -5 & -3 \\ -7 & 2 & 6 \end{vmatrix} = 129$

Thus $\frac{I_1}{-1290} = \frac{-I_2}{1806} = \frac{I_3}{-1161} = \frac{-1}{129}$, giving $\mathbf{I_1 = 10}$ **mA**, $\mathbf{I_2 = 14}$ **mA and** $\mathbf{I_3 = 9}$ **mA.**

C FURTHER PROBLEMS ON THE SOLUTION OF LINEAR EQUATIONS BY DETERMINANTS

In *Problems 1 to 5* simplify and evaluate the determinants given.

1 $\begin{vmatrix} 7 & 5 \\ 21 & 26 \end{vmatrix}$ [77]

2 $\begin{vmatrix} -5 & -4 \\ 10 & 5 \end{vmatrix}$ [15]

3 $\begin{vmatrix} 14 & 14 & 20 \\ 11 & 9 & 13 \\ 15 & 17 & 19 \end{vmatrix}$ [144]

4 $\begin{vmatrix} 8 & -2 & -10 \\ 2 & -3 & -2 \\ 6 & 3 & 8 \end{vmatrix}$ [−328]

5 $\begin{vmatrix} 14 & 42 & 33 \\ 13 & 47 & 36 \\ 17 & 24 & 22 \end{vmatrix}$ [1]

In *Problems 6 to 10* use **matrices** to solve the simultaneous equations given.

6 $3x+4y = 0$
$2x+5y+7 = 0$ [$x = 4; y = -3$]

7 $2p+5q+14.6 = 0$
$3.1p+1.7q+2.06 = 0$ [$p = 1.2; q = -3.4$]

8 $x+2y+3z = 5$
$2x-3y-z = 3$
$-3x+4y+5z = 3$ [$x = 1; y = -1; z = 2$]

9 $3a+4b-3c = 2$
$-2a+2b+2c = 15$ [$a = 2.5; b = 3.5; c = 6.5$]
$7a-5b+4c = 26$

10 $p+2q+3r+7.8 = 0$
$2p+5q-r-1.4 = 0$ [$p = 4.1; q = -1.9; r = -2.7$]
$5p-q+7r-3.5 = 0$

In *Problems 11 to 15* use **determinants** to solve the simultaneous equations given.

11 $3x-5y = -17.6$
$7y-2x-22 = 0$ [$x = -1.2; y = 2.8$]

12 $2.3m-4.4n = 6.84$
$8.5n-6.7m = 1.23$ [$m = -6.4; n = -4.9$]

13 $3x+4y+z = 10$
$2x-3y+5z+9 = 0$
$x+2y-z = 6$ [$x = 1; y = 2; z = -1$]

14 $1.2p-2.3q-3.1r+10.1 = 0$
$4.7p+3.8q-5.3r-21.5 = 0$
$3.7p-8.3q+7.4r+28.1 = 0$ [$p = 1.5; q = 4.5; r = 0.5$]

15 $\frac{x}{2}-\frac{y}{3}+\frac{2z}{5} = -\frac{1}{20}$

$\frac{x}{4}+\frac{2y}{3}-\frac{z}{2} = \frac{19}{40}$ $\left[x = \frac{7}{20};\ y = \frac{17}{40};\ z = -\frac{5}{24}\right]$

$x+y-z = \frac{59}{60}$

16 In two closed loops of an electrical circuit, the currents flowing are given by the simultaneous equations

$I_1+2I_2+4 = 0$
$5I_1+3I_2-1 = 0$

Use matrices to solve for I_1 and I_2 [$I_1 = 2; I_2 = -3$]

17 In a system of forces, the relationship between two forces F_1 and F_2 is given by
$5F_1+3F_2+6 = 0$
$3F_1+5F_2+18 = 0$
Use determinants to solve for F_1 and F_2. [$F_1 = 1.5, F_2 = -4.5$]

18 The relationship between the displacement, s, velocity, v, and acceleration, a, of a piston is given by the equations
$s+2v+2a = 4$
$3s-v+4a = 25$
$3s+2v-a = -4$
Use matrices to determine the values of s, v and a. [$s = 2, v = -3, a = 4$]

19 Kirchhoff's laws are used to determine the current equations in an electrical network and showed that
$i_1+8i_2+3i_3 = -31$
$3i_1-2i_2+i_3 = -5$
$2i_1-3i_2+2i_3 = 6$
Use determinants to solve for i_1, i_2 and i_3. [$i_1 = -5, i_2 = -4, i_3 = 2$]

8 Algebra (5) – Graphs with logarithmic scales

A MAIN POINTS CONCERNED WITH GRAPHS HAVING LOGARITHMIC SCALES

1 (i) Graph paper is available where the scale markings along the horizontal and vertical axes are proportional to the logarithms of the numbers. Such graph paper is called **log-log graph paper**.

Fig 1   1 2 3 4 5 6 7 8 9 10

(ii) A **logarithmic scale** is shown in *Fig 1* where the distances between, say 1 and 2 is proportional to lg 2–lg 1, i.e. 0.3010 of the total distance from 1 to 10. Similarly, the distance between 7 and 8 is proportional to lg 8–lg 7, i.e. 0.057 99 of the total distance from 1 to 10. Thus the distance between markings progressively decreases as the numbers increase from 1 to 10.

(iii) With log-log graph paper the scale markings are from 1 to 9, and this pattern can be repeated several times. The number of times the pattern of markings is repeated on an axis signifies the number of **cycles.** When the vertical axis has, say, 3 sets of values from 1 to 9 and the horizontal axis has 2 sets of values from 1 to 9, then this log-log graph paper is called 'log 3 cycle X 2 cycle' (see *Fig 2*). Many different arrangements are available ranging from 'log 1 cycle X 1 cycle' through to 'log 5 cycle X 5 cycle'.

(iv) To depict a set of values, say, from 0.4 to 161 on an axis of log-log graph paper, 4 cycles are required, from 0.1 to 1, 1 to 10, 10 to 100 and 100 to 1000.

2 (i) **To express graphs of the form $y = ax^n$ in linear form**

Taking logarithms to a base of 10 of both sides of $y = ax^n$ gives:

$$\lg y = \lg (ax^n) = \lg a + \lg x^n$$

i.e. $$\lg y = n \lg x + \lg a$$

which compares with $Y = mX + c$.

Thus, by plotting $\lg y$ vertically against $\lg x$ horizontally, a straight line results, i.e. the equation $y = ax^n$ is reduced to linear form. With log-log graph paper available x and y may be plotted directly, without having to firstly determine their logarithms. (See *Problems 1 to 3*.)

(ii) **To express graphs of the form $y = ab^x$ in linear form**

Taking logarithms to a base of 10 of both sides of $y = ab^x$ gives:

$$\lg y = \lg (ab^x) = \lg a + \lg b^x = \lg a + x \lg b$$

i.e. $$\lg y = (\lg b)x + \lg a$$

which compares with $Y = mX + c$

Thus, by plotting $\lg y$ vertically against x horizontally a straight line results, i.e. the graph $y = ab^x$ is reduced to linear form. In this case, graph paper having a linear horizontal scale and a logarithmic vertical scale may be used. This type of graph paper is called **log-linear graph paper**, and is specified by the number

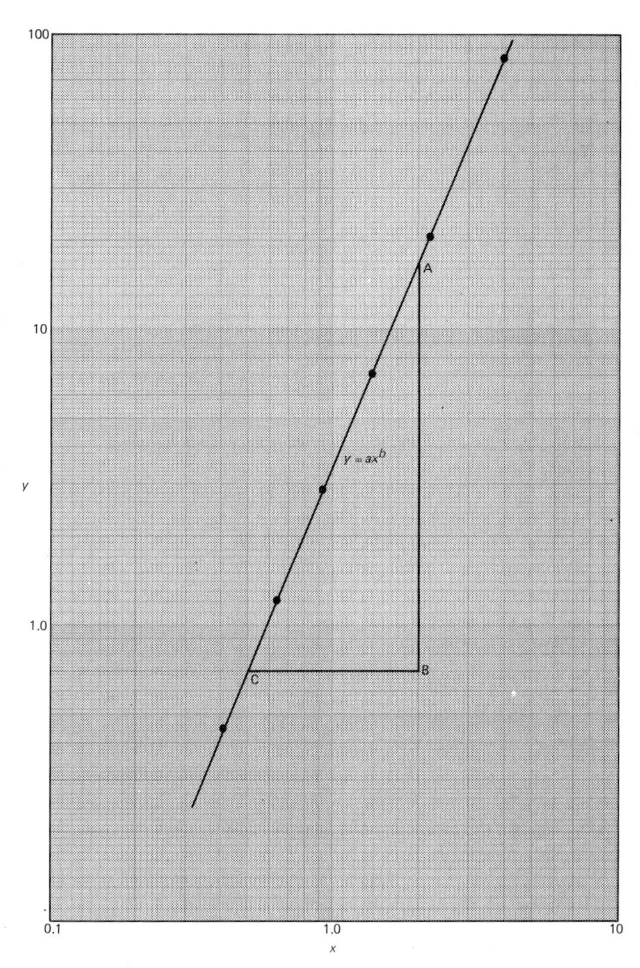

Fig 2 Graph to verify a law of the form $y = ax^b$

of cycles on the logarithmic scale. For example, graph paper having 3 cycles on the logarithmic scale is called 'log 3 cycle × linear' graph paper. (See *Problem 4.*)

(iii) **To express graphs of the form $y = ae^{kx}$ in linear form**

Taking logarithms to a base of e of both sides of $y = ae^{kx}$ gives:

$$\ln y = \ln (ae^{kx}) = \ln a + \ln e^{kx} = \ln a + kx \ln e$$

i.e. $\ln y = kx + \ln a$, (since $\ln e = 1$),

which compares with $Y = mX + c$.

Thus, by plotting $\ln y$ vertically against x horizontally, a straight line results, i.e. the equation $y = ae^{kx}$ is reduced to linear form. Since $\ln y = 2.3026 \lg y$, i.e. $\ln y = (\text{a constant})(\lg y)$, the same log-linear graph paper can be used for Naperian logarithms as for logarithms to a base of 10. (See *Problems 5 and 6*.)

B. WORKED PROBLEMS ON GRAPHS HAVING LOGARITHMIC SCALES

Problem 1 Experimental values of two related quantities x and y are shown below:

x	0.41	0.63	0.92	1.36	2.17	3.95
y	0.45	1.21	2.89	7.10	20.79	82.46

The law relating x and y is believed to be $y = ax^b$, where a and b are constants. Verify that this law is true and determine the approximate values of a and b.
If $y = ax^b$ then $\lg y = b \lg x + \lg a$, from para. 2(i), which is of the form $Y = mX + c$, showing that to produce a straight line graph $\lg y$ is plotted vertically against $\lg x$ horizontally. x and y may be plotted directly on to log-log graph paper as shown in *Fig 2.* The values of y range from 0.45 to 82.46 and 3 cycles are needed, (i.e. 0.1 to 1, 1 to 10 and 10 to 100). The values of x range from 0.41 to 3.95 and 2 cycles are needed (i.e. 0.1 to 1 and 1 to 10). Hence 'log 3 cycle × 2 cycle' is used as shown in *Fig 2* where the axes are marked and the points plotted. Since the points lie on a straight line the law $y = ax^b$ is verified.
To evaluate constants a and b:

Method 1. Any two points on the straight line, say points A and C are selected, and AB and BC are measured (say in centimetres).

$$\text{Then, gradient, } b = \frac{AB}{BC} = \frac{11.5 \text{ units}}{5 \text{ units}} = 2.3$$

Since $\lg y = b \lg x + \lg a$, when $x = 1$, $\lg x = 0$ and $\lg y = \lg a$.
The straight line crosses the ordinate $x = 1.0$ at $y = 3.5$
Hence $\lg a = \lg 3.5$, i.e. $a = \mathbf{3.5}$
Method 2. Any two points on the straight line, say points A and C, are selected. A has co-ordinates (2, 17.25) and C has co-ordinates (0.5, 0.7).
Since $y = ax^b$ then $17.25 = a(2)^b$ (1)
and $0.7 = a(0.5)^b$ (2)
i.e., two simultaneous equations are produced and may be solved for a and b.
Dividing equation (1) by equation (2) to eliminate a gives $\dfrac{17.25}{0.7} = \dfrac{(2)^b}{(0.5)^b} = \left(\dfrac{2}{0.5}\right)^b$
i.e. $24.643 = (4)^b$.
Taking logarithms of both sides gives $\lg 24.643 = b \lg 4$

i.e. $b = \dfrac{\lg 24.643}{\lg 4} = 2.3$, correct to 2 significant figures.

Substituting $b = 2.3$ in equation (1) gives: $17.25 = a(2)^{2.3}$

i.e. $a = \dfrac{17.25}{(2)^{2.3}} = \dfrac{17.25}{4.925} = 3.5$, correct to 2 significant figures.

Hence the law of the graph is $y = 3.5x^{2.3}$

Problem 2 The power dissipated by a resistor was measured for varying values of current flowing in the resistor and the results are as shown:

Current, I amperes	1.4	4.7	6.8	9.1	11.2	13.1
Power, P watts	49	552	1156	2070	3136	4290

Prove that the law relating current and power is of the form $P = RI^n$, where R and n are constants, and determine the law. Hence calculate the power when the current is 12 amperes and the current when the power is 1000 W.

Since $P = RI^n$ then $\lg P = n \lg I + \lg R$, (from para. 2(i)), which is of the form $Y = mX+c$, showing that to produce a straight line graph $\lg P$ is plotted vertically against $\lg I$ horizontally. Power values range from 49 to 4290, hence 3 cycles of log-log graph paper are needed (10 to 100, 100 to 1000 and 1000 to 10 000).

Current values range from 1.4 to 11.2, hence 2 cycles of log-log graph paper are needed (1 to 10 and 10 to 100).

Thus 'log 3 cycles × 2 cycles' is used as shown in *Fig 3* (or, if not available, graph paper having a larger number of cycles per axis can be used). The co-ordinates are plotted and a straight line results which proves that the law relating current and power is of the form $P = RI^n$.

Gradient of straight line $n = \dfrac{AB}{BC} = \dfrac{14 \text{ units}}{7 \text{ units}} = 2$

At point C, $I = 2$ and $P = 100$. Substituting these values into $P = RI^n$ gives: $100 = R(2)^2$. Hence $R = 100/(2)^2 = \mathbf{25}$ which may have been found from the intercept on the $I = 1.0$ axis in *Fig 3*. **Hence the law of the graph is $P = 25I^2$.**
When current $I = 12$, power $P = 25(12)^2 = \mathbf{3600}$ **watts** (which may be read from the graph).
When power $P = 1000$, $1000 = 25I^2$

Hence $\qquad I^2 = \dfrac{1000}{25} = 40$

from which, $I = \sqrt{40} = \mathbf{6.32\ A}$

Problem 3 The pressure p and volume v of a gas are believed to be related by a law of the form $p = cv^n$, where c and n are constants. Experimental values of p and corresponding values of v obtained in a laboratory are:

p pascals	2.28×10^5	8.04×10^5	2.03×10^6	5.05×10^6	1.82×10^7
v m^3	3.2×10^{-2}	1.3×10^{-2}	6.7×10^{-3}	3.5×10^{-3}	1.4×10^{-3}

Verify that the law is true and determine approximate values of c and n.

Since $p = cv^n$, then $\lg p = n \lg v + \lg c$, which is of the form $Y = mX+c$, showing that to produce a straight line graph $\lg p$ is plotted vertically against $\lg v$ horizontally. The co-ordinates are plotted on 'log 3 cycle × 2 cycle' graph paper as shown

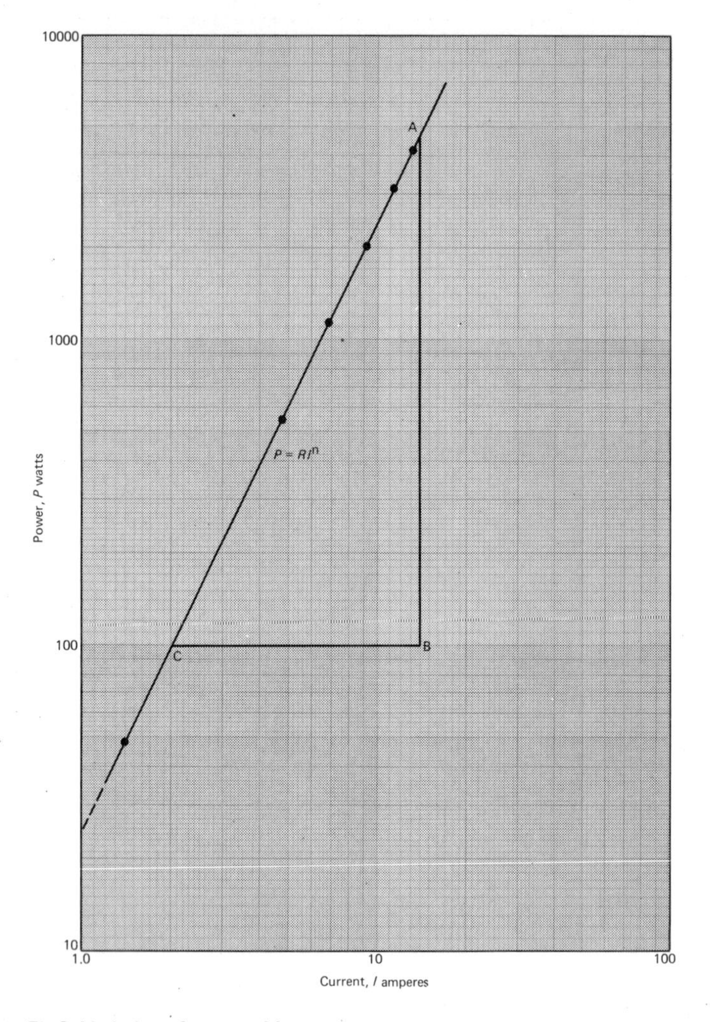

Fig 3 Variation of power with current

in *Fig 4*. With the data expressed in standard form, the axes are marked in standard form also. Since a straight line results the law $p = cv^n$ is verified.

The straight line has a negative gradient and the value of the gradient is given by

$\dfrac{\text{AB}}{\text{BC}} = \dfrac{14 \text{ units}}{10 \text{ units}} = 1.4$. Hence $n = \mathbf{-1.4}$

Selecting any point on the straight line, say point C, having co-ordinates

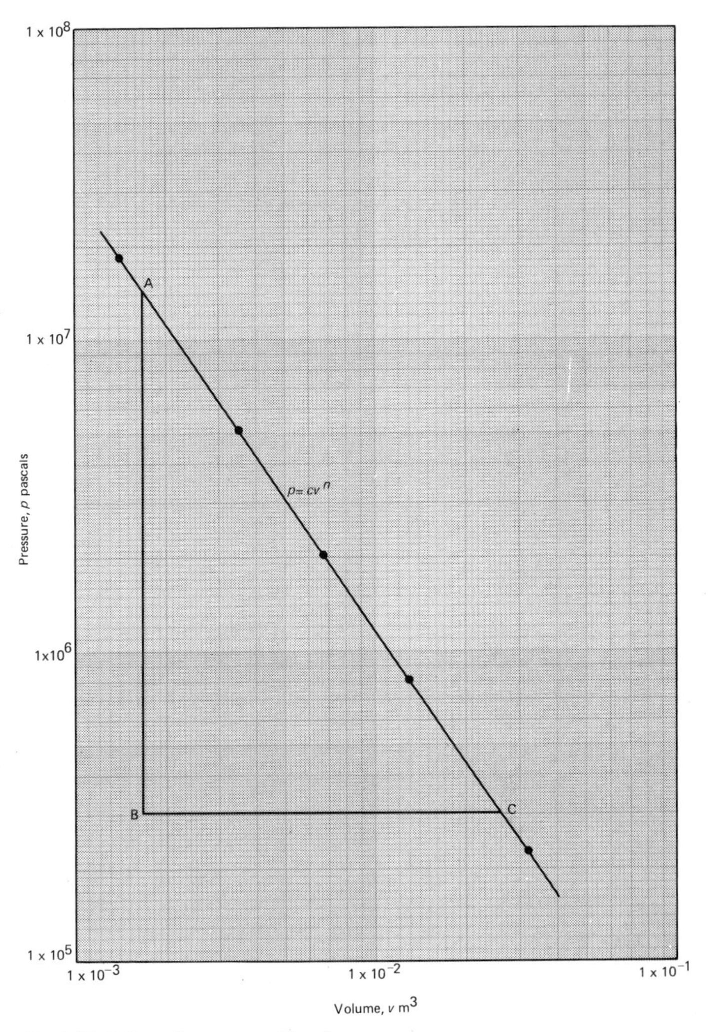

Fig 4 Variation of pressure with volume

$(2.63 \times 10^{-2}, 3 \times 10^5)$, and substituting these values in $p = cv^n$ gives:

$$3 \times 10^5 = c(2.63 \times 10^{-2})^{-1.4}$$

Hence $$c = \frac{3 \times 10^5}{(2.63 \times 10^{-2})^{-1.4}} = \frac{3 \times 10^5}{(0.0263)^{-1.4}} = \frac{3 \times 10^5}{1.63 \times 10^2}$$

$= \mathbf{1840}$, correct to 3 significant figures.

Hence the law of the graph is $p = 1840v^{-1.4}$ or $pv^{1.4} = 1840$

Problem 4 Experimental values of quantities x and y are believed to be related by a law of the form $y = ab^x$, where a and b are constants. The values of x and corresponding values of y are:

x	0.7	1.4	2.1	2.9	3.7	4.3
y	18.4	45.1	111	308	858	1850

Verify the law and determine the approximate values of a and b. Hence evaluate (i) the value of y when x is 2.5 and (ii) the value of x when y is 1200.

Since $y = ab^x$ then $\lg y = (\lg b)x + \lg a$, (from para. 2(ii)), which is of the form $Y = mX + c$, showing that to produce a straight line graph $\lg y$ is plotted vertically against x horizontally. Using log-linear graph paper, values of x are marked on the horizontal scale to cover the range 0.7 to 4.3. Values of y range from 18.4 to 1850 and 3 cycles are needed (i.e. 10 to 100, 100 to 1000 and 1000 to 10 000). Thus 'log 3 cycles X linear' graph paper is used as shown in *Fig 5*. A straight line is drawn through the co-ordinates, hence the law $y = ab^x$ is verified.
Gradient of straight line, $\lg b = AB/BC$. Direct measurement (say in centimetres) is not made with log-linear graph paper since the vertical scale is logarithmic and the horizontal scale is linear.

Hence $\dfrac{AB}{BC} = \dfrac{\lg 1000 - \lg 100}{3.82 - 2.02} = \dfrac{3-2}{1.80} = \dfrac{1}{1.80} = 0.5556$

Here b = antilog $0.5556 (= 10^{0.5556}) = \mathbf{3.6}$, correct to 2 significant figures.
Point A has co-ordinates (3.82, 1000). Substituting these values into $y = ab^x$ gives:

$1000 = a(3.6)^{3.82}$, i.e. $a = \dfrac{1000}{(3.6)^{3.82}} = \mathbf{7.5}$, correct to 2 significant figures.

Hence the law of the graph is $y = 7.5(3.6)^x$
(i) When $x = 2.5$, $y = 7.5(3.6)^{2.5} = \mathbf{184}$
(ii) When $y = 1200$, $1200 = 7.5(3.6)^x$

Hence $(3.6)^x = \dfrac{1200}{7.5} = 160$

Taking logarithms gives $x \lg 3.6 = \lg 160$

i.e. $x = \dfrac{\lg 160}{\lg 3.6} = \dfrac{2.2041}{0.5563} = \mathbf{3.96}$

Problem 5 The data given below is believed to be related by a law of the form $y = ae^{kx}$, where a and b are constants. Verify that the law is true and determine approximate values of a and b. Also determine the value of y when x is 3.8 and the value of x when y is 85.

x	−1.2	0.38	1.2	2.5	3.4	4.2	5.3
y	9.3	22.2	34.8	71.2	117	181	332

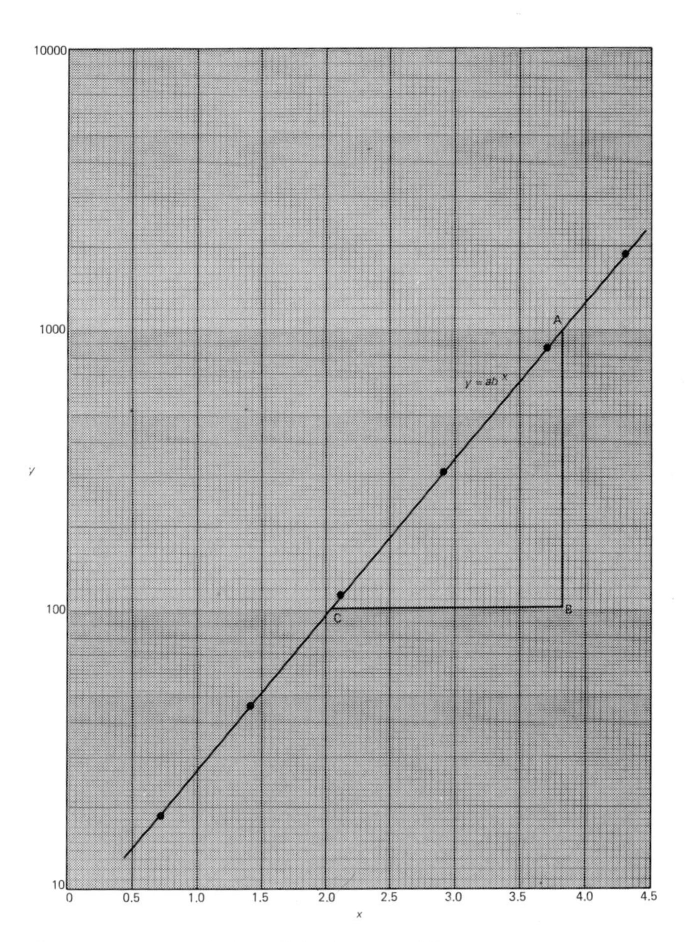

Fig 5 Graph to verify a law of the form $y = ab^x$

Since $y = ae^{kx}$ then $\ln y = kx + \ln a$, (from para. 2(iii)), which is of the form $Y = mX + c$, showing that to produce a straight line graph $\ln y$ is plotted vertically against x horizontally. The value of y ranges from 9.3 to 332 hence 'log 3 cycle × lines' graph paper is used. The plotted co-ordinates are shown in *Fig 6* and since a straight line passes through the points the law $y = ae^{kx}$ is verified.

$$\text{Gradient of straight line, } k = \frac{AB}{BC} = \frac{\ln 100 - \ln 10}{3.12 - (-1.08)} = \frac{2.3026}{4.20}$$

$$= \mathbf{0.55}, \text{ correct to 2 significant figures.}$$

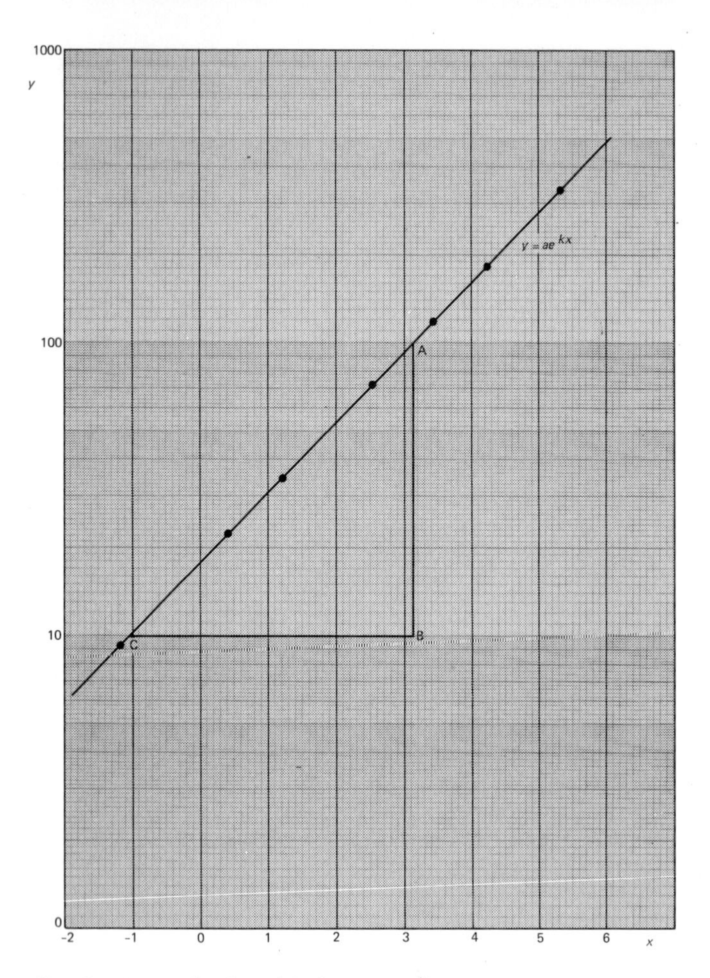

Fig 6 Graph to verify a law of the form $y = ae^{kx}$

Since $\ln y = kx + \ln a$, when $x = 0$, $\ln y = \ln a$, i.e. $y = a$
The vertical axis intercept value at $x = 0$ is 18, hence $\boldsymbol{a = 18}$
The law of the graph is thus $y = \mathbf{18e^{0.55x}}$
When x is 3.8, $y = 18e^{0.55(3.8)} = 18e^{2.09} = 18(8.0849) = \mathbf{146}$

When y is 85, $85 = 18e^{0.55x}$. Hence $e^{0.55x} = \dfrac{85}{18} = 4.7222$

and $\qquad 0.55x = \ln 4.7222 = 1.5523$. Hence $x = \dfrac{1.5523}{0.55} = 2.82$

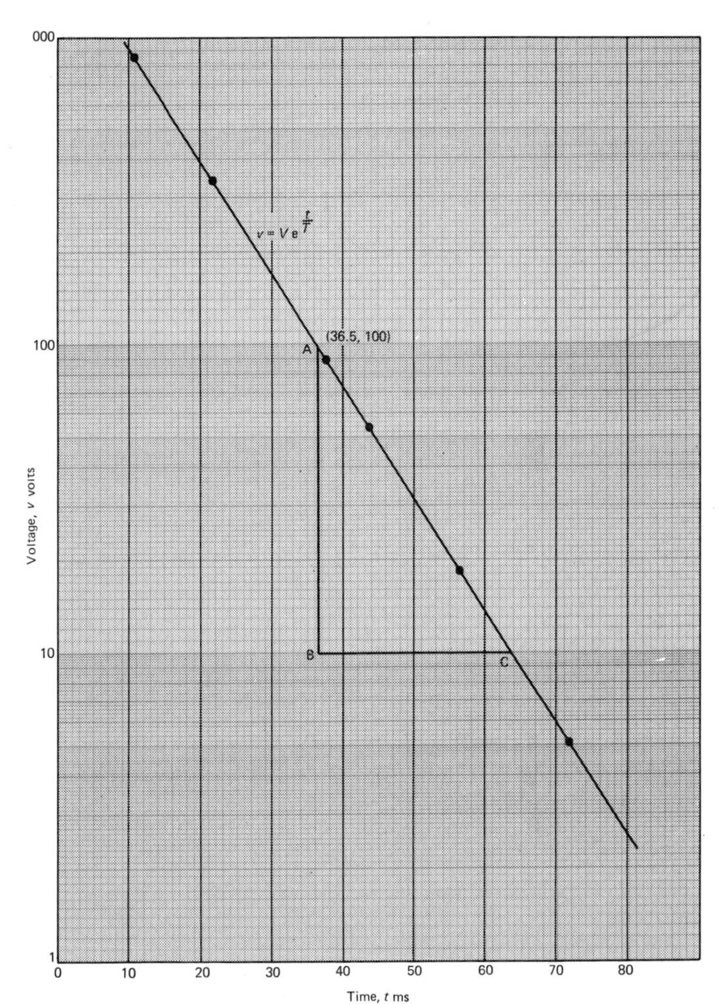

Fig 7 Variation of voltage with time

Problem 6 The voltage, v volts, across an inductor is believed to be related to time, t ms, by the law $v = Ve^{t/T}$, where V and T are constants. Experimental results obtained are:

v volts	883	347	90	55.5	18.6	5.2
t ms	10.4	21.6	37.8	43.6	56.7	72.0

Show that the law relating voltage and time is as stated and determine the approximate values of V and T. Find also the value of voltage after 25 ms and the time when the voltage is 30.0 V.

Since $v = Ve^{t/T}$ then $\ln v = \frac{1}{T}t + \ln V$, from para 2(iii), which is of the form $Y = mX + c$.
Using 'log 3 cycle X linear' graph paper, the points are plotted as shown in *Fig 7*. Since the points are joined by a straight line the law $v = Ve^{t/T}$ is verified.

Gradient of straight line, $\frac{1}{T} = \frac{AB}{BC} = \frac{\ln 100 - \ln 10}{36.5 - 64.2} = \frac{2.3026}{-27.7}$

Hence $T = \frac{-27.7}{2.3026} = $ **−12.0**, correct to 3 significant figures.

Since the straight line does not cross the vertical axis at $t = 0$ in *Fig 7*, the value of V is determined by selecting any point, say A, having co-ordinates (36.5, 100) and substituting these values into $v = Ve^{t/T}$.

Thus $100 = Ve^{\frac{36.5}{-12.0}}$

i.e. $V = \frac{100}{e^{-\frac{36.5}{12.0}}} = $ **2090 volts,** correct to 3 significant figures.

Hence the law of the graph is $v = 2090e^{-\frac{t}{12.0}}$

When time $t = 25$ ms, voltage $v = 2090e^{-\frac{25}{12.0}} = $ **260 V**

When the voltage is 30.0 volts, $30.0 = 2090e^{-\frac{t}{12.0}}$

Hence $e^{-\frac{t}{12.0}} = \frac{30}{2090}$ and $e^{\frac{t}{12.0}} = \frac{2090}{30} = 69.67$

Taking Naperian logarithms gives: $\frac{t}{12.0} = \ln 69.67 = 4.2438$,
from which, time $t = (12.0)(4.2438) = $ **50.9 ms**

C FURTHER PROBLEMS ON GRAPHS HAVING LOGARITHMIC SCALES

1 Quantities x and y are believed to be related by a law of the form $y = ax^n$, where a and n are constants. Experimental values of x and corresponding values of y are:

x	0.8	2.3	5.4	11.5	21.6	42.9
y	8	54	250	974	3028	10410

Show that the law is true and determine the values of a and n. Hence determine the value of y when x is 7.5 and the value of x when y is 5000.

[$a = 12$; $n = 1.8$; 451, 28.5]

2 Show from the following results of voltage V and admittance Y of an electrical circuit that the law connecting the quantities is of the form $V = kY^n$, and determine the values of k and n.

Voltage, V volts	2.88	2.05	1.60	1.22	0.96
Admittance, Y siemens	0.52	0.73	0.94	1.23	1.57

[$k = 1.5; n = -1$]

3 The luminosity I of a lamp varies with the applied voltage V and the relationship between I and V is thought to be $I = kV^n$. Experimental results obtained are:

I candelas	1.92	4.32	9.72	15.87	23.52	30.72
V volts	40	60	90	115	140	160

Verify that the law is true and determine the law of the graph.
Determine also the luminosity when 75 volts is applied across the lamp.

[$I = 0.0012\ V^2$; 6.75 candelas]

4 The head of pressure h and the flow velocity v are measured and are believed to be connected by the law $v = ah^b$, where a and b are constants. The results are as shown below.

h	10.6	13.4	17.2	24.6	29.3
v	9.77	11.00	12.44	14.88	16.24

Verify that the law is true and determine values of a and b.

[$a = 3.0; b = 0.5$]

5 Quantities x and y are believed to be related by a law of the form $y = mn^x$. The values of x and corresponding values of y are:

x	0	0.5	1.0	1.5	2.0	2.5	3.0
y	1.0	3.2	10	31.6	100	316	1000

Verify the law and find the values of m and n. [$m = 1; n = 10$]

6 Experimental values of p and corresponding values of q are shown below.

p	−13.2	−27.9	−62.2	−383.2	−1581	−2931
q	0.30	0.75	1.23	2.32	3.17	3.54

Show that the law relating p and q is $p = ab^q$, where a and b are constants. Determine (i) values of a and b, and state the law, (ii) the value of p when q is 2.0 and (iii) the value of q when p is −2000.

[(i) $a = -8$, $b = 5.3$, $p = -8(5.3)^q$; (ii) −224.7; (iii) 3.31]

7 The activity of a mixture of radioactive isotope is believed to vary according to the law $R = R_0 t^{-c}$, where R_0 and c are constants. Experimental results are shown below.

R	9.72	2.65	1.15	0.47	0.32	0.23
t	2	5	9	17	22	28

Verify that the law is true and determine approximate values of R_0 and c.

[$R_0 = 26.0; c = 1.42$]

8 Experimental values of x and y are measured as follows.

x	0.4	0.9	1.2	2.3	3.8
y	8.35	13.47	17.94	51.32	215.20

The law relating x and y is believed to be of the form $y = a.b^x$, where a and b are constants. Determine the approximate values of a and b. Hence find the value of y when x is 2.0 and the value of x when y is 100.

[$a = 5.7$; $b = 2.6$, 38.53, 3.0]

9 Atmospheric pressure p is measured at varying altitudes h and the results are as shown below:

Altitude, h m	500	1500	3000	5000	8000
pressure, p cm	73.39	68.42	61.60	53.56	43.41

Show that the quantities are related by the law $p = ae^{kh}$, where a and k are constants. Determine the values of a and k and state the law. Find also the atmospheric pressure at 10 000 m.

[$a = 76$; $k = -7 \times 10^{-5}$; $p = 76e^{-7 \times 10^5 h}$; 37.74 cm]

10 At particular times, t minutes, measurements are made of the temperature, θ°C of a cooling liquid and the following results are obtained:

temperature θ°C	92.2	55.9	33.9	20.6	12.5
time t minutes	10	20	30	40	50

Prove that the quantities follow a law of the form $\theta = \theta_0 e^{kt}$, where θ_0 and k are constants, and determine the approximate values of θ_0 and k.

[$\theta_0 = 152$; $k = -0.05$]

11 Determine the law of the form $y = ae^{kx}$ which relates the following values.

y	0.0306	0.285	0.841	5.21	173.2	1181
x	4.0	5.3	9.8	17.4	32.0	40.0

[$y = 0.08e^{0.24x}$]

12 The tension T in a belt passing round a pulley wheel and in contact with the pulley over an angle of θ radians is given by $T = T_0 e^{\mu\theta}$ where T_0 and μ are constants. Experimental results obtained are:

T newtons	47.9	52.8	60.3	70.1	80.9
θ radians	1.12	1.48	1.97	2.53	3.06

Determine approximate values of T_0 and μ. Hence find the tension when θ is 2.25 radians and the value of θ when the tension is 50.0 newtons.

[$T_0 = 35.4$ N; $\mu = 0.27$, 65.0 N, 1.28 radians]

9 Algebra (6) – Simplification of Boolean expressions and cyclic codes

A MAIN POINTS CONCERNED WITH THE SIMPLIFICATION OF BOOLEAN EXPRESSIONS AND CYCLIC CODES

1 A Boolean expression may be used to describe a complex switching circuit or logic system. If the Boolean expression can be simplified, then the number of switches or logic elements can be reduced resulting in a saving in cost. Three principal ways of simplifying Boolean expressions are:
(a) by using the laws and rules of Boolean algebra, (see *Mathematics Checkbook 2,* chapter 19),
(b) by applying de Morgan's laws, and
(c) by using Karnaugh maps.

2 A summary of the principal laws and rules of Boolean algebra are given in *Table 1.* The way in which these laws and rules may be used to simplify Boolean expressions is shown in *Problems 1 and 2.*

TABLE 1 SOME LAWS AND RULES OF BOOLEAN ALGEBRA

Reference	*Name*	*Rule or Law*
1 2	Commutative Laws	$A+B = B+A$ $A.B = B.A$
3 4	Associative Laws	$(A+B)+C = A+(B+C)$ $(A.B).C = A.(B.C)$
5 6	Distributive Laws	$A.(B+C) = A.B+A.C$ $A+(B.C) = (A+B).(A+C)$
7 8 9 10	Sum Rules	$A+0 = A$ $A+1 = 1$ $A+A = A$ $A+\bar{A} = 1$
11 12 13 14	Product Rules	$A.0 = 0$ $A.1 = A$ $A.A = A$ $A.\bar{A} = 0$
15 16 17	Absorption Rules	$A+A.B = A$ $A.(A+B) = A$ $A+\bar{A}.B = A+B$

3 **De Morgan's laws** may be used to simplify **not**-functions having two or more elements. The laws state that:

$$\overline{A+B} = \overline{A}.\overline{B} \text{ and } \overline{A.B} = \overline{A}+\overline{B}$$

and may be verified by using a truth table, (see *Problem 3*). The application of de Morgan's laws in simplifying Boolean expressions is shown in *Problems 4 and 5.*

4 *Karnaugh Maps*

(i) Two-variable Karnaugh maps

A truth table for a two-variable expression is shown in *Table 2(a)*, the '1' in the third row output showing that $Z = A.\overline{B}$. Each of the four possible Boolean expressions associated with a two-variable function can be depicted as shown in *Table 2(b)* in which one cell is allocated to each row of the truth table. A matrix similar to that shown in *Table 2(b)* can be used to depict $Z = A.\overline{B}$, by putting a 1 in the cell corresponding to $A.\overline{B}$ and 0's in the remaining cells. This method of depicting a Boolean expression is called a two-variable **Karnaugh map**, and is shown in *Table 2(c).*

TABLE 2

Inputs A	Inputs B	Output Z	Boolean expression
0	0	0	$\overline{A}.\overline{B}$
0	1	0	$\overline{A}.B$
1	0	1	$A.\overline{B}$
1	1	0	$A.B$

(a)

B \ A	0 ($\overline{A}$)	1 (A)
0($\overline{B}$)	$\overline{A}.\overline{B}$	$A.\overline{B}$
1(B)	$\overline{A}.B$	$A.B$

(b)

B \ A	0	1
0	0	1
1	0	0

(c)

To simplify a two-variable Boolean expression, the Boolean expression is depicted on a Karnaugh map, as outlined above. Any cells on the map having either a common vertical side or a common horizontal side are grouped together to form a **couple.** (This is a coupling together of cells, not just combining two together). The simplified Boolean expression for a couple is given by those variables common to all cells in the couple.

(ii) Three-variable Karnaugh maps

A truth table for a three-variable expression is shown in table 3(a), the 1's in the output column showing that: $Z = \overline{A}.\overline{B}.C+\overline{A}.B.C+A.B.\overline{C}$. Each of the eight possible Boolean expressions associated with a three-variable function can be depicted as shown in *Table 3(b)*, in which one cell is allocated to each row of the truth table. A matrix similar to that shown in *Table 3(b)* can be used to depict: $Z = \bar{A}.\bar{B}.C.+\bar{A}.B.C.+A.B.\bar{C}$, by putting 1's in the cells corresponding to the Boolean terms on the right of the Boolean equation and 0's in the remaining cells. This method of depicting a three-variable Boolean expression is called a three-variable Karnaugh map, and is shown in *Table 3(c)*.

To simplify a three-variable Boolean expression, the Boolean expression is depicted on a Karnaugh map as outlined above. Any cells on the map having common edges either vertically or horizontally are grouped together to form couples of four cells or two cells. During coupling the horizontal lines at the top and bottom of the cells are taken as a common edge, as are the vertical lines on the left and right of the cells. The simplified Boolean expression for a couple is given by those variables common to all cells in the couple.

TABLE 3

Inputs A	B	C	Output Z	Boolean expression
0	0	0	0	$\bar{A}.\bar{B}.\bar{C}$
0	0	1	1	$\bar{A}.\bar{B}.C$
0	1	0	0	$\bar{A}.B.\bar{C}$
0	1	1	1	$\bar{A}.B.C$
1	0	0	0	$A.\bar{B}.\bar{C}$
1	0	1	0	$A.\bar{B}.C$
1	1	0	1	$A.B.\bar{C}$
1	1	1	0	$A.B.C$

(a)

C \ A.B	00 ($\bar{A}.\bar{B}$)	01 ($\bar{A}.B$)	11 ($A.B$)	10 ($A.\bar{B}$)
0($\bar{C}$)	$\bar{A}.\bar{B}.\bar{C}$	$\bar{A}.B.\bar{C}$	$A.B.\bar{C}$	$A.\bar{B}.\bar{C}$
1(C)	$\bar{A}.\bar{B}.C$	$\bar{A}.B.C$	$A.B.C$	$A.\bar{B}.C$

(b)

C \ A.B	00	01	11	10
0	0	0	1	0
1	1	1	0	0

(c)

(iii) Four-variable Karnaugh maps

A truth table for a four-variable expression is shown in Table 4(a), the 1's in the output column showing that:

$$Z = \bar{A}.\bar{B}.C.\bar{D} + \bar{A}.B.C.\bar{D} + A.\bar{B}.C.\bar{D} + A.B.C.\bar{D}.$$

Each of the sixteen possible Boolean expressions associated with a four-variable function can be depicted as shown in *Table 4(b)*, in which one cell is

TABLE 4

Inputs A	Inputs B	Inputs C	Inputs D	Output Z	Boolean expression
0	0	0	0	0	$\bar{A}.\bar{B}.\bar{C}.\bar{D}$
0	0	0	1	0	$\bar{A}.\bar{B}.\bar{C}.D$
0	0	1	0	1	$\bar{A}.\bar{B}.C.\bar{D}$
0	0	1	1	0	$\bar{A}.\bar{B}.C.D$
0	1	0	0	0	$\bar{A}.B.\bar{C}.\bar{D}$
0	1	0	1	0	$\bar{A}.B.\bar{C}.D$
0	1	1	0	1	$\bar{A}.B.C.\bar{D}$
0	1	1	1	0	$\bar{A}.B.C.D$
1	0	0	0	0	$A.\bar{B}.\bar{C}.\bar{D}$
1	0	0	1	0	$A.\bar{B}.\bar{C}.D$
1	0	1	0	1	$A.\bar{B}.C.\bar{D}$
1	0	1	1	0	$A.\bar{B}.C.D$
1	1	0	0	0	$A.B.\bar{C}.\bar{D}$
1	1	0	1	0	$A.B.\bar{C}.D$
1	1	1	0	1	$A.B.C.\bar{D}$
1	1	1	1	0	$A.B.C.D$

(a)

C.D \ A.B	00 ($\bar{A}.\bar{B}$)	01 ($\bar{A}.B$)	11 ($A.B$)	10 ($A.\bar{B}$)
00 ($\bar{C}.\bar{D}$)	$\bar{A}.\bar{B}.\bar{C}.\bar{D}$	$\bar{A}.B.\bar{C}.\bar{D}$	$A.B.\bar{C}.\bar{D}$	$A.\bar{B}.\bar{C}.\bar{D}$
01 ($\bar{C}.D$)	$\bar{A}.\bar{B}.\bar{C}.D$	$\bar{A}.B.\bar{C}.D$	$A.B.\bar{C}.D$	$A.\bar{B}.\bar{C}.D$
11 ($C.D$)	$\bar{A}.\bar{B}.C.D$	$\bar{A}.B.C.D$	$A.B.C.D$	$A.\bar{B}.C.D$
10 ($C.\bar{D}$)	$\bar{A}.\bar{B}.C.\bar{D}$	$\bar{A}.B.C.\bar{D}$	$A.B.C.\bar{D}$	$A.\bar{B}.C.\bar{D}$

(b)

C.D \ A.B	0.0	0.1	1.1	1.0
0.0	0	0	0	0
0.1	0	0	0	0
1.1	0	0	0	0
1.0	1	1	1	1

(c)

allocated to each row of the truth table. A matrix similar to that shown in *Table 4(b)* can be used to depict

$Z = \overline{A}.\overline{B}.C.\overline{D}+\overline{A}.B.C.\overline{D}+A.\overline{B}.C.\overline{D}+A.B.C.\overline{D}$

by putting 1's in the cells corresponding to the Boolean terms on the right of the Boolean equation and 0's in the remaining cells. This method of depicting a four-variable expression is called a four-variable Karnaugh map, and is shown in *Table 4(c)*.

To simplify a four-variable Boolean expression, the Boolean expression is depicted on a Karnaugh map as outlined above. Any cells on the map having common edges either vertically or horizontally are grouped together to form couples of eight cells, four cells or two cells. During coupling, the horizontal lines at the top and bottom of the cells may be considered to be common edges, as are the vertical lines on the left and the right of the cells. The simplified Boolean expression for a couple is given by those variables common to all cells in the couple.

(iv) **Summary of procedure when simplifying a Boolean expression using a Karnaugh map**

(a) Draw a four, eight or sixteen-cell matrix, depending on whether there are two, three or four variables.

(b) Mark in the Boolean expression by putting 1's in the appropriate cells.

(c) Form couples of 8, 4 or 2 cells having common edges, forming the largest groups of cells possible. (Note that a cell containing a 1 may be used more than once when forming a couple. Also note that each cell containing a 1 must be used at least once.)

(d) The Boolean expression for a couple is given by the variables which are common to all cells in the couple.

(See *Problems 6 to 11.*)

5 **Cyclic codes**

In a cyclic code only one bit changes when moving from one number to the next consecutive number and also only one bit changes when moving from the last number in the code back to the first number. One widely used cyclic code is the **Gray code.** A matrix similar to that for a Karnaugh map may be used to generate a cyclic code. When moving from one cell to another, either horizontally or vertically, only one bit changes in the binary number generated. Also, only one bit changes when moving from a cell in the last column to a cell in the first column in the same row and vice versa and from a cell in the bottom row to a cell in the top row in the same column, and vice versa. The techniques of generating a cyclic code are shown in *Problems 12 to 14.*

B WORKED PROBLEMS ON SIMPLIFICATION OF BOOLEAN EXPRESSIONS AND CYCLIC CODES

Problem 1 Simplify $A.\overline{C}+\overline{A}.(B+C)+A.B.(C+\overline{B})$ using the rules of Boolean algebra.

The rules are given in *Table 1.*

	Rule
$A.\overline{C}+\overline{A}.(B+C)+A.B.(C+\overline{B}) = A.\overline{C}+\overline{A}.B+\overline{A}.C+A.B.C+A.B.\overline{B}$	(5)
$= A.\overline{C}+\overline{A}.B+\overline{A}.C+A.B.C+A.0$	(14)

$= A.\overline{C}+\overline{A}.B+\overline{A}.C+A.B.C$	(11)
$= A.(\overline{C}+B.C)+\overline{A}.B+\overline{A}.C$	(5)
$= A.(\overline{C}+B)+\overline{A}.B+\overline{A}.C$	(17)
$= A.\overline{C}+A.B+\overline{A}.B+\overline{A}.C$	(5)
$= A.\overline{C}+B.(A+\overline{A})+\overline{A}.C$	(5)
$= A.\overline{C}+B.1+\overline{A}.C$	(10)
$= \mathbf{A.\overline{C}+B+\overline{A}.C}$	(12)

Problem 2 Simplify the expression $P.\overline{Q}.R+P.Q.(\overline{P}+R)+Q.R(\overline{Q}+P)$, using the rules of Boolean algebra.

The rules are given in *Table 1.*

		Rule
$P.\overline{Q}.R+P.Q.(\overline{P}+R)+Q.R.(\overline{Q}+P)$	$= P.\overline{Q}.R+P.Q.\overline{P}+P.Q.R+Q.R.\overline{Q}+Q.R.P$	(5)
	$= P.\overline{Q}.R+0.Q+P.Q.R+0.R+P.Q.R$	(14)
	$= P.\overline{Q}.R+P.Q.R+P.Q.R$	(7) and (11)
	$= P.\overline{Q}.R+P.Q.R$	(9)
	$= P.R.(Q+\overline{Q})$	(5)
	$= P.R.1$	(10)
	$= \mathbf{P.R}$	(12)

Problem 3 Verify that $\overline{A+B} = \overline{A}.\overline{B}$

A Boolean expression may be verified by using a truth table. In *Table 5*, columns 1 and 2 give all the possible arrangements of the inputs A and B. Column 3 is the **or**-function applied to columns 1 and 2 and column 4 is the **not**-function

TABLE 5

1 A	2 B	3 A+B	4 $\overline{A+B}$	5 $\overline{A}$	6 $\overline{B}$	7 $\overline{A}.\overline{B}$
0	0	0	1	1	1	1
0	1	1	0	1	0	0
1	0	1	0	0	1	0
1	1	1	0	0	0	0

applied to column 3. Columns 5 and 6 are the **not**-function applied to columns 1 and 2 respectively and column 7 is the **and**-function applied to columns 5 and 6. Since columns 4 and 7 have the same pattern of 0's and 1's this verifies that $\overline{A+B} = \overline{A}.\overline{B}$.

Problem 4 Simplify the Boolean expression $(\overline{\overline{A}.B})+(\overline{\overline{A}+B})$ by using de Morgan's laws and the rules of Boolean algebra.

Applying de Morgan's law (see para. 3) to the first term gives:

$\overline{\overline{A}.B} = \overline{\overline{A}}+\overline{B} = A+\overline{B}$, since $\overline{\overline{A}} = A$

Applying de Morgan's law to the second term gives:

$\overline{\overline{A}+B} = \overline{\overline{A}}.\overline{B} = A.\overline{B}$

Thus, $(\overline{\overline{A}.B})+(\overline{\overline{A}+B}) = (A+\overline{B})+A.\overline{B}$

Removing the bracket and reordering gives: $A+A.\overline{B}+\overline{B}$

But, by rule 15, *Table 1*, $A+A.B = A$. It follows that: $A+A.\overline{B} = A$
Thus: $\mathbf{(\overline{A}.B) + (\overline{A}+B) = A+\overline{B}}$

Problem 5 Simplify the Boolean expression $\overline{(A.\overline{B}+C)}.(\overline{A}+\overline{B.\overline{C}})$ by using de Morgan's laws and the rules of Boolean algebra.

Applying de Morgan's laws, (see para. 3), to the first term gives:

$\overline{A.\overline{B}+C} = \overline{A.\overline{B}}.\overline{C} = (\overline{A}+\overline{\overline{B}}).\overline{C} = (\overline{A}+B).\overline{C} = \overline{A}.C+B.\overline{C}$

Applying de Morgan's law to the second term gives:

$\overline{A}+\overline{B.\overline{C}} = \overline{A}+(\overline{B}+C)$

Thus $\overline{(A.\overline{B}+C)}.(\overline{A}+\overline{B.\overline{C}}) = (\overline{A}.C+B.\overline{C}).(\overline{A}+\overline{B}+C)$
$= \overline{A}.\overline{A}.C+\overline{A}.\overline{B}.C+\overline{A}.\overline{C}.C+\overline{A}.B.\overline{C}+B.\overline{B}.\overline{C}+B.\overline{C}.C$

But from *Table 1*, $\overline{A}.\overline{A} = \overline{A}$ and $\overline{C}.C = B.\overline{B} = 0$
Hence the Boolean expression becomes $\overline{A}.C+\overline{A}.\overline{B}.C+\overline{A}.B.\overline{C}$
The first two terms are of the form $P+P.Q = P$, where $\overline{A}.C \equiv P$ and $\overline{B} \equiv Q$, giving $\overline{A}.C+\overline{A}.B.\overline{C}$, i.e., $\overline{A}.(C+B.\overline{C})$
But from *Table 1*, $A+\overline{A}.B = A+B$. Hence, $C+\overline{C}.B = C+B$
Thus: $\mathbf{\overline{(A.\overline{B}+C)}.(\overline{A}+\overline{B.\overline{C}}) = \overline{A}.(B+C)}$

Problem 6 Use the Karnaugh map techniques to simplify the expression $\overline{P}.\overline{Q}+\overline{P}.Q$

Using the procedure given in para. 4(iv),

(a) the two-variable matrix is drawn and is shown in *Table 6*,
(b) The term $\overline{P}.\overline{Q}$ is marked with a 1 in the top left-hand cell, corresponding to $P = 0$ and $Q = 0$. $\overline{P}.Q$ is marked with a 1 in the bottom left-hand cell corresponding to $P = 0$ and $Q = 1$.
(c) The two cells containing 1's have a common horizontal edge and thus a vertical couple, shown by the broken line, can be formed.
(d) The variable common to both cells in the couple is $P = 0$, i.e. $\overline{P}$, thus $\mathbf{\overline{P}.\overline{Q}+\overline{P}.Q = \overline{P}}$

TABLE 6

Q \ P	0	1
0	1	0
1	1	0

Problem 7 **Simplify the expression $\overline{X}.Y.\overline{Z}+\overline{X}.\overline{Y}.Z+X.Y.\overline{Z}+X.\overline{Y}.Z$ by using Karnaugh map techniques.**

Using the procedure given in para. 4(iv),

(a) a three-variable matrix is drawn and is shown in *Table 7*.
(b) The 1's on the matrix correspond to the expression given, i.e., for $\overline{X}.Y.\overline{Z}$, $X = 0$, $Y = 1$ and $Z = 0$ and hence corresponds to the cell in the top row and second column, and so on.
(c) Two couples can be formed, shown by the broken lines. The couple in the bottom row may be formed since the vertical lines on the left and right of th cells are taken as a common edge.
(d) The variables common to the couple in the top row are $Y = 1$ and $Z = 0$,

TABLE 7

Z \ X.Y	0.0	0.1	1.1	1.0
0	0	1	1	0
1	1	0	0	1

that is, $Y.\bar{Z}$ and the variables common to the couple in the bottom row are $Y = 0, Z = 1$, that is, $\bar{Y}.Z$. Hence:

$$\bar{X}.Y.\bar{Z}+\bar{X}.\bar{Y}.Z+X.Y.\bar{Z}+X.\bar{Y}.Z = Y.\bar{Z}+\bar{Y}.Z$$

Problem 8 Use a Karnaugh map technique to simplify the expression $(\overline{\bar{A}.B}).(\overline{\bar{A}+B})$.

Using the procedure given in para. 4(iv), a two-variable matrix is drawn and is shown in *Table 8.*

TABLE 8

B \ A	0	1
0	1	1 2
1		1

TABLE 9

R \ P.Q	0.0	0.1	1.1	1.0
0	3 2	3 2	3 1	3 1
1	4 1	4 2	3 1	4 1

(a)

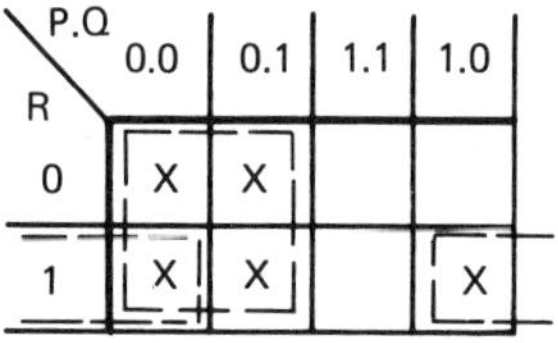

R \ P.Q	0.0	0.1	1.1	1.0
0	X	X		
1	X	X		X

(b)

$\bar{A}.B$ corresponds to the bottom left-hand cell and $(\overline{\bar{A}.B})$ must therefore be all cells except this one, marked with a 1 in *Table 8.* $(\bar{A}+B)$ corresponds to all the cells except the top right-hand cell marked with a 2 in *Table 8.* Hence $(\overline{\bar{A}+B})$ must correspond to the cell marked with a 2. The expression $(\overline{\bar{A}.B}).(\overline{\bar{A}+B})$ corresponds to the cell having both 1 and 2 in it,

i.e. $\quad \mathbf{(\overline{\bar{A}.B}).(\overline{\bar{A}+B}) = A.\bar{B}}$

Problem 9 Simplify $\overline{(P+\bar{Q}.R)}+\overline{(P.Q+\bar{R})}$ using a Karnaugh map technique.

The term $(P+\bar{Q}.R)$ corresponds to the cells marked 1 on the matrix in *Table 9(a)*, hence $\overline{(P+\bar{Q}.R)}$ corresponds to the cells marked 2. Similarly $(P.Q+\bar{R})$ corresponds to the cells marked 3 in *Table 9(a)*, hence $\overline{(P.Q+\bar{R})}$ corresponds to the cells marked 4. The expression $\overline{(P+\bar{Q}.R)}+\overline{(P.Q+\bar{R})}$ corresponds to cells marked with either a 2 or with a 4 and is shown in *Table 9(b)* by X.s. These cells may be coupled as shown by the broken lines. The variables common to the group of four cells is $P = 0$, i.e., $\bar{P}$, and those common to the group of two cells are $Q = 0$, $R = 1$, i.e., $\bar{Q}.R$. Thus:

$$\mathbf{\overline{(P+\bar{Q}.R)}+\overline{(P.Q+\bar{R})} = \bar{P}+\bar{Q}.R}$$

Problem 10 Use Karnaugh map techniques to simplify the expression:
$A.B.\overline{C}.\overline{D}+A.B.C.D+\overline{A}.B.C.D+A.B.C.\overline{D}+\overline{A}.B.C.\overline{D}$

Using the procedure given in para. 4(iv), a four-variable matrix is drawn and is shown in *Table 10.* The 1's marked on the matrix correspond to the expression given. Two couples can be formed and are shown by the broken lines. The four-cell couple has $B = 1$, $C = 1$, i.e. $B.C$ as the common variables to all four cells and the two-cell couple has $A.B.\overline{D}$ as the common variables to both cells. Hence, the expression simplifies to:

$B.C+A.B.\overline{D}$, i.e., $\mathbf{B.(C+A.\overline{D})}$

TABLE 10

C.D \ A.B	0.0	0.1	1.1	1.0
0.0			1	
0.1				
1.1		1	1	
1.0		1	1	

Problem 11 Simplify the expression
$\overline{A}.\overline{B}.\overline{C}.\overline{D}+A.\overline{B}.\overline{C}.\overline{D}+\overline{A}.\overline{B}.C.\overline{D}+A.\overline{B}.C.\overline{D}+A.B.C.D$
by using Karnaugh map techniques.

The Karnaugh map for the expression is shown in *Table 11.* Since the top and bottom horizontal lines are common edges and the vertical lines on the left and right of the cells are common, then the four corner cells form a couple, $\overline{B}.\overline{D}$, (the cells can be considered as if they are stretched to completely cover a sphere, as far as common edges are concerned). The cell $A.B.C.D$ cannot be coupled with any other. Hence the expression simplifies to $\boldsymbol{\overline{B}.\overline{D}+A.B.C.D}$

TABLE 11

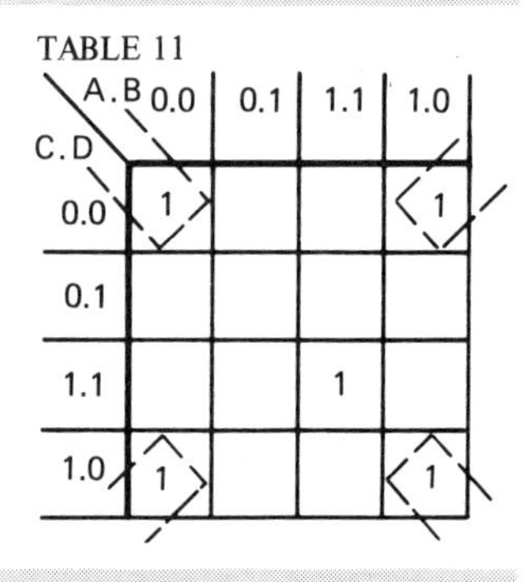

C.D \ A.B	0.0	0.1	1.1	1.0
0.0	1			1
0.1				
1.1			1	
1.0	1			1

Problem 12 Show how the Gray cyclic code can be obtained.

A Karnaugh map matrix, as shown in *Table 12(a)*, is used. Moving one cell at a time either horizontally or vertically generates a cyclic code, care being taken to ensure the last number in the code has a common edge with the first number. The particular movement down the first column, up the second column and so on generates the Gray cyclic code, as shown in *Table 12(b).*

Problem 13 Show how a Gray, excess-three, binary-coded-decimal cyclic code can be obtained.

A Karnaugh map matrix, as shown in *Table 13(a)* is used. Since the code is a BCD, there are only ten states and since it is an 'X's 3' code, the numbers are advanced by three on the natural code. To meet these requirements, the first

TABLE 12

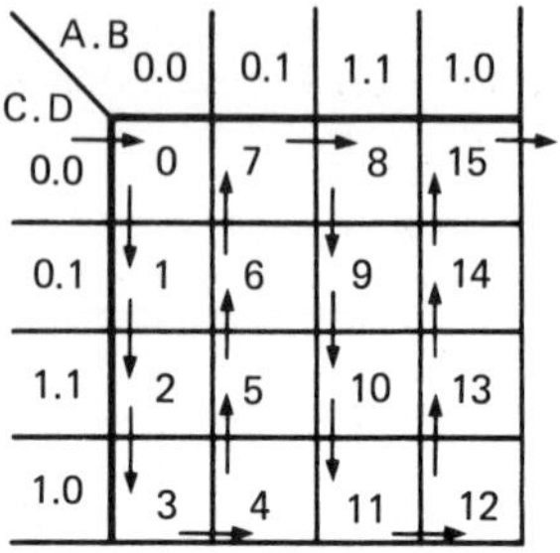

C.D \ A.B	0.0	0.1	1.1	1.0
0.0	0	7	8	15
0.1	1	6	9	14
1.1	2	5	10	13
1.0	3	4	11	12

(a)

Denary number	A	B	C	D
0	0	0	0	0
1	0	0	0	1
2	0	0	1	1
3	0	0	1	0
4	0	1	1	0
5	0	1	1	1
6	0	1	0	1
7	0	1	0	0
8	1	1	0	0
9	1	1	0	1
10	1	1	1	1
11	1	1	1	0
12	1	0	1	0
13	1	0	1	1
14	1	0	0	1
15	1	0	0	0

(b)

TABLE 13

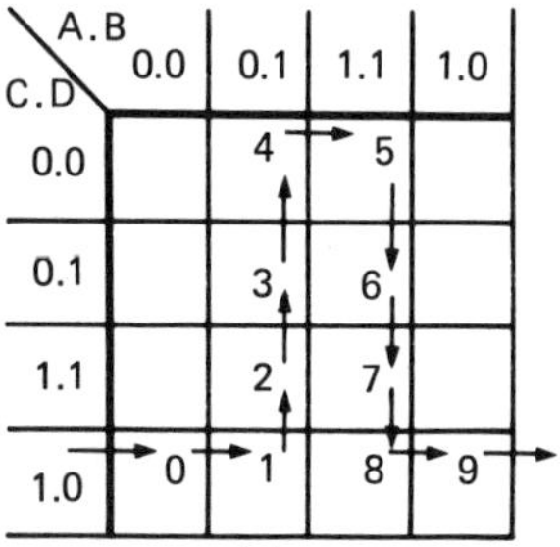

C.D \ A.B	0.0	0.1	1.1	1.0
0.0		4	5	
0.1		3	6	
1.1		2	7	
1.0	0	1	8	9

(a)

Denary number	A	B	C	D
0	0	0	1	0
1	0	1	1	0
2	0	1	1	1
3	0	1	0	1
4	0	1	0	0
5	1	1	0	0
6	1	1	0	1
7	1	1	1	1
8	1	1	1	0
9	1	0	1	0

(b)

TABLE 14

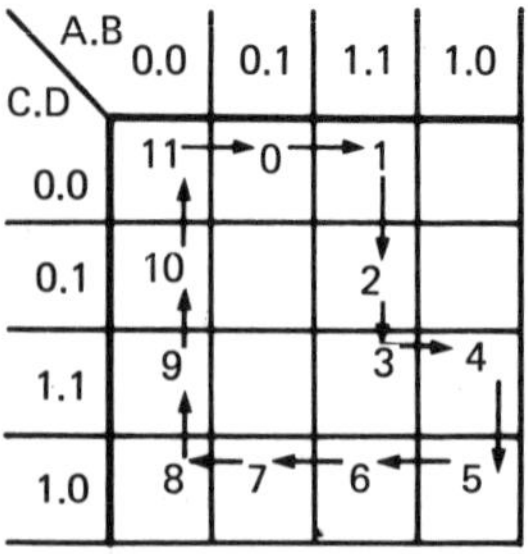

C.D \ A.B	0.0	0.1	1.1	1.0
0.0	11	0	1	
0.1	10		2	
1.1	9		3	4
1.0	8	7	6	5

(a)

Denary number	A	B	C	D
0	0	1	0	0
1	1	1	0	0
2	1	1	0	1
3	1	1	1	1
4	1	0	1	1
5	1	0	1	0
6	1	1	1	0
7	0	1	1	0
8	0	0	1	0
9	0	0	1	1
10	0	0	0	1
11	0	0	0	0

(b)

number in the code starts in the bottom left-hand cell, and the code is as shown in *Table 13(b)*.

Problem 14. Generate a cyclic code having twelve states starting with 0100, 1100, 1101, ...

There is no unique solution to this problem, but one possible solution is shown in *Table 14.* The first three given states are marked on the Karnaugh matrix as 0, 1, 2. Moving one cell at a time either horizontally or vertically, and completing the numbering with number 11, (since there are twelve states starting at 0), in a cell having a common edge with the cell containing 0, one possible solution is as shown in *Table 14(a).* The corresponding cyclic code is obtained by reading the appropriate *ABCD* states (i.e. 0 or 1 values) from the edges of the matrix and is shown in *Table 14(b).*

C FURTHER PROBLEMS ON THE SIMPLIFICATION OF BOOLEAN EXPRESSIONS AND CYCLIC CODES

In *Problems 1 to 5*, use the rules of Boolean algebra, given in *Table 1*, to simplify the expressions given.

1 $\bar{F}.\bar{G}+F.\bar{G}+\bar{G}.(F+\bar{F})$ $[\bar{G}]$
2 $F.\bar{G}+F.(G+\bar{G})+F.G$ $[F]$
3 $\bar{P}.\bar{Q}.\bar{R}+\bar{P}.Q.R+P.\bar{Q}.\bar{R}$ $[\bar{Q}.\bar{R}+\bar{P}.Q.R]$
4 $R.(P.Q+P.\bar{Q})+\bar{R}.(\bar{P}.\bar{Q}+\bar{P}.Q)$ $[P.R+\bar{P}.\bar{R}]$
5 $\bar{R}.(\bar{P}.\bar{Q}+P.Q+P.\bar{Q})+P.(Q.R+\bar{Q}.R)$ $[P+\bar{Q}.\bar{R}]$
6 Use a truth table to verify that $\overline{A.B} = \bar{A}+\bar{B}$
7 Verify by using a truth table that $\bar{A}+\bar{B}+\bar{C} = \overline{A.B.C}$
8 Show that $\overline{\bar{A}+\bar{B}+\bar{C}} = A.B.C$

In *Problems 9 to 13* use de Morgan's laws and the rules of Boolean algebra given in *Table 1* to simplify the expressions given.

9 $(\bar{A}.\bar{B}).\overline{(\bar{A}.B)}$ $[\bar{A}.\bar{B}]$
10 $\overline{(A+\overline{\bar{B}.\bar{C}})}+(\overline{A.B}+C)$ $[\bar{A}+\bar{B}+C]$
11 $\overline{(\bar{A}.B+B.\bar{C})}.\overline{A.\bar{B}}$ $[\bar{A}.\bar{B}+A.B.C]$
12 $(\overline{A.\bar{B}}+\overline{B.\bar{C}})+\overline{(\bar{A}.B)}$ $[1]$
13 $(\overline{P.\bar{Q}}+\overline{\bar{P}.R}).(\bar{P}.\overline{Q.R})$ $[\bar{P}.(\bar{Q}+\bar{R})]$

In *Problems 14 to 27* use Karnaugh map techniques to simplify the expressions given.

14 $\bar{X}.Y+X.Y$ $[Y]$
15 $\bar{X}.\bar{Y}+X.\bar{Y}+X.Y$ $[X+\bar{Y}]$
16 $\bar{X}.\bar{Y}+\bar{X}.Y+X.Y$ $[\bar{X}+Y]$
17 $(\bar{P}.\bar{Q}).\overline{(\bar{P}.Q)}$ $[\bar{P}.\bar{Q}]$
18 $\overline{(P.\bar{Q}.R)}+\overline{(P+\bar{Q}+R)}$ $[\bar{P}+Q+\bar{R}]$
19 $\bar{P}.\bar{Q}.\bar{R}+\bar{P}.Q.\bar{R}+P.Q.\bar{R}$ $[\bar{R}.(\bar{P}+Q)]$
20 $\bar{P}.\bar{Q}.\bar{R}+P.Q.\bar{R}+P.Q.R+P.\bar{Q}.R$ $[P.(Q+R)+\bar{P}.\bar{Q}.\bar{R}]$
21 $(\overline{X.Y}+Z)+\overline{(X+\overline{\bar{Y}.\bar{Z}})}$ $[\bar{X}+\bar{Y}+Z]$
22 $\overline{(\bar{X}.Z+X.Y.Z)}.\overline{(X.\bar{Y})}$ $[\bar{Z}.(\bar{X}+Y)]$
23 $(\overline{\bar{X}.Y}+\overline{\bar{X}.Z})+(X.\overline{Y.Z})$ $[X+\bar{Y}+\bar{Z}]$
24 $\bar{A}.\bar{B}.\bar{C}.\bar{D}+\bar{A}.B.\bar{C}.\bar{D}+\bar{A}.B.\bar{C}.D$ $[\bar{A}.\bar{C}.(B+\bar{D})]$
25 $\bar{A}.\bar{B}.C.D+\bar{A}.\bar{B}.C.\bar{D}+A.\bar{B}.C.\bar{D}$ $[\bar{B}.C.(\bar{A}+\bar{D})]$
26 $\bar{A}.B.\bar{C}.D+A.B.\bar{C}.D+A.B.C.D+A.\bar{B}.\bar{C}.D+A.\bar{B}.C.D$ $[D.(A+B.\bar{C})]$

27 $\overline{A}.\overline{B}.\overline{C}.D+A.B.\overline{C}.\overline{D}+A.\overline{B}.\overline{C}.\overline{D}+A.B.C.\overline{D}+A.\overline{B}.C.\overline{D}$ [$A.\overline{D}+\overline{A}.\overline{B}.\overline{C}.D$]

In *Problems 28 to 30*, generate cyclic codes having the number of states indicated and starting with the groups of bits shown. There are no unique solutions, but one possible solution is shown in each case.

28 10 states: 0100, 1100, 1101, ...

[0100, 1100, 1101, 0101, 0001, 0011, 0111, 1111, 1110, 0110]

29 12 states: 0000, 0100, 0101, 0001, ...

[0000, 0100, 0101, 0001, 0011, 0111, 1111, 1011, 1010, 1110, 0110, 0010]

30 10 states: 1011, 0011, 0010, ...

[1011, 0011, 0010, 0110, 0111, 0101, 0100, 1100, 1101, 1111]

10 Algebra (7) – Logic circuits

A MAIN POINTS CONCERNED WITH LOGIC CIRCUITS

1 In practice, logic gates are used to perform the **and**, **or** and **not**-functions introduced in *Mathematics 2 Checkbook*, chapter 19, and used in chapter 9 of this book. Logic gates can be made from switches, magnetic devices or fluidic devices, but most logic gates in use are electronic devices. Various logic gates are available. For example, the Boolean expression $(A.B.C)$ can be produced using a three-input, **and**-gate and $(C+D)$ by using a two-input **or**-gate. The principal gates in common use are introduced in paras. 2 to 6. The term 'gate' is used in the same

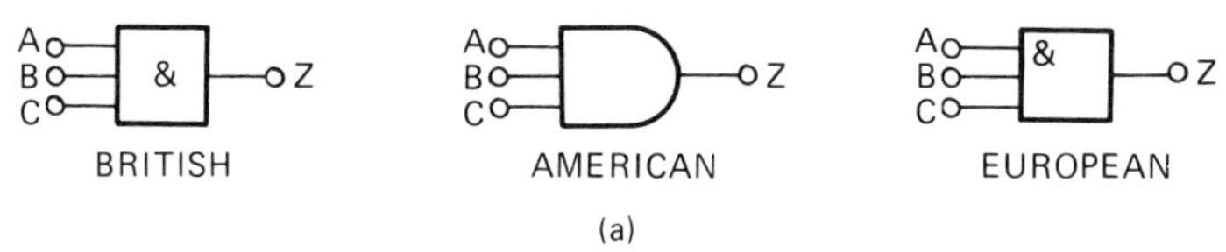

(a)

INPUTS A	B	C	OUTPUT Z=A.B.C
0	0	0	0
0	0	1	0
0	1	0	0
0	1	1	0
1	0	0	0
1	0	1	0
1	1	0	0
1	1	1	1

(b)

Fig 1

sense as a normal gate, the open state being indicated by a binary '1' and the closed state by a binary '0'. A gate will only open when the requirements of the gate are met and, for example, there will only be a '1' output on a two-input **and**-gate when both the inputs to the gate are at a '1' state.

2 **The and-gate**

The different symbols used for a three-input, **and**-gate are shown in *Fig 1(a)* and the truth table is shown in *Fig 1(b).* This shows that there will only be a '1' output when A is 1 and B is 1 and C is 1, written as:

$Z = A.B.C$

3 **The or-gate**

The different symbols used for a three-input **or**-gate are shown in *Fig 2(a)* and the truth table is shown in *Fig 2(b).* This shows that there will be a '1' output when A is 1, or B is 1, or C is 1, or any combination of A, B or C is 1, written as $Z = A+B+C$.

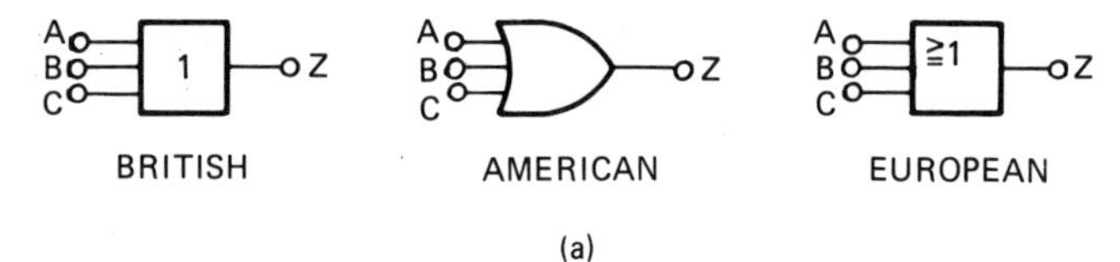

(a)

INPUTS A	B	C	OUTPUT Z=A+B+C
0	0	0	0
0	0	1	1
0	1	0	1
0	1	1	1
1	0	0	1
1	0	1	1
1	1	0	1
1	1	1	1

Fig 2

(b)

4 **The invert-gate or not-gate**

The different symbols used for an **invert**-gate are shown in *Fig 3(a)* and the truth table is shown in *Fig 3(b).* This shows that a '0' input gives a '1' output and vice versa, i.e., it is an 'opposite to' function. The invert of A is written $\overline{A}$ and is called 'not-A'.

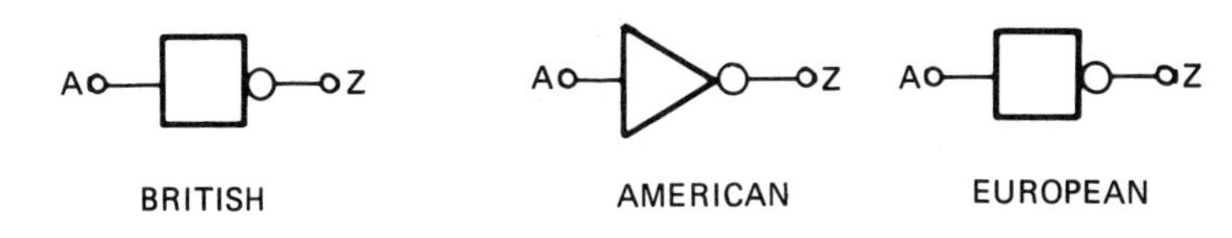

(a)

Fig 3

INPUT A	OUTPUT $Z=\overline{A}$
0	1
1	0

(b)

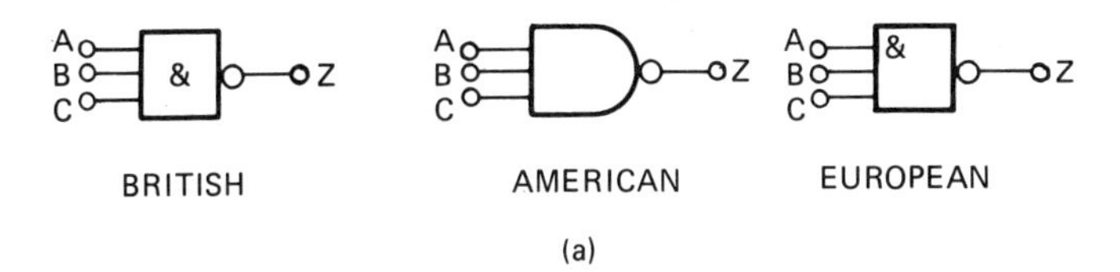

(a)

INPUTS A	B	C	A.B.C.	OUTPUT $Z=\overline{A.B.C.}$
0	0	0	0	1
0	0	1	0	1
0	1	0	0	1
0	1	1	0	1
1	0	0	0	1
1	0	1	0	1
1	1	0	0	1
1	1	1	1	0

Fig 4 (b)

5 **The nand-gate**

The different symbols used for a **nand**-gate are shown in *Fig 4(a)* and the truth table is shown in *Fig 4(b)*. This gate is equivalent to an **and-gate** and an **invert-gate** in series (not–and = nand) and the output is written as: $Z = \overline{A.B.C}$.

6 **The nor-gate**

The different symbols used for a **nor**-gate are shown in *Fig 5(a)* and the truth table is shown in *Fig 5(b)*. This gate is equivalent to an **or**-gate and an **invert**-gate in series, (not–or = nor), and the output is written as: $Z = \overline{A+B+C}$.

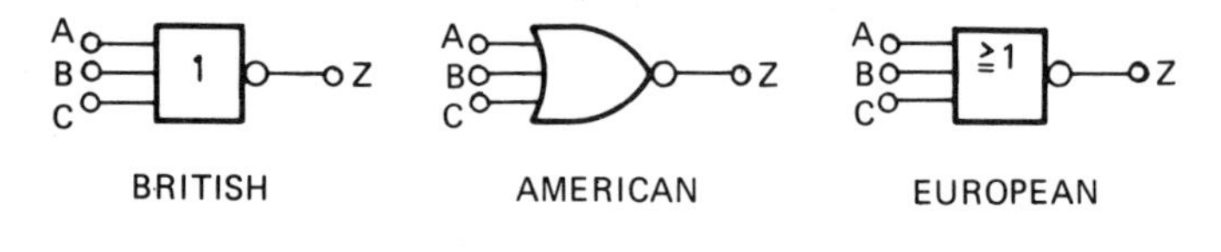

(a)

INPUTS				OUTPUT
A	B	C	A+B+C	$Z=\overline{A+B+C}$
0	0	0	0	1
0	0	1	1	0
0	1	0	1	0
0	1	1	1	0
1	0	0	1	0
1	0	1	1	0
1	1	0	1	0
1	1	1	1	0

Fig 5

(b)

7 **Combinational logic networks**

In most logic circuits, more than one gate is needed to give the required output. Except for the **invert**-gate, logic gates generally have two, three or four inputs and are confined to one function only, thus, for example, a two-input, **or**-gate or a four-input **and**-gate can be used when designing a logic circuit. The way in which logic gates are used to generate a given output is shown in *Problems 1 to 4*.

8 **Universal logic gates**

The function of any of the five logic gates in common use can be obtained by using either **nand**-gates or **nor**-gates and when used in this manner, the gate selected is called a **universal gate**. The way in which a universal **nand**-gate is used to produce the **invert, and, or** and **nor**-functions is shown in *Problem 5*. The way in which a universal **nor**-gate is used to produce the **invert, or, and** and **nand**-functions is shown in *Problem 6.*

B WORKED PROBLEMS ON LOGIC CIRCUITS

Problem 1 Devise a logic system to meet the requirements of: $Z = A.\overline{B}+C$

With reference to *Fig 6* an **invert**-gate, shown as (1), gives $\overline{B}$. The **and**-gate,

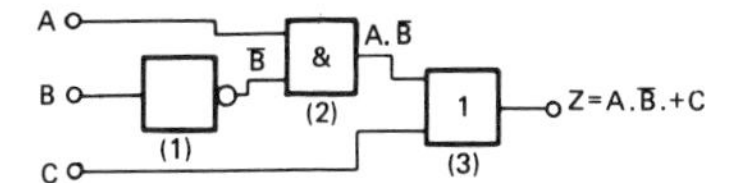

Fig 6

shown as (2), has inputs of A and $\overline{B}$, giving $A.\overline{B}$. The **or**-gate, shown as (3), has inputs of $A.\overline{B}$ and C, giving $\boldsymbol{Z = A.\overline{B}+C}$.

Problem 2 Devise a logic system to meet the requirements of $(P+\overline{Q}).(\overline{R}+S)$

The logic system is shown in *Fig 7*. The given expression shows that two **invert**-functions are needed to give $\overline{Q}$ and $\overline{R}$ and these are shown as gates (1) and (2).

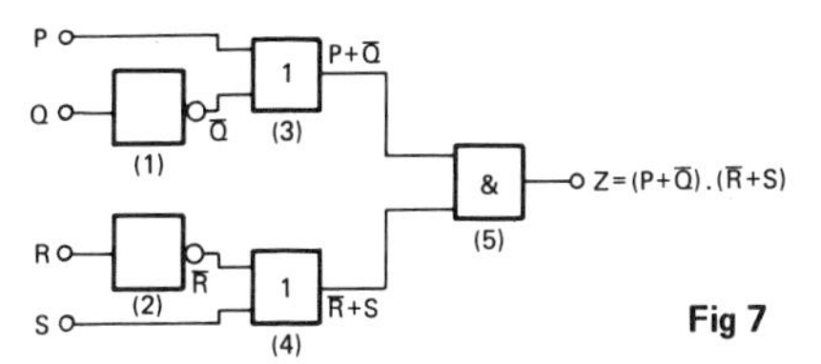

Fig 7

Two **or**-gates, shown as (3) and (4), give $(P+\overline{Q})$ and $(\overline{R}+S)$ respectively. Finally, an **and**-gate, shown as (5), gives the required output, $Z = (P+\overline{Q}).(\overline{R}+S)$.

Problem 3 Devise a logic circuit to meet the requirements of the output given in *Table 1*, using as few gates as possible.

The '1' outputs in rows 6, 7 and 8 of *Table 1* show that the Boolean expression is: $Z = A.\overline{B}.C+A.B.\overline{C}+A.B.C$. The logic circuit for this expression can be built using

TABLE 1

Inputs			*Output*
A	B	C	Z
0	0	0	0
0	0	1	0
0	1	0	0
0	1	1	0
1	0	0	0
1	0	1	1
1	1	0	1
1	1	1	1

three, 3-input **and**-gates and one, 3-input **or**-gate, together with two **invert**-gates. However, the number of gates required can be reduced by using the techniques introduced in chapter 9, resulting in the cost of the circuit being reduced. Any of the techniques in chapter 9 can be used, and in this case, the rules of Boolean algebra given in chapter 9, *Table 1* are used.

$$
\begin{aligned}
Z &= A.\overline{B}.C+A.B.\overline{C}+A.B.C \\
&= A.[\overline{B}.C+B.\overline{C}+B.C] \\
&= A.[\overline{B}.C+B(\overline{C}+C)] \\
&= A.[\overline{B}.C+B] \\
&= A.[B+\overline{B}.C] = A.[B+C]
\end{aligned}
$$

A, B, C; 1; B+C; &; Z = A.(B+C)

Fig 8

The logic circuit to give this simplified expression is shown in *Fig 8.*

Problem 4 Simplify the expression:
$Z=\overline{P}.\overline{Q}.\overline{R}.\overline{S}+\overline{P}.\overline{Q}.\overline{R}.S+\overline{P}.Q.\overline{R}.\overline{S}+\overline{P}.Q.\overline{R}.S+P.\overline{Q}.\overline{R}.\overline{S}$
and devise a logic circuit to give this output.

The given expression is simplified using the Karnaugh map techniques introduced in chapter 9. Two couples are formed as shown in *Fig 9(a)* and the simplified

R.S \ P.Q	0.0	0.1	1.1	1.0
0.0	1	1		1
0.1	1	1		
1.1				
1.0				

(a)

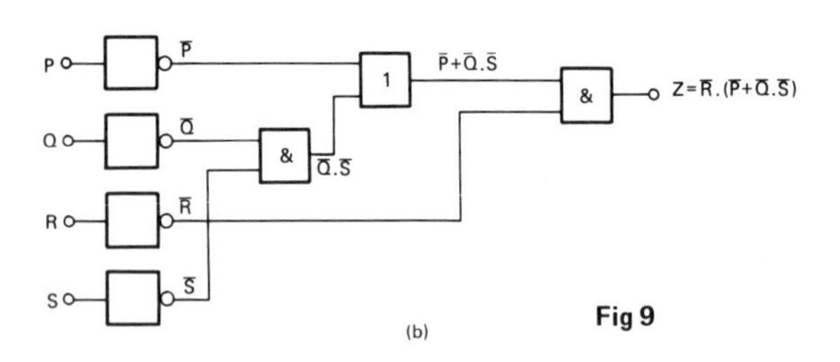

(b)

Fig 9

expression becomes: $Z=\overline{R}.(\overline{P}+\overline{Q}.\overline{S})$. The logic circuit to produce this expression is shown in *Fig 9(b).*

Problem 5 Show how **invert, and, or** and **nor**-functions can be produced using **nand**-gates only.

A single input to a **nand**-gate gives the **invert**-function, as shown in *Fig 10(a).* When two **nand**-gates are connected, as shown in *Fig 10(b),* the output from the first gate is $\overline{A.B.C}$ and this is inverted by the second gate, giving

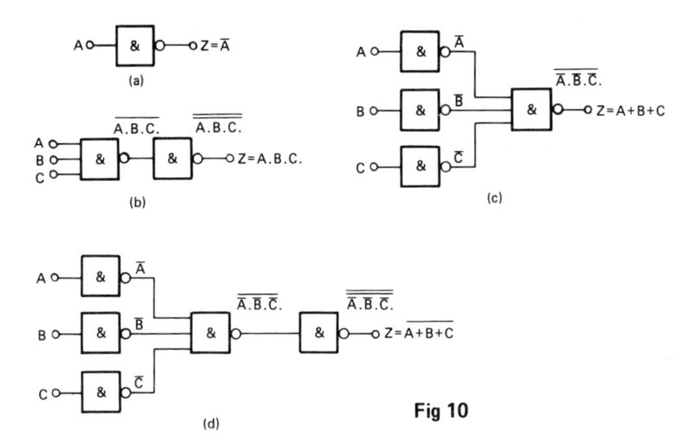

Fig 10

$Z = \overline{\overline{A.B.C}} = A.B.C$, i.e., the **and**-function is produced. When $\overline{A}$, $\overline{B}$ and $\overline{C}$ are the inputs to a **nand**-gate, the output is $\overline{\overline{A}.\overline{B}.\overline{C}}$.

By de Morgan's law, $\overline{\overline{A}.\overline{B}.\overline{C}} = \overline{\overline{A}}+\overline{\overline{B}}+\overline{\overline{C}} = A+B+C$, i.e., a **nand**-gate is used to produce the **or**-function. The logic circuit is shown in *Fig 10(c)*. If the output from the logic circuit in *Fig 10(c)* is inverted by adding an additional **nand**-gate, the output becomes the invert of an **or**-function, i.e., the **nor**-function, as shown in *Fig 10(d)*.

Problem 6 Show how **invert, or, and** and **nand**-functions can be produced by using **nor**-gates only.

A single input to a **nor**-gate gives the **invert**-function, as shown in *Fig 11(a)*. When two **nor**-gates are connected, as shown in *Fig 11(b)*, the output from the

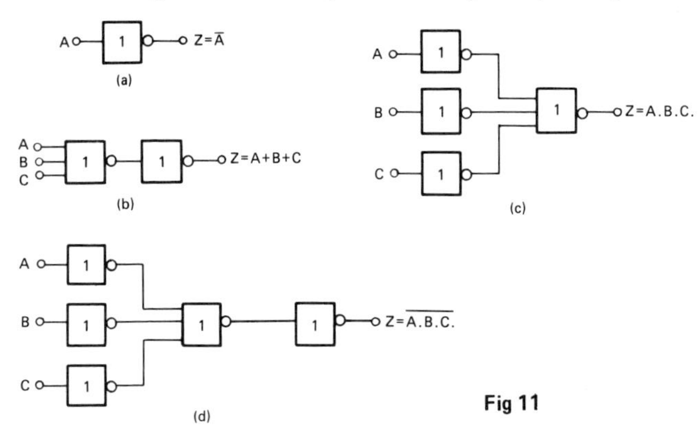

Fig 11

first gate is $\overline{A+B+C}$ and this is inverted by the second gate, giving $Z = \overline{\overline{A+B+C}} = A+B+C$, i.e. the **or**-function is produced. Inputs of $\overline{A}$, $\overline{B}$ and $\overline{C}$ to a **nor**-gate give an output of $\overline{\overline{A}+\overline{B}+\overline{C}}$.

By de Morgan's law: $\overline{\overline{A}+\overline{B}+\overline{C}} = \overline{\overline{A}}.\overline{\overline{B}}.\overline{\overline{C}} = A.B.C$, i.e. the **nor**-gate can be used to produce the **and**-function. The logic circuit is shown in *Fig 11(c).* When the output of the logic circuit, shown in *Fig 11(c)*, is inverted by adding an additional **nor**-gate, the output then becomes the invert of an **or**-function, i.e., the **nor**-function as shown in *Fig 11(d).*

Problem 7 Design a logic circuit, using **nand**-gates having not more than three inputs, to meet the requirements of the Boolean expression: $Z = \overline{A}+\overline{B}+C+\overline{D}$

When designing logic circuits, it is often easier to start at the output of the circuit. The given expression shows there are four variables joined by **or**-functions. From the principles introduced in *Problem 5*, if a four-input **nand**-gate is used to give the expression given, the inputs are $\overline{\overline{A}}$, $\overline{\overline{B}}$, $\overline{C}$ and $\overline{\overline{D}}$, that is $A, B, \overline{C}$ and D. However, the problem states that three-inputs are not to be exceeded, so two of the variables are joined, i.e., the inputs to the three-input **nand**-gate, shown as gate (1)

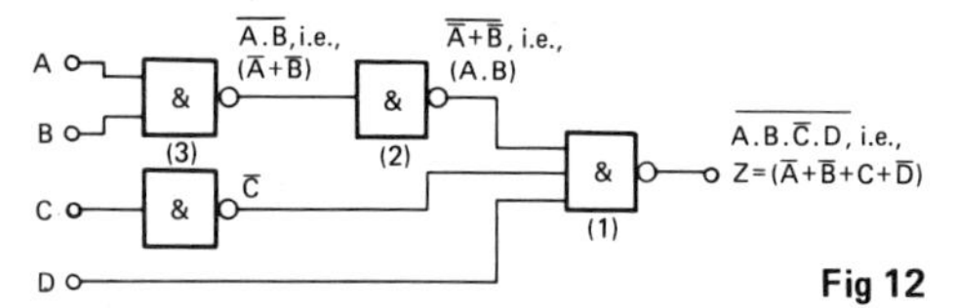

Fig 12

in *Fig 12*, is $A.$ B, $\overline{C}$ and D. From *Problem 5*, the **and**-function is generated by using two **nand**-gates connected in series, as shown by gates (2) and (3) in *Fig 12.* The logic circuit required to produce the given expression is as shown in *Fig 12.*

Problem 8 Use **nor**-gates only to design a logic circuit to meet the requirements of the expression: $Z = \overline{D}.(\overline{A}+B+\overline{C})$

It is usual in logic circuit design to start the design at the output. From *Problem 6*, the **and**-function between $\overline{D}$ and the terms in the bracket can be produced by using inputs of $\overline{\overline{D}}$ and $\overline{\overline{A}+B+\overline{C}}$ to a **nor**-gate, i.e., by de Morgan's law, inputs of

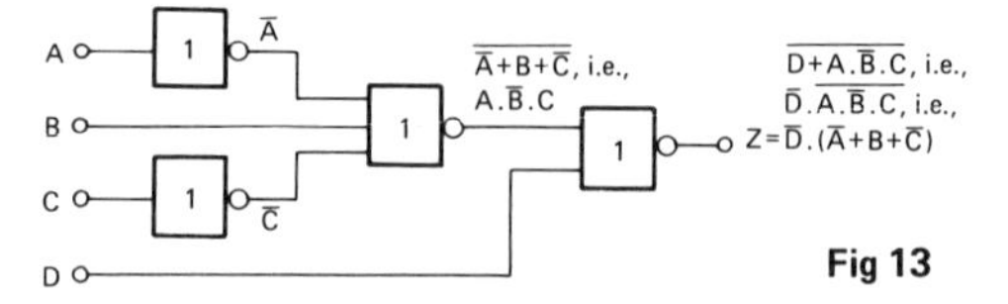

Fig 13

D and $A.\overline{B}.C.$ Again, with reference to *Problem 6*, inputs of $\overline{A}.B$ and $\overline{C}$ to a **nor**-gate give an output of $\overline{\overline{A}+B+\overline{C}}$, which by de Morgan's law is $A.\overline{B}.C$. The logic circuit to produce the required expression is as shown in *Fig 13.*

Problem 9 An alarm indicator in a grinding mill complex should be activated if (a) the power supply to all mills is off and (b) the hopper feeding the mills is less than 10% full, and (c) if less than two of the three grinding mills are in action. Devise a logic system to meet these requirements.

Let variable A represent the power supply on to all the mills, then $\overline{A}$ represents the power supply off. Let B represent the hopper feeding the mills being more than 10% full, then $\overline{B}$ represents the hopper being less than 10% full. Let C, D and E represent the three mills respectively being in action, then $\overline{C}$, $\overline{D}$ and $\overline{E}$ represent the three mills respectively not being in action. The required expression to activate the alarm is:

$Z = \overline{A}.\overline{B}.(\overline{C}+\overline{D}+\overline{E})$.

There are three variables joined by **and**-functions in the output, indicating that a three-input **and**-gate is required, having inputs of $\overline{A}$, $\overline{B}$ and $(\overline{C}+\overline{D}+\overline{E})$. The term

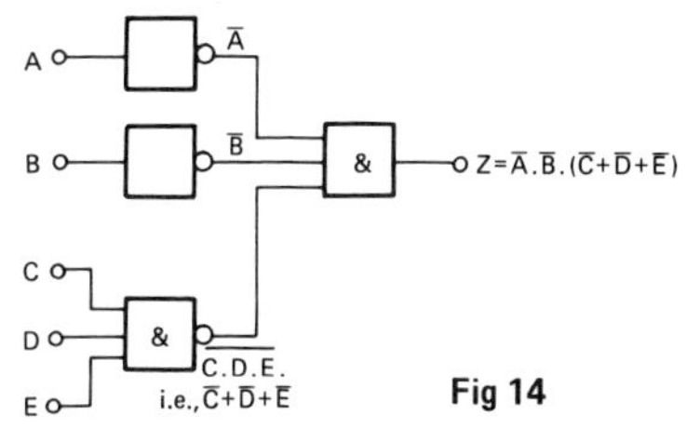

Fig 14

$(\overline{C}+\overline{D}+\overline{E})$ is produced by a three-input **nand**-gate. When variables C, D and E are the inputs to a **nand**-gate, the output is $\overline{C.D.E}$, which, by de Morgan's law is $\overline{C}+\overline{D}+\overline{E}$. Hence the required logic circuit is as shown in *Fig 14*.

C WORKED PROBLEMS ON LOGIC CIRCUITS

In *Problems 1 to 5*, devise logic systems to meet the requirements of the Boolean expressions given.

1 $Z = \overline{A}+B.C$ [See *Fig 15(a)*]
2 $Z = A.\overline{B}+B.\overline{C}$ [See *Fig 15(b)*]
3 $Z = A.B.\overline{C}+\overline{A}.\overline{B}.C$ [See *Fig 15(c)*]
4 $Z = (\overline{A}+B).(\overline{C}+D)$ [See *Fig 15(d)*]
5 $Z = A.\overline{B}+B.\overline{C}+C.\overline{D}$ [See *Fig 15(e)*]

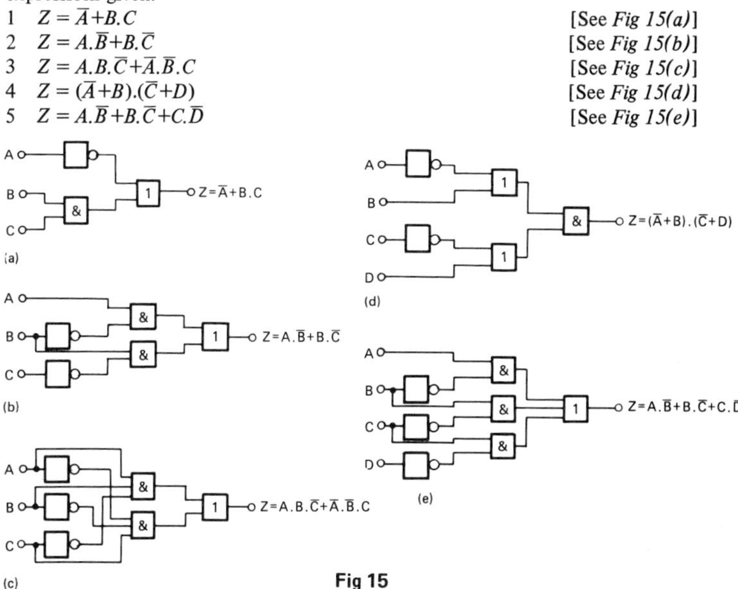

Fig 15

In *Problems 6 to 8*, simplify the expression given in the truth table and devise a logic circuit to meet the requirements stated.

6 Column 4 of table 2. [$A.B+C$, see *Fig 16(a)*]

7 Column 5 of table 2. [$A.\overline{B}+B.C$, see *Fig 16(b)*]

8 Column 6 of table 2. [$A.C+B$, see *Fig 16(c)*]

TABLE 2

1 A	2 B	3 C	4 Z_1	5 Z_2	6 Z_3
0	0	0	0	0	0
0	0	1	1	0	0
0	1	0	0	0	1
0	1	1	1	1	1
1	0	0	0	1	0
1	0	1	1	1	1
1	1	0	1	0	1
1	1	1	1	1	1

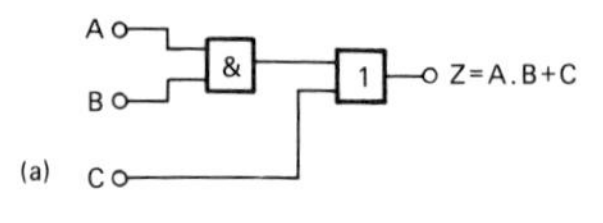

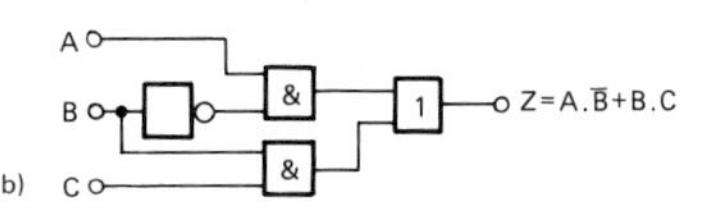

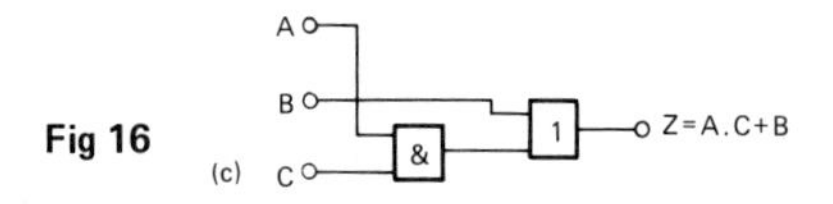

Fig 16

In *Problems 9 to 15*, simplify the Boolean expressions given and devise logic circuits to give the requirements of the simplified expressions.

9 $\overline{P}.\overline{Q}+\overline{P}.Q+P.Q$ [$\overline{P}+Q$, see *Fig 17(a)*]

10 $\overline{P}.\overline{Q}.\overline{R}+P.Q.\overline{R}+P.\overline{Q}.\overline{R}$ [$\overline{R}.(P+\overline{Q})$, see *Fig 17(b)*]

11 $P.\overline{Q}.R+P.\overline{Q}.\overline{R}+\overline{P}.\overline{Q}.\overline{R}$ [$\overline{Q}.(P+\overline{R})$, see *Fig 17(c)*]

12 $\overline{P}.\overline{Q}.R+P.Q.\overline{R}+P.Q.R+P.\overline{Q}.R$ [$\overline{P}.\overline{Q}.R+P.(Q+R)$, see *Fig 17(d)*]

13 $\overline{A}.\overline{B}.\overline{C}.\overline{D}+A.\overline{B}.\overline{C}.\overline{D}+\overline{A}.\overline{B}.C.\overline{D}+\overline{A}.B.C.\overline{D}+A.\overline{B}.C.\overline{D}$

[$\overline{D}.(\overline{A}.C+\overline{B})$, see *Fig 18(a)*]

14 $\overline{A}.\overline{B}.C.\overline{D}+\overline{A}.B.\overline{C}.D+A.B.\overline{C}.D+\overline{A}.B.C.D+A.B.C.D$

[$\overline{A}.\overline{B}.C.\overline{D}+B.D$, see *Fig 18(b)*]

15 $(\overline{\overline{P}.Q.R}).(\overline{P+Q.R})$ [$\overline{P}.(\overline{Q}+\overline{R})$, see *Fig 18(c)*]

Fig 17

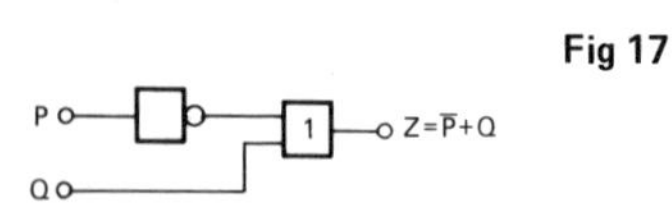

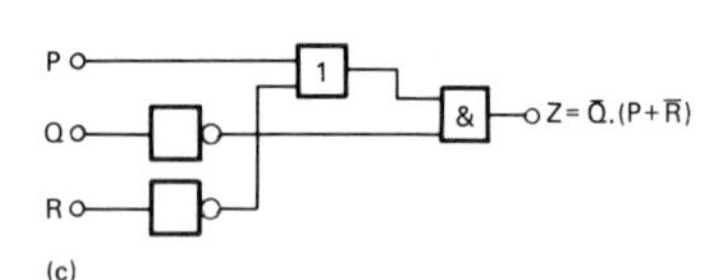

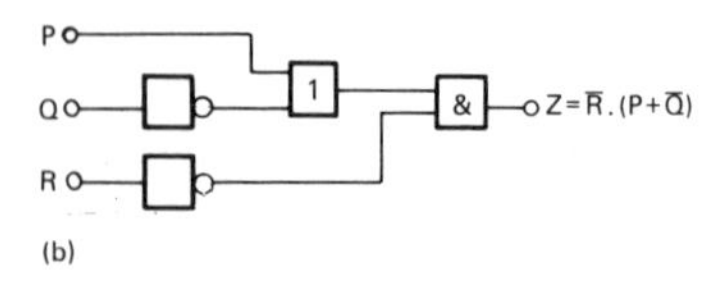

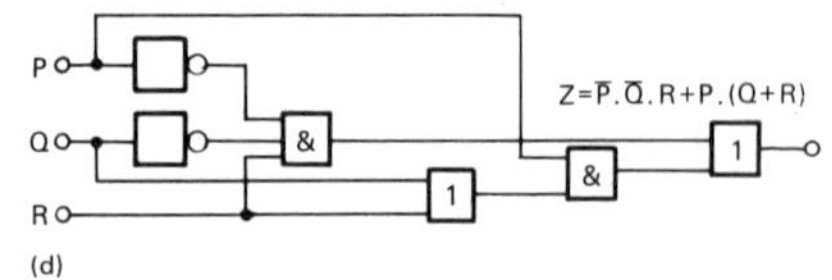

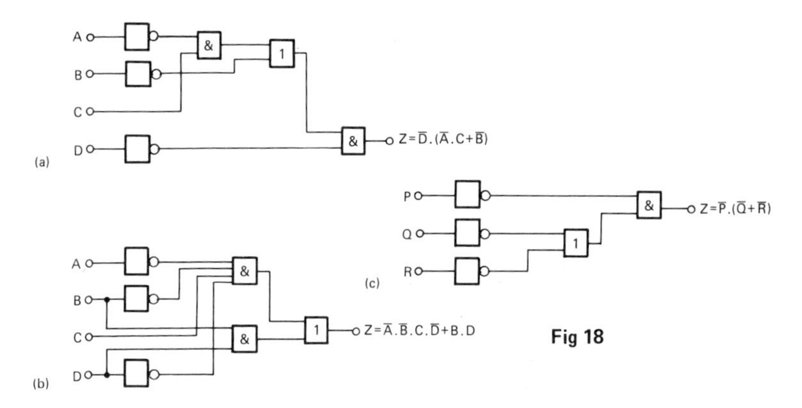

Fig 18

In *Problems 16 to 18*, use **nand**-gates only to devise the logic systems stated.

16 $Z = A+B.C$ [See *Fig 19(a)*]

17 $Z = A.\overline{B}+B.\overline{C}$ [See *Fig 19(b)*]

18 $Z = A.B.\overline{C}+\overline{A}.\overline{B}.C$ [See *Fig 19(c)*]

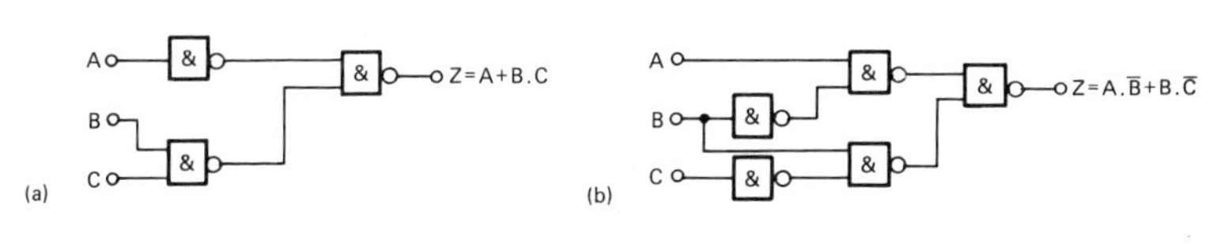

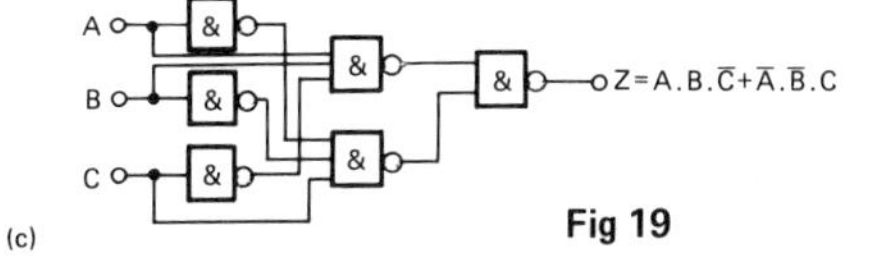

Fig 19

In *Problems 19 to 21*, use **nor**-gates only to devise the logic systems stated.

19 $Z = (\overline{A}.B).(\overline{C}+D)$ [See *Fig 20(a)*]

20 $Z = A.\overline{B}+B.\overline{C}+C.\overline{D}$ [See *Fig 20(b)*]

21 $Z = \overline{P}.Q+P.(Q+R)$ [See *Fig 20(c)*]

22 In a chemical process, three of the transducers used are P, Q and R, giving output signals of either 0 or 1. Devise a logic system to give a 1 output when:
(a) P and Q and R all have 0 output, or when:
(b) P is 0 and (Q is 1 or R is 0). [$\overline{P}.(Q+\overline{R})$, see *Fig 21(a)*]

23 Lift doors should close, (Z), if:
(a) the master switch, (A), is on and either
(b) a call, (B), is received from any other floor, or
(c) the doors, (C), have been open for more than 10 seconds, or
(d) the selector push within the lift (D), is pressed for another floor.
Devise a logic circuit to meet these requirements.
[$Z = A.(B+C+D)$, see *Fig 21(b)*]

24 A water tank feeds three separate processes. When any two of the processes are in operation at the same time, a signal is required to start a pump to maintain

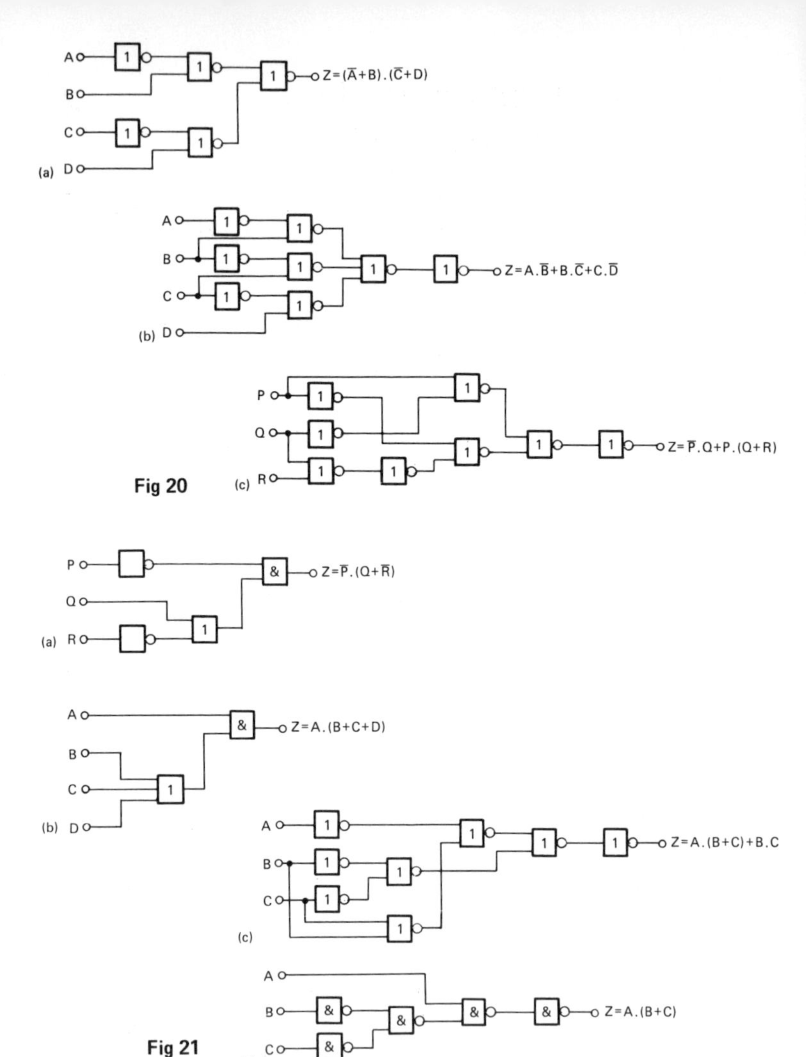

the head of water in the tank. Devise a logic circuit using **nor**-gates only to give the required signal. [$Z = A.(B+C)+B.C$, see *Fig 21(c)*]

25 A logic signal is required to give an indication when:

(a) the supply to an oven is on, and
(b) the temperature of the oven exceeds 210°C, or
(c) the temperature of the oven is less than 190°C.

Devise a logic circuit using **nand**-gates only to meet these requirements.

[$Z = A.(B+C)$, see *Fig 21(d)*]

11 Trigonometry (1) – The solution of three-dimensional triangulation problems

A. MAIN POINTS CONCERNED WITH THREE-DIMENSIONAL TRIANGULATION PROBLEMS

1 A **scalene triangle** is one in which all its sides and angles are unequal.

2 To **solve a triangle** means to find the values of all unknown sides and angles. If a triangle is right-angled, trigonometric ratios and the theorem of Pythagoras may be used for its solution. For a scalene triangle, the sine rule and the cosine rule are used for its solution.

3 For the scalene triangle shown in *Fig 1*:

(i) the **sine rule** states:

$$\frac{a}{\sin A} = \frac{b}{\sin B} = \frac{c}{\sin C}$$

(ii) the **cosine rule** states:

$$a^2 = b^2 + c^2 - 2bc \cos A$$

or $$b^2 = a^2 + c^2 - 2ac \cos B$$

or $$c^2 = a^2 + b^2 - 2ab \cos C$$

Fig 1

4 **The angle between a line and a plane** is defined as the angle between the line and its projection on the plane. In *Fig 2*, the line PQ meets a plane at P. If QR is constructed perpendicular to the plane then the projection of PQ on the plane is PR. The angle between the line PQ and the plane in *Fig 2* is θ.

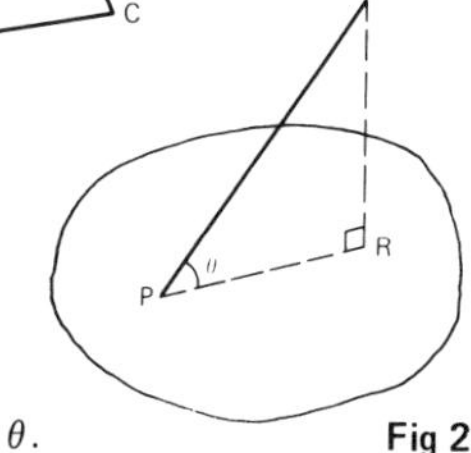

Fig 2

5 *Fig 3* shows two planes ABCD and ABEF intersecting along the line AB. **The angle between the planes** is defined as the angle between any two straight lines drawn on each plane which meet at, and are perpendicular to, the line of intersection of the planes. In *Fig 3*, PQ and PR are both perpendicular to AB; thus the angle between the two intersecting planes is $\angle$RPQ.

6 Three-dimensional triangulation problems rely on the ability to (i) visualise the problem and (ii) solve triangles. A clearly labelled sketch is thus usually invaluable. Determining the location, speed and direction of moving objects, such as ships and aircraft, usually involves the solution of three-dimensional problems.

7 (i) If, in *Fig 4(a)*, BC represents horizontal ground and AB a vertical flagpole, then **the angle of elevation** of the top of the flagpole, A, from the point C is the angle that the imaginary straight line AC must be raised (or elevated) from the horizontal CB, i.e. angle θ.

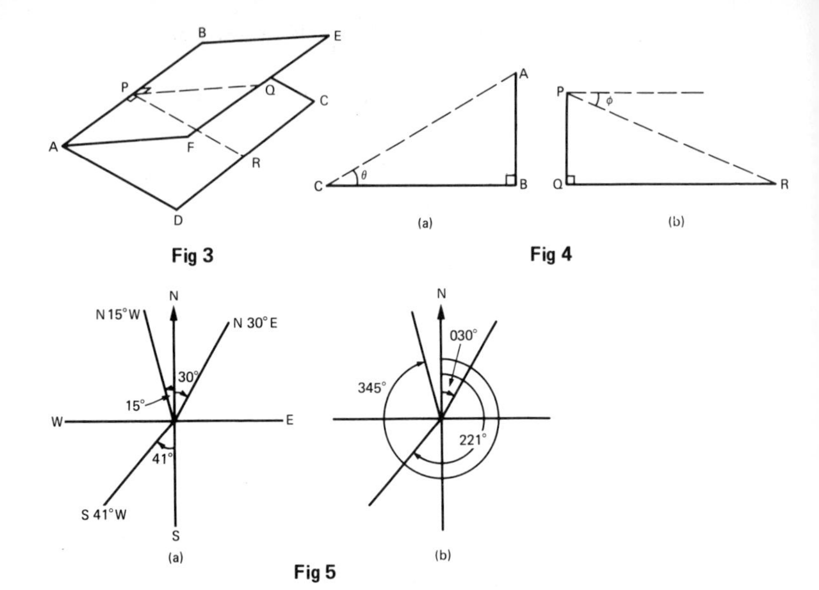

Fig 3

Fig 4

Fig 5

(ii) If, in *Fig 4(b)*, PQ represents a vertical cliff and R a ship at sea, then **the angle of depression** of the ship from point P is the angle through which the imaginary straight line PR must be lowered (or depressed) from the horizontal to the ship, i.e. angle ϕ. (Note, $\angle$PRQ is also ϕ – alternate angles between parallel lines.)

8 **Bearings** provide a method of specifying the position of an object with respect to the points of the compass. There are two methods of stating bearings:

(i) A bearing of N 30° E means an angle of 30° measured from the north towards east as shown in *Fig 5(a)*. Similarly, bearings of S 41° W and N 15° W are also shown.

(ii) The angle denoting the bearing is measured from north in a clockwise direction, north being considered as 0°. Three figures are always stated. Bearings of 030°, 221° and 345° are shown in *Fig 5(b)* and are equivalent to N 30° E, S 41° W and N 15° W respectively.

B WORKED PROBLEMS ON THREE-DIMENSIONAL TRIGONOMETRY

Problem 1 In triangle DEF, $D = 53°$, $E = 69°$ and EF = 125 mm. Solve the triangle.

Triangle DEF is shown in *Fig 6*. Since the angles in a triangle add up to 180° then $F = 180° - 53° - 69° = \mathbf{58°}$

Applying the sine rule: $\dfrac{125}{\sin 53°} = \dfrac{e}{\sin 69°}$, from which, $e = \dfrac{125 \sin 69°}{\sin 53°} = \mathbf{146.1\ mm}$

Also, $\frac{125}{\sin 53°} = \frac{f}{\sin 58°}$, from which, $f = \frac{125 \sin 58°}{\sin 53°}$ = **132.7 mm**

Problem 2 Solve the triangle ABC given AC = 7.0 cm, AB = 9.0 cm and $C = 48°$.

Triangle ABC is shown in *Fig 7*

Applying the sine rule: $\frac{9.0}{\sin 48°} = \frac{7.0}{\sin B}$ from which, $\sin B = \frac{7.0 \sin 48°}{9.0} = 0.5780$

Hence $B = \arcsin 0.5780 = 35°\ 19'$ or $144°\ 41'$

When $B = 35°\ 19'$, $A = 180° - 48° - 35°\ 19' = 96°\ 41'$

When $B = 144°\ 41'$, $A = 180° - 48° - 144°\ 41' = -12°\ 41'$. This latter solution is impossible and is thus neglected. (Note that if both values of B had been

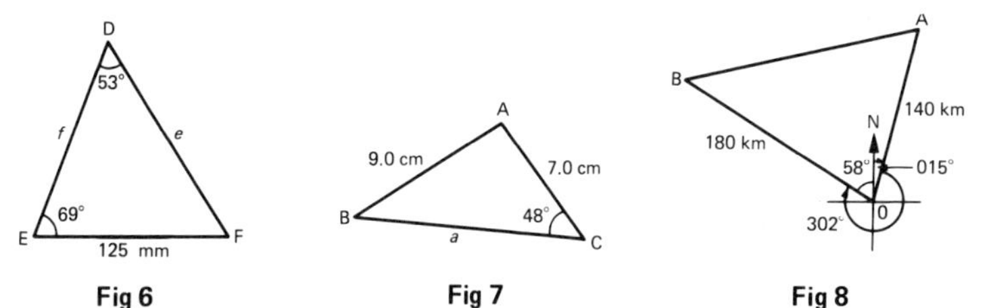

Fig 6 Fig 7 Fig 8

feasible then two different solutions of the triangle are produced. This is known as the ambiguous case.)

Applying the sine rule: $\frac{a}{\sin 96°\ 41'} = \frac{9.0}{\sin 48°}$,

from which $a = \frac{9.0 \sin 96°\ 41'}{\sin 48°} = 12.0$ cm

Hence A = 96° 41′, B = 35° 19′ and BC = 12.0 cm.

Problem 3 A ship A sails at a steady speed of 35 km/h from a port on a bearing of 015°. At the same time another ship B leaves the port at a steady speed of 45 km/h on a bearing of 302°. Determine their distance apart after 4 hours.

In 4 hours ship A travels $4 \times 35 = 140$ km (shown as OA in *Fig 8*), and ship B travels $4 \times 45 = 180$ km (shown as OB).

The distance the two ships are apart after 4 hours is given by AB in *Fig 8.*

$\angle AOB = 15° + 58° = 73°$. Applying the cosine rule:

$$AB^2 = OA^2 + OB^2 - 2(OA)(OB) \cos AOB$$
$$= (140)^2 + (180)^2 - [2(140)(180) \cos 73°]$$
$$= 19\,600 + 32\,400 - 14\,735.5 = 37\,264.5$$

and $AB = \sqrt{(37\,264.5)}$ = **193.0 km.**

Problem 4 Three forces acting on a fixed point are represented by the three sides of a triangle of dimensions 7.2 cm, 9.6 cm and 11.0 cm. Determine the angles between the lines of action of the three forces.

The triangle representing the forces is shown in *Fig 9* and is labelled ABC. The largest angle is C since 11.0 cm is the longest side.
Since $c^2 = a^2 + b^2 - 2ab \cos C$ from the cosine rule,

$$\text{then } \cos C = \frac{a^2+b^2-c^2}{2ab} = \frac{(7.2)^2+(9.6)^2-(11.0)^2}{2(7.2)(9.6)} = 0.1664$$

$$C = \arccos 0.1664 = \mathbf{80^\circ\ 25'}.$$

Since C is positive the triangle is acute-angled. If the value of $\cos C$ had been negative C would have been obtuse (i.e. lie between 90° and 180°).

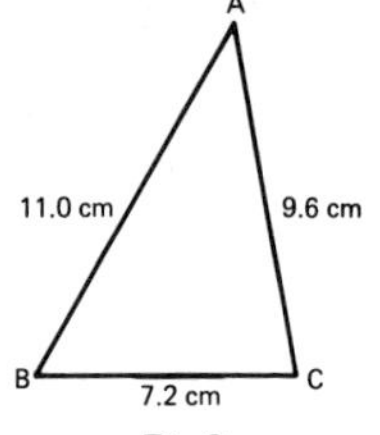

Fig 9

Applying the sine rule: $\dfrac{11.0}{\sin 80^\circ\ 25'} = \dfrac{7.2}{\sin A}$, from which, $\sin A = \dfrac{7.2 \sin 80^\circ\ 25'}{11.0}$

$= 0.6454$

Hence $A = \arcsin 0.6454 = \mathbf{40^\circ\ 12'}$

Also, $\dfrac{11.0}{\sin 80^\circ\ 25'} = \dfrac{9.6}{\sin B}$, from which, $\sin B = \dfrac{9.6 \sin 80^\circ\ 25'}{11.0} = 0.8605$

Hence $B = \arcsin 0.8605 = \mathbf{59^\circ\ 23'}$.

Problem 5 A surveyor, standing S 65° W of a tower measures the angle of elevation of the top of the tower as 46° 30′. From a point S 67° E from the tower the elevation of the top is 37° 15′. Determine the height of the tower if the distance between the two observations is 75 m.

In *Fig 10*, AB represents the tower and C and D the initial and final observation positions.

From triangle ABC, $\tan 46^\circ\ 30' = \dfrac{\text{AB}}{\text{BC}}$

from which $\text{BC} = \dfrac{\text{AB}}{\tan 46^\circ\ 30'} = \text{AB} \cot 46^\circ\ 30'$

Similarly, for triangle ABD, $\text{BD} = \text{AB} \cot 37^\circ\ 15'$.
Applying the cosine rule to triangle BCD gives:

$$75^2 = (\text{AB} \cot 46^\circ\ 30')^2 + (\text{AB} \cot 37^\circ\ 15')^2 - [2(\text{AB} \cot 46^\circ\ 30')(\text{AB} \cot 37^\circ\ 15') \cos (65^\circ + 67^\circ)]$$

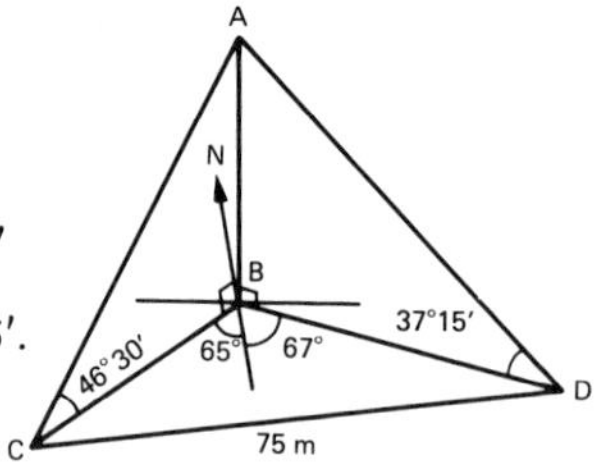

Fig 10

i.e. $5625 = \text{AB}^2(0.9005) + \text{AB}^2(1.7294) - \text{AB}^2(-1.6701)$
i.e. $5625 = \text{AB}^2(0.9005 + 1.7294 + 1.6701) = 4.3000\ \text{AB}^2$

Hence height of tower, $\mathbf{AB} = \sqrt{\dfrac{5625}{4.3000}} = \mathbf{36.2\ m}$

Problem 6 A pyramid has a rectangular base 8 cm by 6 cm and each sloping edge is 13 cm. Determine (a) the perpendicular height of the pyramid and (b) the angle a sloping edge makes with the base.

The pyramid is shown in *Fig 11*, where diagonals AC and BD intersect at F.
(a) Length of diagonal $\text{AC} = \sqrt{(8^2+6^2)} = 10$ cm.

Length $\text{AF} = \frac{1}{2}\text{AC} = \frac{1}{2}(10) = 5$ cm

From triangle AEF, $EF = \sqrt{(AE^2 - AF^2)}$
$= \sqrt{(13^2 - 5^2)} = 12$ cm.

Hence the perpendicular height of the pyramid is 12 cm.

(b) The angle which a line makes with a plane is the angle which it makes with its projection on the plane. In *Fig 11*, AF is the projection of AE on to the plane ABCD. From triangle AEF, $\cos A = \frac{AF}{AE} = \frac{5}{13}$,

from which

$A = \arccos \frac{5}{13} = 67° \ 23'$.

Hence each sloping edge makes an angle of 67° 23′ with the base.

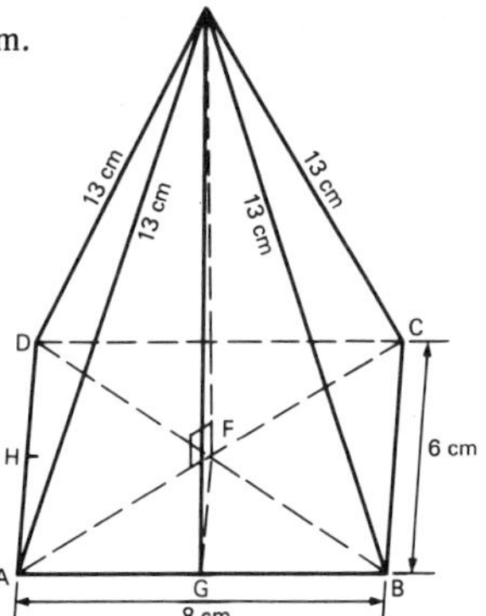

Fig 11

Problem 7 For the pyramid shown in *Fig 11*, determine (a) the angle face AEB makes with the base, and (b) the angle the plane EGH makes with the base, where G and H are the mid-points of AB and AD respectively.

(a) Since G is the mid-point of AB then FG is perpendicular to AB. Also, EG is the perpendicular height of triangle AEB. Thus the angle between the two intersecting lines EG and GF is the angle between the face AEB and the base. In triangle EFG, EF = 12 cm (from *Problem 6*) and $FG = \frac{6}{2} = 3$ cm.

Hence $\angle EGF = \arctan \frac{12}{3} = 75° \ 58'$.

Hence the angle plane AEB makes with the base is 75° 58′.

(b) To determine the angle between plane EGH and the base ABCD it is necessary to establish a perpendicular to the common edge GH in each of the planes. Triangle FGH is shown in *Fig 12(a)*, where FJ is the perpendicular to HG. FJ = 3 sin FGH and $\angle FGH = \arctan \frac{4}{3} = 53° \ 8'$.

Hence FJ = 3 sin 53° 8′ = 2.40 cm.

From triangle EFJ shown in *Fig 12(b)*,

$\angle EJF = \frac{12}{2.4} = 78° \ 41'$.

Hence the angle between plane EGH and the base is 78° 41′.

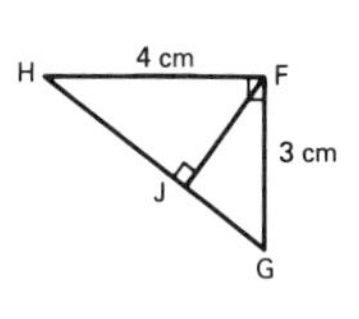

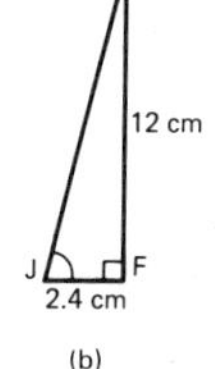

Fig 12 (a) (b)

Problem 8 A 10 cm cube is shown in *Fig 13*, where X and Y are the mid-points of sides AB and BC respectively. Determine (a) the angle between EC and CH, and (b) the angle between FX and CX.

(a) Diagonal $AC = \sqrt{(10^2 + 10^2)} = 14.14$ cm, which is the length of all the diagonals of the cube.
From the right-angled triangle, EAC, $EC = \sqrt{(AC^2 + EA^2)} = \sqrt{[(14.14)^2 + (10)^2}$
$= 17.32$ cm.

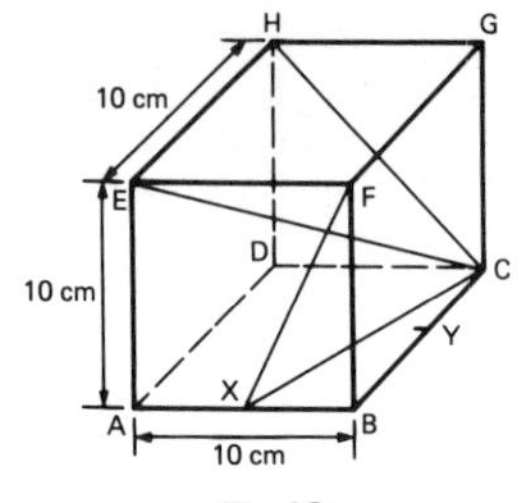

Fig 13

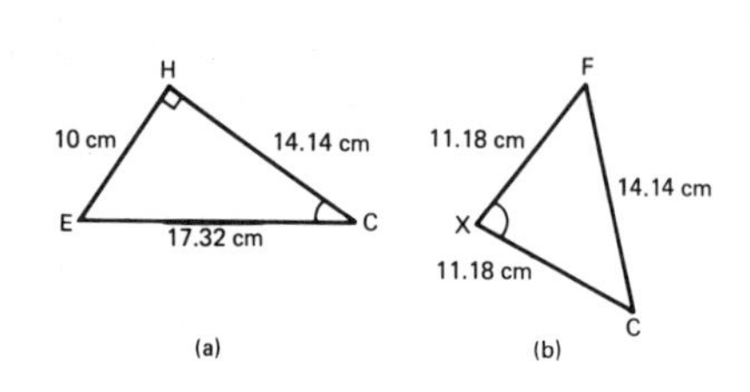

Fig 14

Triangle EHC is shown separately in *Fig 14(a)*, from which,

$$\angle ECH = \arctan \frac{10}{14.14} = 35° \ 16'.$$

Hence the angle between EC and CH is 35° 16′.

(b) From the right-angled triangle BFX,
$FX = \sqrt{(BF^2 + XB^2)} = \sqrt{(10^2 + 5^2)} = 11.18$ cm. Similarly, CX = 11.18 cm.
Triangle FXC is shown separately in *Fig 14(b)*. Applying the cosine rule to triangle FXC gives: $14.14^2 = 11.18^2 + 11.18^2 - 2(11.18)(11.18) \cos X$,

$$\text{from which, } \cos X = \frac{11.18^2 + 11.18^2 - 14.14^2}{2(11.18)(11.18)} = 0.2000$$

and $\quad$ FXC = arccos 0.2000 = 78° 28′.

Hence the angle between FX and CX is 78° 28′.

Problem 9 For the cube shown in *Fig 13*, determine (a) the angle between EC and plane ABCD, and (b) the angle between the planes HXY and the plane ABCD.

(a) The projection of the line EC on to the plane ABCD is given by the diagonal AC, where AC and EA are perpendicular to each other. Triangle ECA is shown in *Fig 15(a)*, and

$$\angle ECA = \arctan \frac{10}{14.14} = 35° \ 16'.$$

Hence the angle between EC and plane ABCD is 35° 16′.

(b) From *Fig 15(b)*, XY is the common edge between planes HXY and ABCD. From the right-angled triangle XBY (see *Fig 13*), $XY = \sqrt{(5^2 + 5^2)} = 7.071$ cm.
Let Z be the mid-point of XY in *Fig 15(b)*, then XZ

$$= \frac{7.071}{2} = 3.536 \text{ cm.}$$

DX = 11.18 cm (which is the same as CX of *Problem 8(b)*).
$DZ = \sqrt{(DX^2 - XZ^2)} = \sqrt{[(11.18)^2 - (3.536)^2]} = 10.61$ cm.
Both DZ and HZ are perpendicular to XY. Thus the angle between planes HXY and ABCD is given by ∠HZD.

$$\angle HZD = \arctan \frac{DH}{DZ}$$

$$= \arctan \frac{10}{10.61} = \mathbf{43° \ 18'}.$$

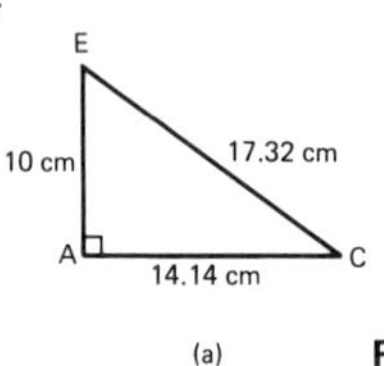

Fig 15

Problem 10 A ship is observed from the top of a 130 m high cliff on a bearing of 205° at an angle of depression of 9°. 5 minutes later the same ship is seen on a bearing of 280° at an angle of depression of 10°. Calculate the speed of the ship in km/h.

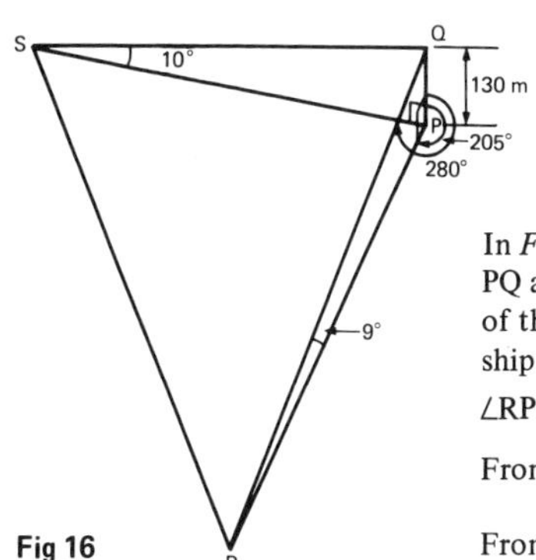

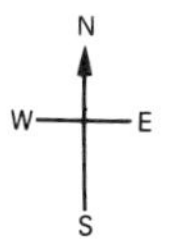

Fig 16

In *Fig 16*, Q represents the top of the cliff PQ and R and S the initial and final positions of the ship. To calculate the speed of the ship distance RS is required.

$\angle RPS = 280° - 205° = 75°$.

From triangle PQR, $PR = \dfrac{130}{\tan 9°} = 820.8$ m.

From triangle PQS, $PS = \dfrac{130}{\tan 10°} = 737.3$ m.

Applying the cosine rule to triangle PRS gives:

$RS^2 = (820.8)^2 + (737.3)^2 - [2(820.8)(737.3) \cos 75°]$

from which, RS = 950.8 m or 0.9508 km.

Speed of ship $= \dfrac{\text{distance travelled}}{\text{time taken}} = \dfrac{0.9508 \text{ km}}{5/60 \text{ h}} = \mathbf{11.41 \text{ km/h}}$.

Problem 11 An aeroplane is sighted due east from a radar station at an elevation of 40° and a height of 8000 m, and later at an elevation of 35° and height 5500 m in a direction S 20° E. If it is descending uniformly find the angle of descent. Determine also the speed of the aeroplane in km/h if the time between the two observations is 45 seconds.

In *Fig 17*, R represents the radar station and P and S the initial and final positions of the aeroplane.

From the right-angled triangle PQR, $RQ = \dfrac{8000}{\tan 40°} = 9534$ m.

From the right-angled triangle STR, $RT = \dfrac{5500}{\tan 35°} = 7855$ m.

Angle QRT = 90° − 20° = 70°.
Applying the cosine rule to triangle QRT gives:

$$QT^2 = RQ^2 + RT^2 - 2(RQ)(RT) \cos QRT$$
$$= (9534)^2 + (7855)^2 - [2(9534)(7855) \cos 70°]$$

from which, QT = 10 068 m.
The angle of descent is shown as angle θ in *Fig 18*, from which,

$$\tan \theta = \frac{PX}{SX} = \frac{PX}{QT} = \frac{8000-5500}{10\,068} = \frac{2500}{10\,068} = 0.2483$$

Hence the angle of descent, θ = arctan 0.2483 = **13° 57′**.

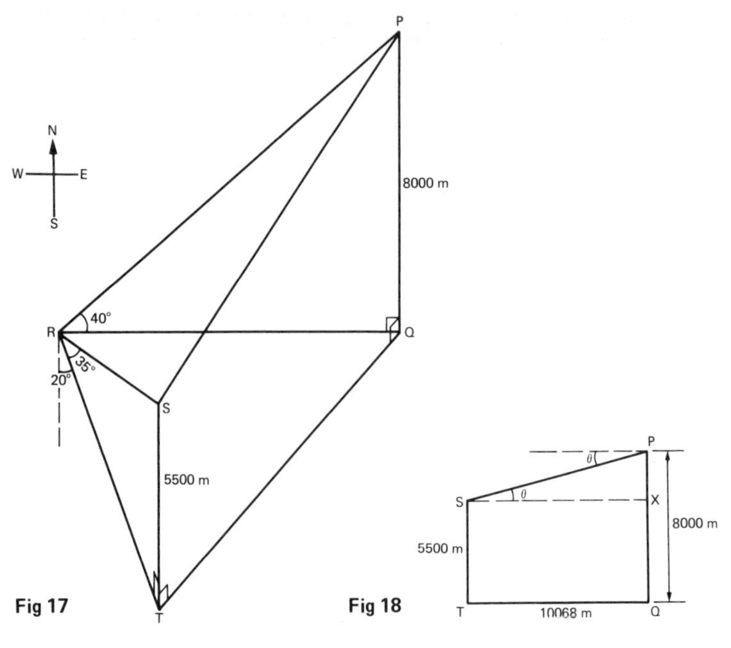

Fig 17

Fig 18

From triangle PXS of *Fig 18*, $\sin\theta = \dfrac{PX}{SP}$, from which, $SP = \dfrac{PX}{\sin\theta}$

$$= \frac{2500}{\sin 13^\circ\, 57'} = 10\ 370 \text{ m}$$

$$\text{Speed of aeroplane} = \frac{\text{distance SP}}{\text{time taken}} = \frac{10\ 370 \text{ m}}{45 \text{ s}} = \frac{10.37 \text{ km}}{\left(\dfrac{45}{60 \times 60}\right)\text{h}} = \mathbf{829.6\ km/h}$$

C FURTHER PROBLEMS ON THREE-DIMENSIONAL TRIGONOMETRY

1 Solve the following triangles ABC:

(a) $A = 30°; B = 72°; b = 3.40$ cm.
(b) $A = 115°; C = 25°; a = 20.0$ mm.
(c) $b = 5.5$ m; $c = 7.2$ m; $C = 80°$

[(a) $a = 1.787$ cm; $c = 3.497$ cm; $C = 78°$
(b) $B = 40°; b = 14.18$ mm; $c = 9.326$mm
(c) $B = 48° 47'; A = 51° 13'; a = 5.70$ m]

2 Solve the following triangles PQR:

(a) $p = 3.9$ cm; $q = 3.2$ cm; $Q = 35°$
(b) $q = 45$ mm; $r = 35$ mm; $R = 36°$
(c) $P = 55°; q = 12$ cm; $r = 17$ cm.

[(a) $P = 44° 21'; R = 100° 39'; r = 5.48$ cm
OR $P = 135° 39'; R = 9° 21'; r = 0.906$ cm
(b) $Q = 49° 5'; P = 94° 55'; p = 59.3$ mm
OR $Q = 130° 55'; P = 13° 5'; p = 13.5$ mm
(c) $p = 14.1$ cm; $Q = 44° 12'; R = 80° 48'$]

3 Solve the following triangles XYZ:

(a) $X = 104°; y = 3.4$ m; $z = 4.3$ m
(b) $x = 9.0$ cm; $y = 7.5$ cm; $z = 6.7$ cm
(c) $x = 20$ mm; $y = 32$ mm; $z = 43$ mm

[(a) $x = 6.09$ m, $Y = 32° 48', Z = 43° 12'$
(b) $X = 78° 26', Y = 54° 44', Z = 46° 50'$
(c) $Z = 109° 24', Y = 44° 35', X = 26° 1'$]

4 A jib crane is shown in *Fig 19.* If the tie rod AC is 7.50 m long and AB is 4.20 m long determine (a) the length of jib BC, and (b) the angle between the jib and the tie rod. [(a) 10.49 m; (b) 19° 9′]

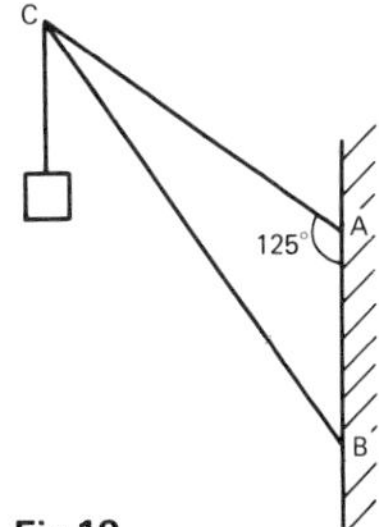

Fig 19

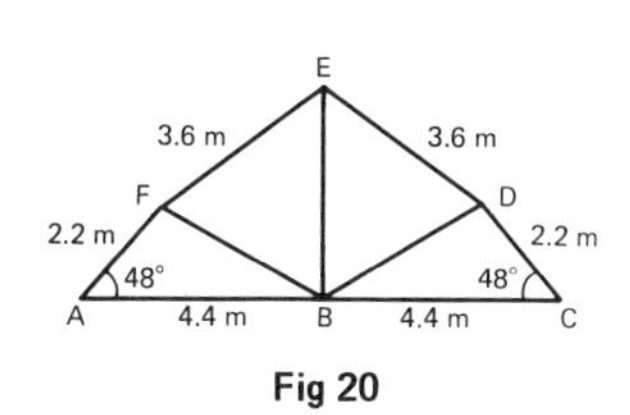

Fig 20

5 Determine the lengths of members BF and EB of the roof truss shown in *Fig 20.* [3.35 m; 3.73 m]

6 An 8.4 m wide room has a span roof which slopes at 37° on one side and 43° on the other. Determine the lengths of the roof slopes. [5.82 m; 5.13 m]

7 A vertical aerial PQ, 9.2 m high stands on ground which is inclined 13° to the horizontal. A stay connects the top of the aerial P to a point R on the ground 12 m downhill from Q, the foot of the aerial. Determine (a) the length of the stay and (b) the angle the stay makes with the ground. [(a) 16.7 m; (b) 32° 28′]

8 A reciprocating engine mechanism is shown in *Fig 21*. The crank AB is 12.5 cm long and the connecting rod BC is 34.0 cm long. For the position shown determine (a) the length of AC and (b) the angle between the crank and the connecting rod. (c) Find also how far C moves when angle CAB changes from 50° to 150°, B moving in an anti-clockwise direction. [(a) 40.66 cm; (b) 113° 39′; (c) 18.06 cm]

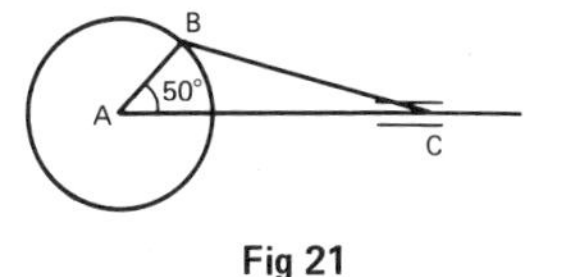

Fig 21

9 *Fig 22.* Determine the angle between plane PRS and QRS in *Fig 22.* [32° 2′]

10 A surveyor standing due west of a building finds the angle of elevation of the top to be 50°. He moves due north 30 m and finds the elevation of the top of the building to be 46°. Calculate the height of the building. [62.76 m]

11 The base of a right pyramid of vertex X is a rectangle ABCD. AX is 12 cm, AB is 10 cm and BC is 8 cm. Calculate (a) the perpendicular height of the pyramid, and (b) the angle edge AX makes with the base. [(a) 10.15 cm; (b) 57° 45′]

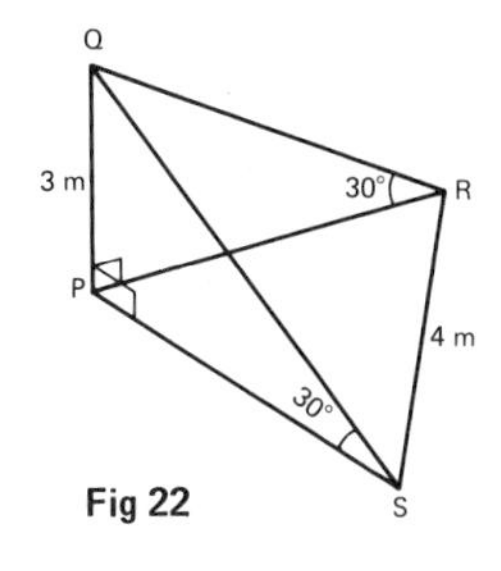

Fig 22

12 In the pyramid of *Problem 11*, P and Q are the mid-points of AB and BC respectively. Determine (a) the angles which the faces XAB and XBC make with the base, and (b) the angle plane XPQ makes with base ABCD. [(a) 68° 29′, 63° 46′; (b) 72° 54′]

13 A marquee is in the form of a regular octagonal pyramid with each side of its

base 5 m in length. If the perpendicular height of the marquee is 12 m find (a) the angle between a sloping face and the base, and (b) the angle of inclination of an edge of the sloping face to the base. [(a) 63° 18′; (b) 61° 26′]

14 An aircraft is sighted due west from a radar station at an elevation of 50° at a height of 8000 m, and later at an elevation of 39° and height 6000 m on a bearing of 160°. If it is descending uniformly, find the angle of descent and the speed of the aircraft, in km/h, if the time between the two observations is 1 minute. [9° 48′; 704.8 km/h]

15 For the cuboid shown in *Fig 23* determine the angles between the line AX and the planes ABCD, ADSP and ABQP. [6° 40′; 37° 48′; 31° 49′]

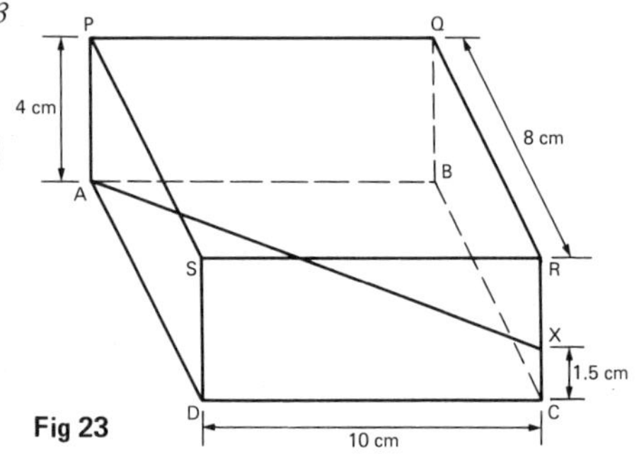

Fig 23

16 A 16 cm high right circular cone of vertex X has a base diameter of 8 cm. If point A is on the circumference of the base find the inclination of AX to the base. B is also a point on the circumference 6 cm from A (i.e. AB is a chord of the circular base). Determine the inclination of plane XAB to the base. [75° 58′; 80° 37′]

17 From a ship situated on a bearing 210° from a vertical lighthouse, the angle of elevation of the top of the lighthouse is 15°. Later, when the ship is on a bearing of 170° from the light house, the angle of elevation is 11°. Determine the height of the lighthouse,if the distance travelled by the ship between the two observations is 120 m. [36.22 m]

12 Trigonometry (2) – Graphs of trigonometric functions

A MAIN POINTS CONCERNED WITH GRAPHS OF TRIGONOMETRIC FUNCTIONS

1 In *Fig 1*, OR represents a vector that is free to rotate anticlockwise about 0 at a velocity of ω rads/s. A rotating vector is called a **phasor**. After a time t seconds OR will have turned through an angle ωt radians (shown as angle TOR in *Fig 1*). If ST is constructed perpendicular to OR, then $\sin \omega t = ST/OT$, i.e. $\mathbf{ST = OT \sin \omega t}$. If all such vertical components are projected on to a graph of y against ωt, a **sine wave** results of maximum value OR.

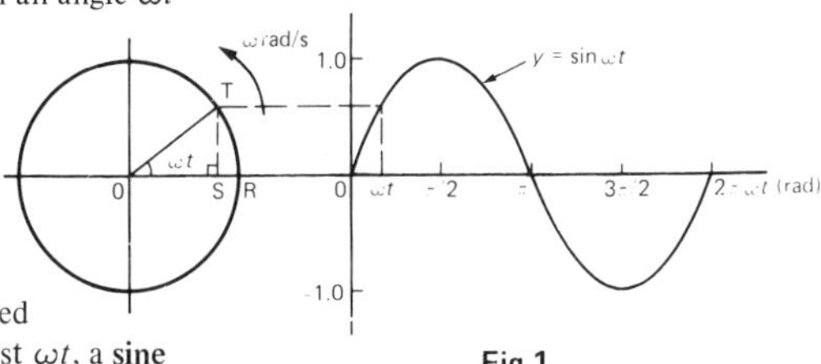

Fig 1

2 If phasor OR of *Fig 1* makes one revolution (i.e. 2π radians) in T seconds, then the angular velocity, $\omega = 2\pi/T$ rads/s, from which, $T = 2\pi/\omega$ **seconds.** T is known as the **periodic time.**

3 The number of complete cycles occurring per second is called the **frequency**, f.

$$\text{Frequency} = \frac{\text{number of cycles}}{\text{second}} = \frac{1}{T} = \frac{\omega}{2\pi} \text{ Hz, i.e. } f = \frac{\omega}{2\pi} \text{ Hz.}$$

Hence **angular velocity,** $\omega = 2\pi f$ **rads/s.**

4 *Fig 2* shows graphs of $y = \sin \omega t$, for $\omega = 1/2, 1, 2$ and 3 rads/s. The graphs are plotted either from a calculated table of values or by the rotating vector approach.

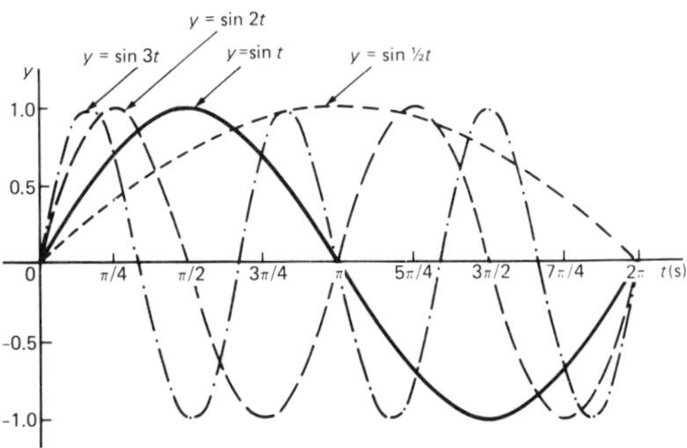

Fig 2

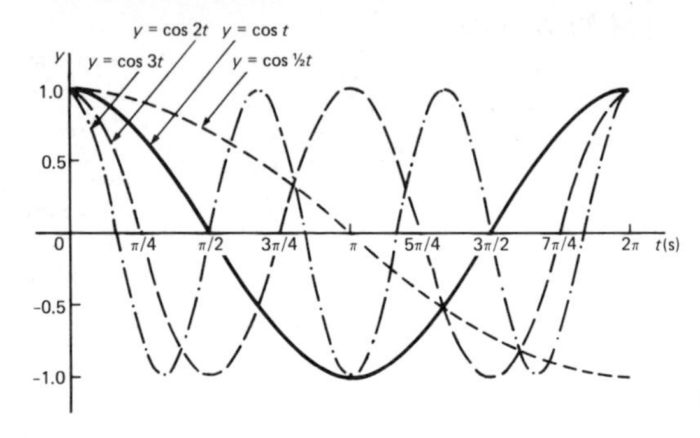

Fig 3

5 *Fig 3* shows graphs of $y = \cos \omega t$, for $\omega = 1/2$, 1, 2 and 3 rads/s.

6 Each of the waveforms shown in *Figs 2 and 3* repeat themselves after periodic time T $(= 2\pi/\omega$ seconds) and such functions are known as **periodic functions.** It is noted from the graphs that in 2π seconds:

$\sin \frac{1}{2}t$ and $\cos \frac{1}{2}t$ complete $\frac{1}{2}$ cycle,
$\sin t$ and $\cos t$ complete 1 cycle,
$\sin 2t$ and $\cos 2t$ complete 2 cycles, and
$\sin 3t$ and $\cos 3t$ complete 3 cycles.

7 *Fig 4* shows graphs of $y = \sin^2 t$ and $y = \cos^2 t$ which may be obtained by drawing up a table of values. Both are periodic functions of periodic time π seconds and both contain only positive values.

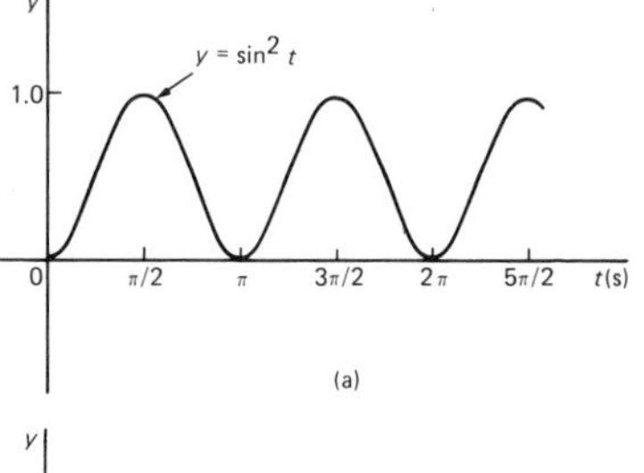

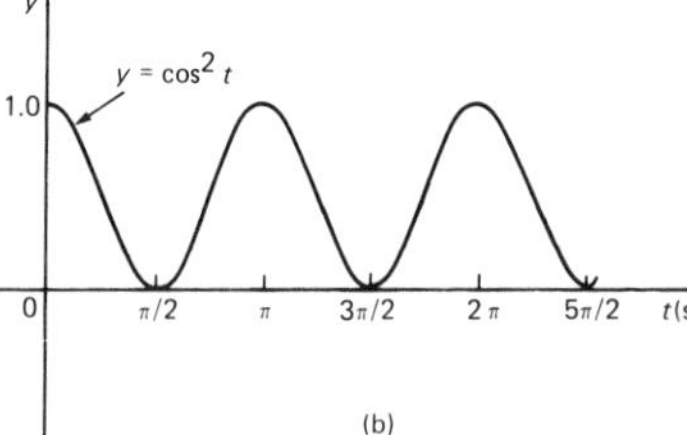

Fig 4

8 (i) *Fig 5(a)* shows a graph of $y = \sin^2 3t$, which has a periodic time of $\pi/3$ seconds.

(ii) *Fig 5(b)* shows a graph of $y = \cos^2 2t$, which has a periodic time of $\pi/2$ seconds.

(iii) In general, if $y = \sin^2 \omega t$ or $y = \cos^2 \omega t$ then the periodic time is given by π/ω seconds.

(iv) Graphs of the form $y = \sin^2 \omega t$ and $y = \cos^2 \omega t$ are **not** sine waves and **cannot** be produced by the rotating vector approach.

The general form of a sine wave, $R \sin (\omega t \pm \alpha)$.

9 **Amplitude** is the name given to the maximum or peak value of a sine wave. Each of the graphs shown in *Figs 2 and 3* have an amplitude of 1. However, if $y = 4 \sin \omega t$, the maximum value, and thus amplitude, is 4.

10 **Lagging and leading angles**

(i) A sine or cosine curve may not always start at zero. To show this a periodic function is represented by $y = \sin(\omega t \pm \alpha)$ or $y = \cos(\omega t \pm \alpha)$, where α is a phase displacement compared with $y = \sin \omega t$ or $y = \cos \omega t$.

(ii) By drawing up a table of values, a graph of $y = \sin(\omega t - \frac{\pi}{3})$ may be plotted as shown in *Fig 6*. If $y = \sin \omega t$ is assumed to start at zero, then $y = \sin(\omega t - \pi/3)$ starts $\pi/3$ radians later (i.e. has a zero value $\pi/3$ rads later). Thus $y = \sin(\omega t - \pi/3)$ is said to **lag** $y = \sin \omega t$ by $\pi/3$ rads.

(iii) By drawing up a table of values, a graph of $y = \cos(\omega t + \pi/4)$ may be plotted as shown in *Fig 7*. If $y = \cos \omega t$ is assumed to start at zero, then $y = \cos(\omega t + \pi/4)$ starts $\pi/4$ radians earlier (i.e. has a value of 1, $\pi/4$ rad earlier). Thus $y = \cos(\omega t + \pi/4)$ is said to **lead** $y = \cos \omega t$ by $\pi/4$ rad.

(iv) Generally, a graph of $y = \sin(\omega t - \alpha)$ lags $y = \sin \omega t$ by angle α, and a graph of $y = \sin(\omega t + \alpha)$ leads $y = \sin \omega t$ by angle α.

(v) A cosine curve is the same shape as a sine curve but starts $\pi/2$ rad earlier, i.e. leads by $\pi/2$ rad. Hence $\mathbf{\cos \omega t = \sin(\omega t + \pi/2)}$.

11 In terms of time, $y = R \sin(\omega t - \alpha)$ lags $y = R \sin \omega t$ by α/ω seconds, and $y = R \sin(\omega t + \alpha)$ leads $y = R \sin \omega t$ by α/ω seconds.

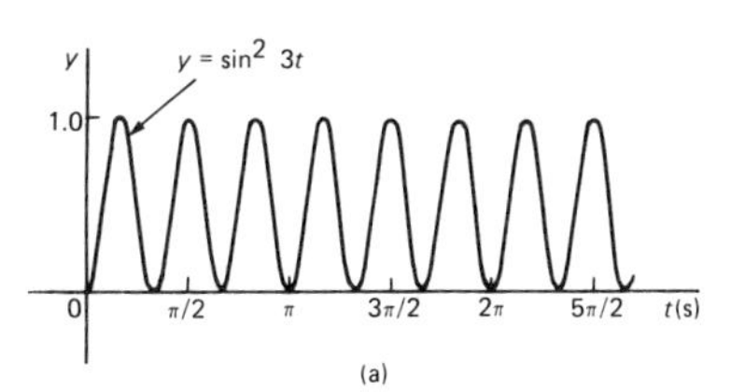

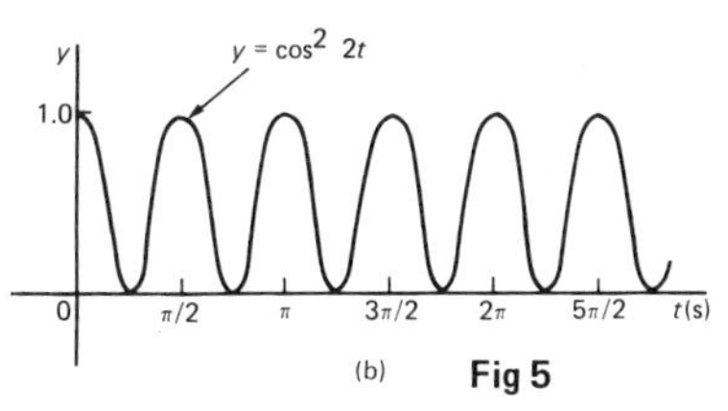

Fig 5

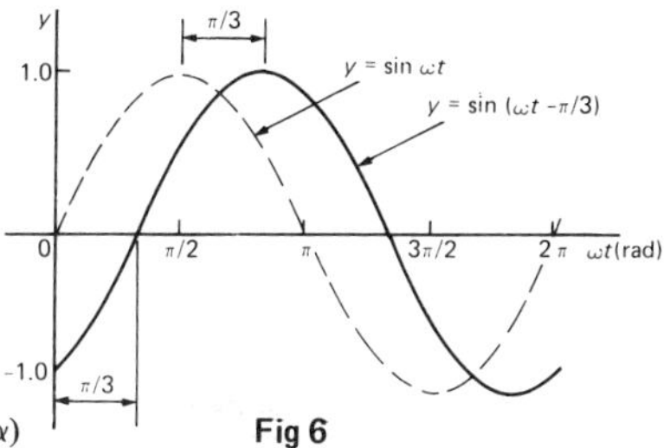

Fig 6

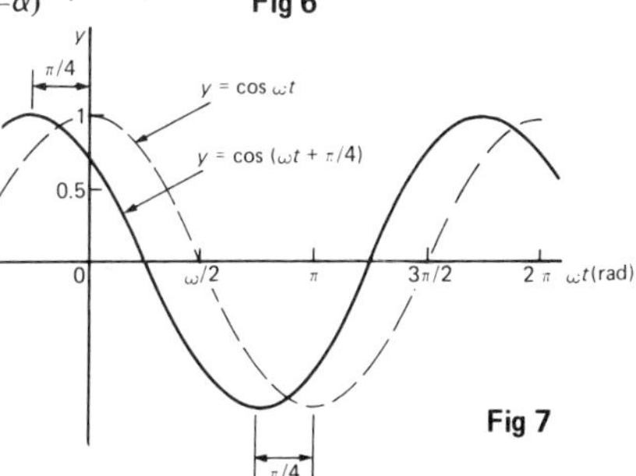

Fig 7

12 **Summary**

Given a general sinusoidal periodic function $y = R \sin(\omega t \pm \alpha)$, then: R = amplitude, ω = angular velocity, $2\pi/\omega$ = periodic time, T, $\omega/2\pi$ = frequency, f, and α = angle of lead or lag (compared with $y = R \sin \omega t$).

B WORKED PROBLEMS ON GRAPHS OF TRIGONOMETRIC FUNCTIONS

Problem 1 The instantaneous value of a sinusoidal voltage is given by $v = 100 \sin 50 \pi t$ volts. Determine (a) the amplitude; (b) the periodic time and (c) the frequency.

$v = 100 \sin 50\pi t$ may be compared with $v = R \sin \omega t$.

(a) Amplitude, or peak value, = **100 V.**

(b) Angular velocity, $\omega = 50\pi$ rads/s.

Hence periodic time, $T = \frac{2\pi}{\omega} = \frac{2\pi}{50\pi} = \frac{1}{25}$ **or 0.04 s.**

(c) Frequency, $f = \frac{\omega}{2\pi} = \frac{50\pi}{2\pi} =$ **25 Hz.**

Problem 2 Determine the amplitude, periodic time and frequency of the curve $y = 40 \cos^2 \pi t/2$

Amplitude = **40**

Angular velocity, $\omega = \pi/2$ rads/s. Periodic time, $T = \pi/\omega$ (from para 8(iii))

i.e. $T = \frac{\pi}{\pi/2} =$ **2 seconds.**

Frequency, $f = \frac{1}{T} = \frac{1}{2}$ **Hz.**

Problem 3 An alternating current is given by $i = 15 \sin (100\pi t - 0.37)$ amperes. Find (a) the maximum value, (b) the periodic time; (c) the frequency; (d) the phase angle (in degrees and minutes) of the oscillation compared with $i = 15 \sin 100\pi t$, and (e) the time interval between $i = 15 \sin 100\pi t$ and $i = 15 \sin (100\pi t - 0.37)$.

$i = 15 \sin (100\pi t - 0.37)$ may be compared with $i = R \sin (\omega t - \alpha)$.

(a) Maximum value = amplitude = **15 A.**

(b) Angular velocity, $\omega = 100\pi$ rads/s. Periodic time $T = \frac{2\pi}{\omega} = \frac{2\pi}{100\pi} = \frac{1}{50}$

= **0.02 or 20 ms.**

(c) Frequency, $f = \frac{\omega}{2\pi} = \frac{100\pi}{2\pi} =$ **50 Hz.**

(d) Phase angle, $\alpha = 0.37$ rads $= \left(0.37 \times \frac{180}{\pi}\right)^\circ = 21.2^\circ$ or

or **21° 12′ lagging** $15 \sin 100\pi t$.

(e) Time interval between $15 \sin 100\pi t$ and $15 \sin (100\pi t - 0.37) = \frac{0.37}{100\pi}$ s

or **1.178 ms.**

Problem 4 Sketch (a) $y = 4 \sin 3t$, and (b) $y = 3 \cos 2t$, from $t = 0$ to $t = 2$ seconds, showing important points.

(a) $y = 4 \sin 3t$ has amplitude = 4, angular velocity, $\omega = 3$ rads/s and

periodic time, $T = \frac{2\pi}{\omega} = \frac{2\pi}{3}$ s.

A sketch of $y = 4 \sin 3t$ is shown in *Fig 8(a).*

(b) $y = 3 \cos 2t$ has amplitude = 3, angular velocity, $\omega = 2$ rads/s, and

periodic time, $T = \frac{2\pi}{\omega} = \frac{2\pi}{2} = \pi$ s.

A sketch of $y = 3 \cos 2t$ is shown in *Fig 8(b).*

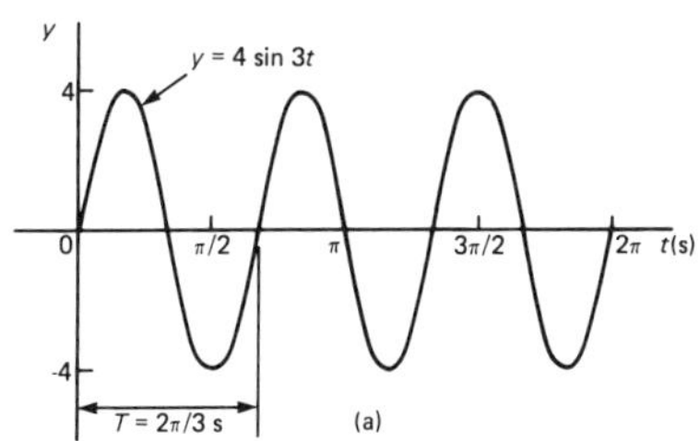

Fig 8

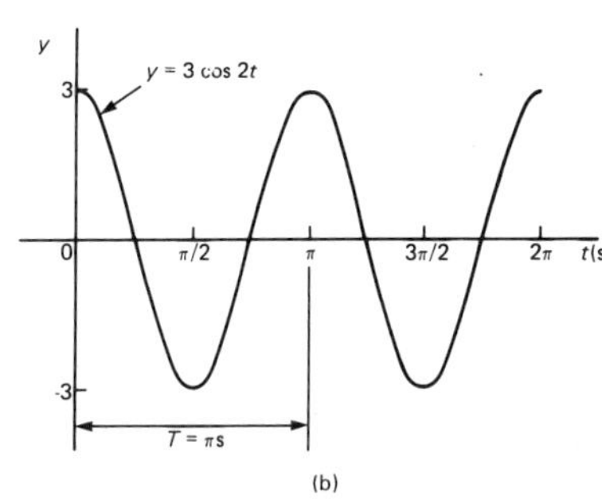

Problem 5 Sketch (a) $y = 5 \sin (t + \frac{\pi}{3})$, and (b) $y = 2 \cos (t - \frac{\pi}{6})$, from $t = 0$ to $t = 2\pi$ seconds, showing important points.

(a) $y = 5 \sin (t + \frac{\pi}{3})$ has amplitude = 5,
angular velocity, $\omega = 1$ rads/s,
periodic time, $T = 2\pi/\omega = 2\pi$ s.
$y = 5 \sin (t+(\pi/3))$ leads $y = 5 \sin t$ by $\pi/3$ seconds.
A sketch of $y = 5 \sin (t+(\pi/3))$ is shown in *Fig 9(a)*.

(b) $y = 2 \cos (t - \frac{\pi}{6})$ has amplitude = 2, angular velocity, $\omega = 1$ rads/s, periodic time, $T = 2\pi/\omega = 2\pi$ s.
$y = 2 \cos (t-\pi/6)$ lags $2 \cos t$ by $\pi/6$ seconds. A sketch of $y = 2 \cos (t-\pi/6)$ is shown in *Fig 9(b)*.

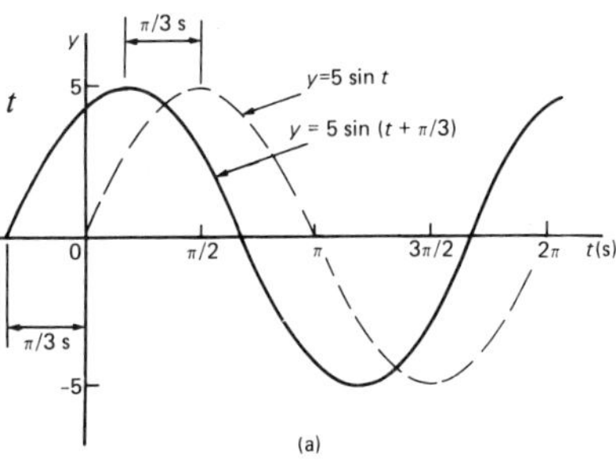

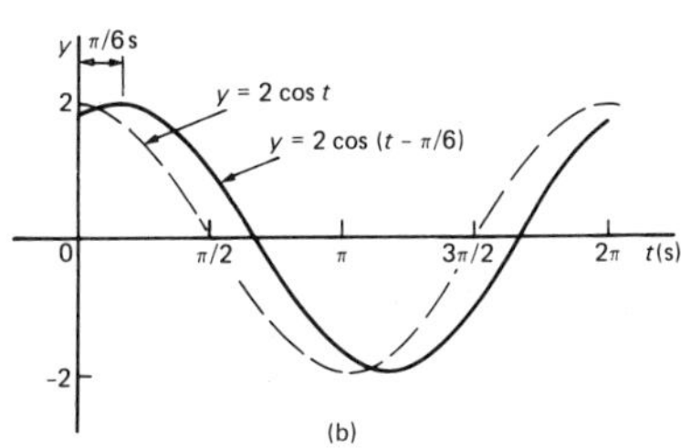

Fig 9

Problem 6 Sketch (a) $y = 3 \sin (2t - \frac{\pi}{4})$, and (b) $y = 8 \cos (3t + \frac{\pi}{3})$, from $t = 0$ to $t = 2\pi$ seconds, showing important points.

(a) $y = 3 \sin (2t - \frac{\pi}{4})$ has a maximum value of 5 angular velocity, w = 2 rads/s,

periodic time, $T = \frac{2\pi}{\omega} = \frac{2\pi}{2}$ $= \pi$ s. $y = 3 \sin (2t - \frac{\pi}{4})$ lags $y = 3 \sin 2t$ by $\frac{\frac{\pi}{4} \text{ rads}}{2 \text{ rads/s}}$ $= \frac{\pi}{8}$ s (see para. 11).

A sketch of $y = 3 \sin (2t - \frac{\pi}{4})$ is shown in *Fig 10(a)*.

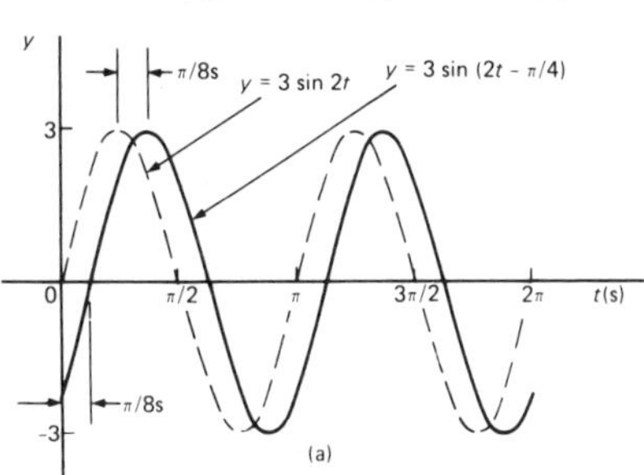

(b) $y = 8 \cos (3t + \frac{\pi}{3})$ has a maximum value of 7, angular velocity, $\omega = 3$ rads/s,

periodic time, $T = \frac{2\pi}{\omega} = \frac{2\pi}{3}$ s.

$y = 8 \cos (3t + \frac{\pi}{3})$ leads $y = 8 \cos 3t$ by $\frac{\frac{\pi}{3} \text{ rads}}{3 \text{ rads/s}} = \frac{\pi}{9}$ s.

A sketch of $y = 8 \cos (3t + \frac{\pi}{3})$ is shown in *Fig 10(b)*.

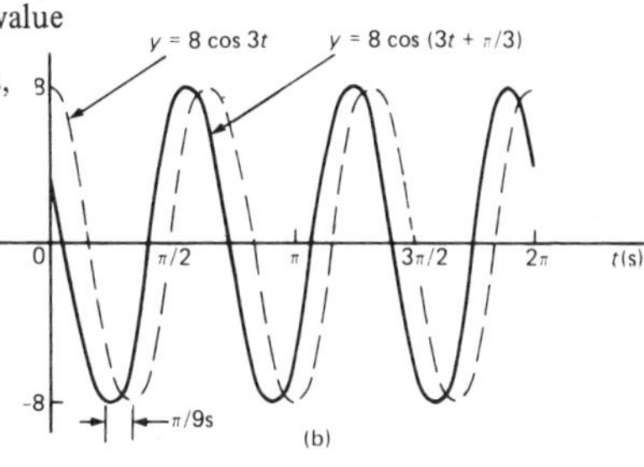

Fig 10

Problem 7 Sketch (a) $y = 5 \sin^2 2t$ and (b) $y = 7 \cos^2 3t$, from $t = 0$ to $t = 2\pi$ seconds, showing important points.

(a) $y = 5 \sin^2 2t$ has a maximum value of 5, angular velocity, $\omega = 2$ rads/s,

periodic time, $T = \frac{\pi}{\omega} = \frac{\pi}{2}$ s. A sketch of $y = 5 \sin^2 2t$ is shown in *Fig 11(a)*.

(b) $y = 7 \cos^2 3t$ has a maximum value of 7, angular velocity, $\omega = 3$ rads/s,

periodic time, $T = \frac{\pi}{\omega} = \frac{\pi}{3}$ s. A sketch of $y = 7 \cos^2 3t$ is shown in *Fig 11(b)*.

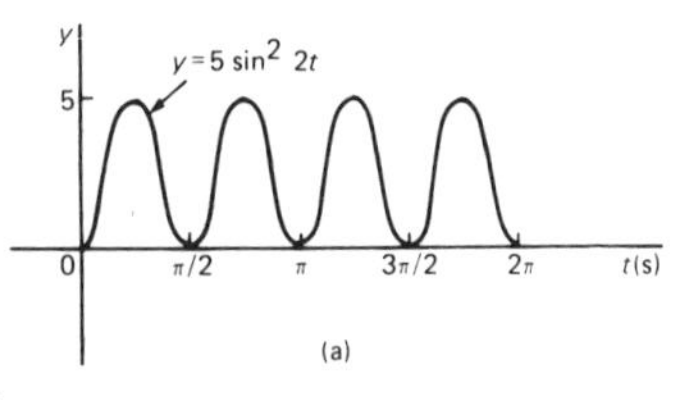

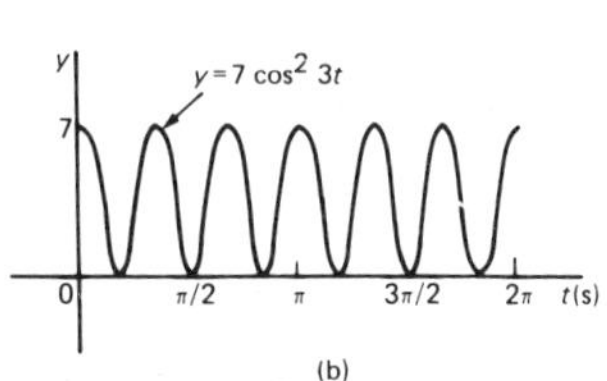

Fig 11

Problem 8 An alternating voltage has a periodic time of 20 ms and a maximum value of 50 V. When time t is zero, the voltage is 12 V. Express the instantaneous voltage v in the form $v = R \sin(\omega t \pm \alpha)$.

Amplitude R = maximum value = 50 V.

Periodic time, $T = 20 \text{ ms} = 0.02 \text{ s} = \frac{2\pi}{\omega}$.

Hence angular velocity, $\omega = \frac{2\pi}{0.02} = 100\pi$ rads/s.

Thus $v = 50 \sin(100\,\pi t+\alpha)$. (Note, α is leading, i.e., positive, since voltage has a positive value at $t = 0$.)
When $t = 0$, $v = 12$ V. Thus $12 = 50 \sin \alpha$, from which $\alpha = \arcsin \frac{12}{50} = 13.89° = 0.242$ rads.

Hence $v = 50 \sin(100\,\pi t+0.242)$ V.

Problem 9 At any time t seconds an alternating current is given by $i = 20 \sin 100t$ amperes. Determine the times in the first cycle when the current is -8 A.

If $i = -8$ A then $-8 = 20 \sin 100t$, from which $\sin 100t = \left(-\frac{8}{20}\right)$

and $100t = \arcsin\left(-\frac{8}{20}\right) = 203°\,35'$ or $336°\,25'$,

i.e. $100t = 3.553$ rads or 5.872 rads.

Hence $t = \frac{3.553}{100} = 0.03553$ or $t = \frac{5.872}{100} = 0.05872$,

i.e., **the times in the first cycle when the current is -8 amperes are 35.53 ms and 58.72 ms.**

Problem 10 An alternating voltage is given by $v = 200 \sin(200\,\pi t+0.26)$ V. Find (a) the amplitude, periodic time, frequency and phase angle (in degrees and minutes), (b) the value of voltage when time t is zero, (c) the value of voltage when time t is 4 ms, (d) the time when the voltage first reaches 100 V and (e) the time when the voltage is first a maximum.
Sketch one cycle of the waveform.

(a) $v = 200 \sin(200\,\pi t+0.26)$ V. Thus amplitude = **200 V.**

Angular velocity, $\omega = 200\pi$ rads/s. Periodic time, $T = \frac{2\pi}{\omega} = \frac{2\pi}{200\pi} = \frac{1}{100}$

= **0.01 s or 10 ms.**

Frequency, $f = \frac{1}{T}$ = **100 Hz.**

Phase angle = 0.26 rads = $(0.26 \times \frac{180}{\pi})°$ = 14.90° = $14°\,54'$ leading $200 \sin(200\,\pi t)$.

(In terms of time $v = 200 \sin(200\,\pi t+0.26)$ leads $v = 200 \sin 200\,\pi t$ by $\frac{0.26}{200\pi}$ s = **0.41 ms.**)

(b) When time $t = 0$, $v = 200 \sin(0+0.26) = 200 \sin 14°\,54'$ = **51.43 V.**

(c) When time $t = 4$ ms, $v = 200 \sin [200\pi(4/10^3)+0.26] = 200 \sin 2.7733$
$= 200 \sin 158.9° = \mathbf{72.0\ V}$.

(d) When $v = 100$ V, $100 = 200 \sin (200\ \pi t+0.26)$

i.e. $\frac{100}{200} = \sin (200\ \pi t+0.26)$, from which,

$$(200\ \pi t+0.26) = \arcsin\left(\frac{100}{200}\right) = 30° \text{ or } 0.5236 \text{ rads.}$$

i.e. $200\ \pi t+0.26 = 0.5236$.

Hence time, $t = \frac{0.5236-0.26}{200\pi} = \mathbf{0.42\ ms.}$

(e) When the voltage is first a maximum, $v = 200$ volts.
Hence $200 = 200 \sin (200\ \pi t+0.26)$
i.e., $\sin (200\ \pi t+0.26) = 1$, from which, $(200\ \pi t+0.26)$
$= \arcsin 1 = 90° = \frac{\pi}{2} = 1.5708$ rads.

Hence time, $t = \frac{1.5708-0.26}{200\pi} = \mathbf{2.09\ ms.}$

One cycle of the waveform is shown in *Fig 12.*

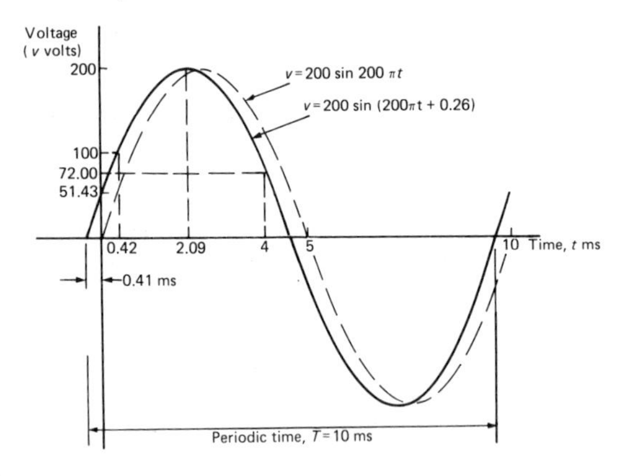

Fig 12

C FURTHER PROBLEMS ON GRAPHS OF TRIGONOMETRIC FUNCTIONS

1 State the amplitude, periodic time and frequency of the following waveforms:
(a) $5 \sin 400\ \pi t$; (b) $120 \sin 314.2\ t$; (c) $2 \cos \frac{1}{2} t$; (d) $4 \sin (2t+0.8)$

[(a) 5, 5ms, 200 Hz; (b) 120, 20 ms, 50 Hz;
(c) 2, 12.57s, 0.0796 Hz; (d) 4, 3.142s, 0.318 Hz.]

In *Problems 2 to 4* find the amplitude, periodic time, frequency and phase angle (in degrees and minutes) for each of the alternating quantities shown.

2 $i = 25 \sin (50\ \pi t+0.29)$ [25, 0.04 s, 25 Hz, leading $25 \sin 50\ \pi t$ by $16°\ 37'$]

3 $x = 30 \sin (628.3t-0.416)$

[30, 0.01 s, 100 Hz, lagging $30 \sin 628.3t$ by $23°\ 50'$]

4 $y = 85 \sin(40t - 0.50)$

[85, 0.157 s, 6.37 Hz, lagging 85 sin $40t$ by 28° 39′]

5 An instantaneous value of an alternating voltage is given by $v = 340 \sin(157.1t + 0.52)$ V. Find (a) the maximum value, (b) the periodic time, (c) the frequency, (d) the phase angle (in degrees and minutes) and (e) the time interval between $v = 340 \sin 157.1\,t$ and $v = 340 \sin(157.1t + 0.52)$.

[(a) 340 V; (b) 0.04 s; (c) 25 Hz; (d) 29° 48′ leading; (e) 3.31 ms]

6 Determine the maximum value, periodic time and frequency of the curves: (a) $y = 4 \sin^2 2t$; (b) $y = 50 \cos^2 \pi t$

[(a) 4, 1.571 s, 0.637 Hz; (b) 50, 1 s, 1 Hz]

In *Problems 7 and 8* sketch the given curves from $t = 0$ to $t = 2\pi$ seconds showing important points.

7 (a) $y = 3 \sin 2t$; (b) $y = 10 \cos 3t$;

(c) $y = 16 \sin(t - \frac{\pi}{6})$; (d) $y = 5 \cos(t + \frac{\pi}{4})$.

8 (a) $y = 7 \sin(3t + \frac{\pi}{2})$; (b) $y = 12 \cos(2t - \frac{\pi}{6})$;

(c) $y = 4 \sin^2 \frac{1}{2}t$; (d) $y = 6 \cos^2 2t$.

9 An oscillating mechanism has a maximum displacement of 2.5 m and a frequency of 100 Hz. At time $t = 0$, the displacement is zero. Express the instantaneous displacement x in the form $x = R \sin(\omega t \pm \alpha)$.

[$x = 2.5 \sin(200\pi t)$ metres]

10 An alternating current has a periodic time of 20 ms and a maximum value of 5 A. When time t is zero the current is −3 A. Express the instantaneous current in the form $i = R \sin(\omega t \pm \alpha)$. [$i = 5 \sin(100\pi t - 0.644)$A]

11 At any time t seconds an alternating voltage is given by $v = 250 \sin 50t$ V. Determine the times in the first cycle when the current is (a) 100 volts and (b) −180 volts. [(a) 8.23 ms and 54.60 ms; (b) 78.91 ms and 109.6 ms]

12 An alternating current at any time t seconds is given by $i = 12 \sin(100\pi t - 0.724)$ A. Determine (a) the amplitude, periodic time, frequency and phase angle, stating whether it is leading or lagging $12 \sin(100\pi t)$, (b) the value of current at time $t = 0$, (c) the value of current at time, $t = 6$ ms, (d) the times in the first cycle when the current is 8 A, (e) the times in the first cycle when the current is −10 A, and (f) the time when the current is first a maximum. Sketch the curve showing important points.

[(a) 12 A, 0.02 s, 50 Hz, 41° 29′ lagging;
(b) −7.949 A; (c) 11.01 A;
(d) 4.627 ms and 9.982 ms;
(e) 15.44 ms and 19.17 ms; (f) 7.305 ms.]

13 Trigonometry (3) – Compound angles

A MAIN POINTS CONCERNED WITH COMPOUND ANGLES

1 Angles such as $(A+B)$ or $(A-B)$ are called **compound angles** since they are the sum or difference of two angles, A and B.

2 The **compound angle formulae** for sines and cosines of the sum and difference of two angles A and B are:

$$\sin(A+B) = \sin A \cos B + \cos A \sin B$$
$$\sin(A-B) = \sin A \cos B - \cos A \sin B$$
$$\cos(A+B) = \cos A \cos B - \sin A \sin B$$
$$\cos(A-B) = \cos A \cos B + \sin A \sin B$$

(Note, $\sin(A+B)$ is **not** equal to $(\sin A + \sin B)$, and so on.)

3 The formulae stated in para. 2 may be used to derive two further compound-angle formulae:

$$\tan(A+B) = \frac{\tan A + \tan B}{1-\tan A \tan B}$$

$$\tan(A-B) = \frac{\tan A - \tan B}{1+\tan A \tan B}$$

4 The compound-angle formulae are true for all values of A and B, and by substituting values of A and B into the formulae they may be shown to be true. (See *Problems 1 to 10*.)

5 (i) $R\sin(\omega t+\alpha)$ represents a sine wave of maximum value R, periodic time $2\pi/\omega$, frequency $\omega/2\pi$ and leading $R\sin\omega t$ by angle α. (See Chapter 12).

(ii) $R\sin(\omega t+\alpha)$ may be expanded using the compound-angle formula for $\sin(A+B)$, where $A = \omega t$ and $B = \alpha$.

Hence $R\sin(\omega t+\alpha) = R[\sin\omega t\cos\alpha+\cos\omega t\sin\alpha]$
$= R\sin\omega t\cos\alpha+R\cos\omega t\sin\alpha$
$= (R\cos\alpha)\sin\omega t+(R\sin\alpha)\cos\omega t$

(iii) If $a = R\cos\alpha$ and $b = R\sin\alpha$, where a and b are constants, then
$R\sin(\omega t+\alpha) = a\sin\omega t+b\cos\omega t$
i.e., a sine and cosine function of the same frequency when added produce a sine wave of the same frequency (as shown in *Mathematics 2 checkbook*, chapter 3).

(iv) Since $a = R\cos\alpha$, then $\cos\alpha = a/R$, and since $b = R\sin\alpha$, then $\sin\alpha = b/R$.

If the values of a and b are known then the values of R and α may be calculated. The relationship between constants a, b, R and α are shown in *Fig 1*.

From *Fig 1*, by Pythagoras' theorem:

$\boldsymbol{R = \sqrt{(a^2 + b^2)}}$ and from trigonometric ratios: $\alpha =$ **arctan** $\boldsymbol{b/a}$.

(See *Problems 11 to 16.*)

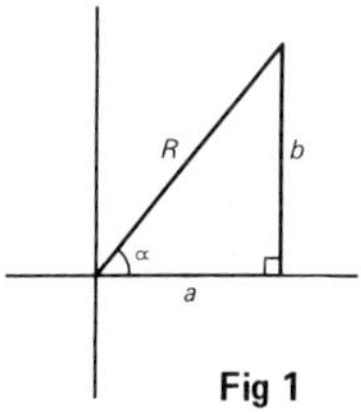

Fig 1

6 Double angles

(i) If, in the compound-angle formula for $\sin(A+B)$, we let $B = A$ then
$\boldsymbol{\sin 2A = 2 \sin A \cos A}$.
Also, for example, $\sin 4A = 2 \sin 2A \cos 2A$
and $\sin 8A = 2 \sin 4A \cos 4A$, and so on.

(ii) If, in the compound-angle formula for $\cos(A+B)$, we let $B = A$
then $\boldsymbol{\cos 2A = \cos^2 A - \sin^2 A}$.
Since $\cos^2 A + \sin^2 A = 1$, then $\cos^2 A = 1 - \sin^2 A$, and $\sin^2 A = 1 - \cos^2 A$, and two further formula for $\cos 2A$ can be produced.

Thus $\cos 2A = \cos^2 A - \sin^2 A = (1 - \sin^2 A) - \sin^2 A$

i.e. $\boldsymbol{\cos 2A = 1 - 2 \sin^2 A}$

and $\cos 2A = \cos^2 A - \sin^2 A = \cos^2 A - (1 - \cos^2 A)$

i.e. $\boldsymbol{\cos 2A = 2 \cos^2 A - 1}$.

Also, for example,

$\cos 4A = \cos^2 2A - \sin^2 2A$ or $1 - 2 \sin^2 2A$ or $2 \cos^2 2A - 1$

and $\cos 6A = \cos^2 3A - \sin^2 3A$ or $1 - 2 \sin^2 3A$ or $2 \cos^2 3A - 1$, and so on.

(iii) If, in the compound-angle formula for $\tan(A+B)$, we let $B = A$

then $$\boldsymbol{\tan 2A = \frac{2 \tan A}{1 - \tan^2 A}}$$

Also, for example, $\tan 4A = \dfrac{2 \tan 2A}{1 - \tan^2 2A}$ and $\tan 5A = \dfrac{2 \tan \frac{5}{2}A}{1 - \tan^2 \frac{5}{2}A}$,

and so on.

(See *Problems 17 to 20.*)

7 Changing products of sines and cosines into sums or differences

(i) $\sin(A+B) + \sin(A-B) = 2 \sin A \cos B$ (from the formula in para. 2)

i.e. $$\boldsymbol{\sin A \cos B = \frac{1}{2}[\sin(A+B) + \sin(A-B)]} \quad (1)$$

(ii) $\sin(A+B) - \sin(A-B) = 2 \cos A \sin B$

i.e. $$\boldsymbol{\cos A \sin B = \frac{1}{2}[\sin(A+B) - \sin(A-B)]} \quad (2)$$

(iii) $\cos(A+B) + \cos(A-B) = 2 \cos A \cos B$

i.e. $$\boldsymbol{\cos A \cos B = \frac{1}{2}[\cos(A+B) + \cos(A-B)]} \quad (3)$$

(iv) $\cos(A+B) - \cos(A-B) = -2 \sin A \sin B$

i.e. $$\boldsymbol{\sin A \sin B = -\frac{1}{2}[\cos(A+B) - \cos(A-B)]} \quad (4)$$

(See *Problems 21 to 25.*)

8 Changing sums or differences of sines and cosines into products

In the compound-angle formula let $(A+B) = X$ and $(A-B) = Y$

Solving the simultaneous equations gives $A = \dfrac{X+Y}{2}$ and $B = \dfrac{X-Y}{2}$

Thus $\sin(A+B) + \sin(A-B) = 2 \sin A \cos B$

becomes $\sin X+\sin Y = 2\sin\left(\frac{X+Y}{2}\right)\cos\left(\frac{X-Y}{2}\right)$ (5)

Similarly, $\sin X-\sin Y = 2\cos\left(\frac{X+Y}{2}\right)\sin\left(\frac{X-Y}{2}\right)$ (6)

$\cos X+\cos Y = 2\cos\left(\frac{X+Y}{2}\right)\cos\left(\frac{X-Y}{2}\right)$ (7)

$\cos X-\cos Y = -2\sin\left(\frac{X+Y}{2}\right)\sin\left(\frac{X-Y}{2}\right)$ (8)

(See *Problems 26 to 30.*)

B WORKED PROBLEMS ON COMPOUND ANGLES

Problem 1 Show that the compound-angle formula for (a) $\sin(A+B)$ and (b) $\cos(A-B)$ are true, correct to 3 significant figures, when $A=45°$ and $B=60°$.

(a) $\sin(A+B) = \sin A\cos B+\cos A\sin B$
When $A = 45°$ and $B = 60°$, $\sin(45°+60°) = \sin 45°\cos 60°+\cos 45°\sin 60°$
$= (0.7071)(0.5000)+(0.7071)(0.8660) = 0.3536+0.6123$
i.e. $\sin 105° = 0.9659$
$\sin 105°$ ($\equiv \sin 75°$) $= 0.9659$ from tables or calculator.
Thus the formula for $\sin(A+B)$ is true when $A = 45°$ and $B = 60°$.

(b) $\cos(A+B) = \cos A\cos B-\sin A\sin B$.
When $A = 45°$ and $B = 60°$, $\cos(45°+60°) = \cos 45°\cos 60°-\sin 45°\sin 60°$
$= (0.7071)(0.5000)-(0.7071)(0.8660) = 0.3536-0.6123$
i.e. $\cos 105° = -0.2587$
$\cos 105°$ ($\equiv -\cos 75°$) $= -0.2588$ from tables or calculator.
Thus the formula for $\cos(A+B)$ is true when $A = 45°$ and $B = 60°$, the small error being due to 'rounding-off'.

Problem 2 Show that the compound-angle formula for (a) $\sin(A-B)$ and (b) $\cos(A-B)$ are true when $A = 211°$ and $B = 124°$.

(a) $\sin(A-B) = \sin A\cos B-\cos A\sin B$
When $A = 211°$ and $B = 124°$, $\sin(211°-124°) = \sin 211°\cos 124° - \cos 211°\sin 124°$
$= (-0.5150)(-0.5592)-(-0.8572)(0.8290) = 0.2880+0.7106$
i.e. $\sin 87° = 0.9986$. From tables or calculator $\sin 87° = -0.9986$
Thus the formula for $\sin(A-B)$ is true when $A = 211°$ and $B = 124°$.

(b) $\cos(A-B) = \cos A\cos B+\sin A\sin B$
When $A = 211°$ and $B = 124°$, $\cos(211°-124°)$
$= \cos 211°\cos 124°+\sin 211°\sin 124°$
$= (-0.8572)(-0.5592)+(-0.5150)(0.8290) = 0.4793-0.4269$
i.e. $\cos 87° = 0.0524$
From tables or calculator, $\cos 87° = 0.0524$
Thus the formula for $\cos(A-B)$ is true when $A = 211°$ and $B = 124°$.

Problem 3 Verify that when $A = 162°$ and $B = 55°$ the compound-angle formulae for (a) $\tan(A+B)$ and (b) $\tan(A-B)$ are true, correct to 3 significant figures.

(a) $\tan(A+B) = \dfrac{\tan A+\tan B}{1-\tan A \tan B}$

When $A = 162°$ and $B = 55°$, $\tan(162°+55°) = \dfrac{\tan 162°+\tan 55°}{1-\tan 162° \tan 55°}$

$$= \frac{(-0.3249)+(1.4281)}{1-(-0.3249)(1.4281)}$$

i.e. $\tan 217° = \dfrac{1.1032}{1.4640} = 0.7536$

From tables or calculator, $\tan 217°$ (= $+\tan 37°$) = 0.7536

Thus the formula for $\tan(A+B)$ is true when $A = 162°$ and $B = 55°$.

(b) $\tan(A-B) = \dfrac{\tan A-\tan B}{1+\tan A \tan B}$

When $A = 162°$ and $B = 55°$, $\tan(162°-55°) = \dfrac{\tan 162°-\tan 55°}{1+\tan 162° \tan 55°}$

$$= \frac{(-0.3249)-(1.4281)}{1+(-0.3249)(1.4281)}$$

i.e. $\tan 107° = \dfrac{-1.7530}{0.5360} = -3.2705 = -3.271$ correct to 3 significant figures

From tables or calculator, $\tan 107°$ (= $-\tan 73°$) = $-3.2709 = -3.271$ correct to 3 significant figures.

Thus the formula for $\tan(A-B)$ is true when $A = 162°$ and $B = 55°$.

Problem 4 Expand and simplify the following expressions: (a) $\sin(\pi+\alpha)$; (b) $-\cos(90°+\beta)$; (c) $\sin(A-B)-\sin(A+B)$.

(a) $\sin(\pi+\alpha) = \sin\pi\cos\alpha + \cos\pi\sin\alpha$ (from the formula for $\sin(A+B)$)

$= (0)(\cos\alpha)+(-1)\sin\alpha = \mathbf{-\sin\alpha}$.

(b) $-\cos(90°+\beta) = -[\cos 90° \cos\beta - \sin 90° \sin\beta]$

$= -[(0)(\cos\beta)-(1)\sin\beta] = \mathbf{\sin\beta}$

(c) $\sin(A-B)-\sin(A+B) = [\sin A\cos B-\cos A\sin B]-[\sin A\cos B+\cos A\sin B]$

$= \mathbf{-2\cos A\sin B}$.

Problem 5 Prove that $\cos(y-\pi)+\sin(y+\frac{\pi}{2}) = 0$.

$\cos(y-\pi) = \cos y\cos\pi+\sin y\sin\pi = (\cos y)(-1)+(\sin y)(0) = -\cos y$

$\sin(y+\frac{\pi}{2}) = \sin y\cos\frac{\pi}{2}+\cos y\sin\frac{\pi}{2} = (\sin y)(0)+(\cos y)(1) = \cos y$

Hence $\cos(y-\pi)+\sin(y+\frac{\pi}{2}) = (-\cos y)+(\cos y) = 0$.

Problem 6 Show that $\tan(x+\frac{\pi}{4})\tan(x-\frac{\pi}{4}) = -1$.

$\tan(x+\frac{\pi}{4}) = \dfrac{\tan x+\tan\frac{\pi}{4}}{1-\tan x\tan\frac{\pi}{4}}$ (from the formula for $\tan(A+B)$),

$= \dfrac{\tan x+1}{1-(\tan x)(1)} = \left(\dfrac{1+\tan x}{1-\tan x}\right)$, since $\tan\frac{\pi}{4} = 1$

$$\tan\left(x-\frac{\pi}{4}\right) = \frac{\tan x - \tan\frac{\pi}{4}}{1+\tan x \tan\frac{\pi}{4}} = \frac{\tan x - 1}{1+\tan x}$$

Hence $\tan\left(x+\frac{\pi}{4}\right)\tan\left(x-\frac{\pi}{4}\right) = \left(\frac{1+\tan x}{1-\tan x}\right)\left(\frac{\tan x-1}{1+\tan x}\right) = \frac{\tan x-1}{1-\tan x} = \frac{-(1-\tan x)}{1-\tan x} = -1$

Problem 7 Given that $\sin A = 12/13$ and $\cos B = 8/17$, where A and B are acute angles, determine without using trigonometric tables the values of (a) $\sin(A+B)$; (b) $\cos(A-B)$ and (c) $\tan(A-B)$ each correct to 3 significant figures.

$\sin A = \frac{12}{13} = \frac{\text{opposite side}}{\text{hypotenuse}}$, as shown in *Fig 2(a).*

From *Fig 2(a),* side $p = \sqrt{(13^2-12^2)} = 5$, by Pythagoras' theorem.

Hence $\cos A = \frac{5}{13}$ and $\tan A = \frac{12}{5}$

Similarly, $\cos B = \frac{8}{17} = \frac{\text{adjacent side}}{\text{hypotenuse}}$

as shown in *Fig 2(b).*
From *Fig 2(b)*, side $q = \sqrt{(17^2-8^2)} = 15$, by Pythagoras' theorem.

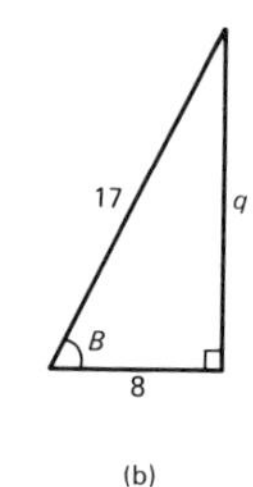

Fig 2

Hence $\sin B = \frac{15}{17}$ and $\tan B = \frac{15}{8}$.

(a) $\sin(A+B) = \sin A \cos B + \cos A \sin B = \left(\frac{12}{13}\right)\left(\frac{8}{17}\right) + \left(\frac{5}{13}\right)\left(\frac{15}{17}\right)$

$= \frac{96}{221} + \frac{75}{221} = \frac{171}{221}$ **or 0.774**

(b) $\cos(A-B) = \cos A \cos B + \sin A \sin B = \left(\frac{5}{13}\right)\left(\frac{8}{17}\right) + \left(\frac{12}{13}\right)\left(\frac{15}{17}\right)$

$= \frac{40}{221} + \frac{180}{221} = \frac{220}{221}$ **or 0.995**

(c) $\tan(A-B) = \frac{\tan A - \tan B}{1+\tan A \tan B} = \frac{\frac{12}{5}-\frac{15}{8}}{1+\left(\frac{12}{5}\right)\left(\frac{15}{8}\right)} = \frac{0.525}{5.5} = \mathbf{0.095}$

Problem 8 If $\sin P = 0.8142$ and $\cos Q = 0.4432$ evaluate, correct to 3 decimal places (a) $\sin(P-Q)$; (b) $\cos(P+Q)$ and (c) $\tan(P+Q)$, using the compound-angle formulae.

Since $\sin P = 0.8142$ then $P = \arcsin 0.8142 = 54.51°$.
Thus $\cos P = \cos 54.51° = 0.5806$ and $\tan P = \tan 54.51° = 1.4025$
Since $\cos Q = 0.4432$, $Q = \arccos 0.4432 = 63.69°$
Thus $\sin Q = \sin 63.69° = 0.8964$ and $\tan Q = \tan 63.69° = 2.0225$
(a) $\sin(P-Q) = \sin P \cos Q - \cos P \sin Q$
$= (0.8142)(0.4432)-(0.5806)(0.8964) = 0.3609-0.5204 = \mathbf{-0.160}$
(b) $\cos(P+Q) = \cos P \cos Q - \sin P \sin Q$
$= (0.5806)(0.4432)-(0.8142)(0.8964) = 0.2573-0.7298 = \mathbf{-0.473}$

(c) $\tan(P+Q) = \dfrac{\tan P+\tan Q}{1-\tan P\tan Q} = \dfrac{(1.4025)+(2.0225)}{1-(1.4025)(2.0225)} = \dfrac{3.4250}{-1.8366} = \mathbf{-1.865}$

Problem 9 Determine, in surd form, the value of sin 75° without using trigonometric tables.

$\sin 75° = \sin(45°+30°) = \sin 45° \cos 30°+\cos 45° \sin 30°$, from the formula for $\sin(A+B)$,

$$= \left(\frac{1}{\sqrt{2}}\right)\left(\frac{\sqrt{3}}{2}\right) + \left(\frac{1}{\sqrt{2}}\right)\left(\frac{1}{2}\right) = \frac{\sqrt{3}}{2\sqrt{2}} + \frac{1}{2\sqrt{2}} = \mathbf{\frac{\sqrt{3}+1}{2\sqrt{2}}}$$

Problem 10 Solve the equation $4\sin(x-20°) = 5\cos x$ for values of x between 0° and 90°.

$4\sin(x-20°) = 4[\sin x \cos 20°-\cos x \sin 20°]$, from the formula for $\sin(A-B)$
$= 4[\sin x\,(0.9397)-\cos x\,(0.3420)]$
$= 3.7588\sin x-1.3680\cos x$

Since $4\sin(x-20°) = 5\cos x$ then $3.7588\sin x-1.3680\cos x = 5\cos x$

Rearranging gives: $3.7588\sin x = 5\cos x+1.3680\cos x = 6.3680\cos x$

and $\dfrac{\sin x}{\cos x} = \dfrac{6.3680}{3.7588} = 1.6942$

i.e. $\tan x = 1.6942$, and $x = \arctan 1.6942 = 59.449°$ or **59° 27′**.

[*Check:* LHS $= 4\sin(59.449°-20°) = 4\sin 39.449° = 2.542$
RHS $= 5\cos x = 5\cos 59.449° = 2.542$]

Problem 11 Find an expression for $3\sin\omega t+4\cos\omega t$ in the form $R\sin(\omega t+\alpha)$ and sketch graphs of $3\sin\omega t$, $4\cos\omega t$ and $R\sin(\omega t+\alpha)$ on the same axes.

Let $3\sin\omega t+4\cos\omega t = R\sin(\omega t+\alpha)$
then $3\sin\omega t+4\cos\omega t = R[\sin\omega t\cos\alpha+\cos\omega t\sin\alpha]$
$= (R\cos\alpha)\sin\omega t+(R\sin\alpha)\cos\omega t$

Equating coefficients of $\sin\omega t$ gives: $3 = R\cos\alpha$, from which, $\cos\alpha = \dfrac{3}{R}$

Equating coefficients of $\cos\omega t$ gives: $4 = R\sin\alpha$, from which, $\sin\alpha = \dfrac{4}{R}$

There is only one quadrant where both $\sin\alpha$ and $\cos\alpha$ are positive, and this is the first, as shown in *Fig 3*. From *Fig 3*, by Pythagoras' theorem:
$R = \sqrt{(3^2+4^2)} = 5$.

From trigonometric ratios: $\alpha = \arctan\dfrac{4}{3} = 53°\ 8'$ or 0.927 radians.

Hence $3\sin\omega t+4\cos\omega t = 5\sin(\omega t+0.927)$.

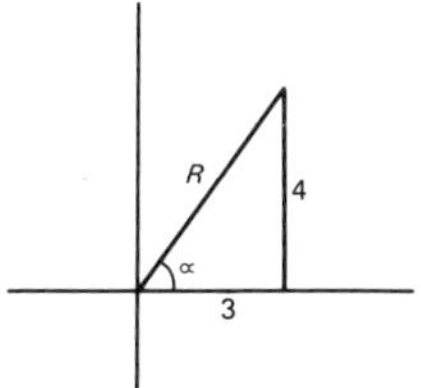

Fig 3

A sketch of $3\sin\omega t$, $4\cos\omega t$ and $5\sin(\omega t+0.927)$ is shown in *Fig 4*.
(It was shown in Mathematics 2 checkbook, chapter 3, that two periodic functions of the same frequency could be combined by
(a) plotting the functions graphically and combining ordinates at intervals, or

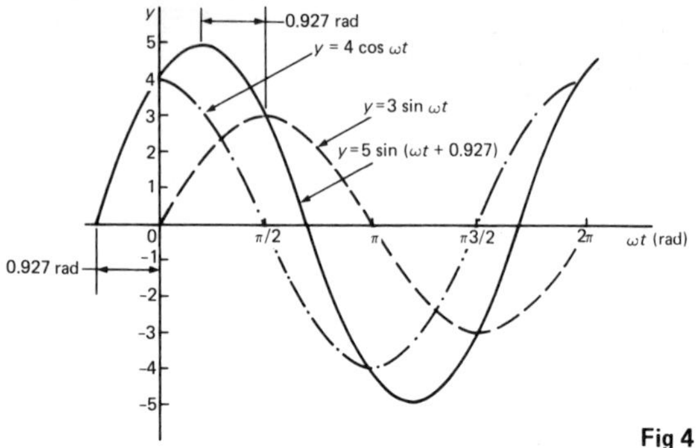

Fig 4

(b) by resolution of phasors by drawing or calculation.
This problem, together with the following, demonstrate a third method of combining waveforms.)

Problem 12 Express $4.6 \sin \omega t - 7.3 \cos \omega t$ in the form $R \sin(\omega t + \alpha)$.

Let $4.6 \sin \omega t - 7.3 \cos \omega t = R \sin(\omega t + \alpha)$,
then $4.6 \sin \omega t - 7.3 \cos \omega t = R[\sin \omega t \cos \alpha + \cos \omega t \sin \alpha]$
$= (R \cos \alpha) \sin \omega t + (R \sin \alpha) \cos \omega t$

Equating coefficients of $\sin \omega t$ gives: $4.6 = R \cos \alpha$, from which, $\cos \alpha = \dfrac{4.6}{R}$

Equating coefficients of $\cos \omega t$ gives $-7.3 = R \sin \alpha$, from which $\sin \alpha = \dfrac{-7.3}{R}$

There is only one quadrant where cosine is positive and sine is negative, i.e., the fourth quadrant, as shown in *Fig 5*. By Pythagoras' theorem: $R = \sqrt{[(4.6)^2 + (-7.3)^2]} = 8.628$

From trigonometric ratios:

$\alpha = \arctan\left(\dfrac{-7.3}{4.6}\right) = -57.78°$ or -1.008 radians.

Hence $4.6 \sin \omega t - 7.3 \cos \omega t = 8.628 \sin(\omega t - 1.008)$.

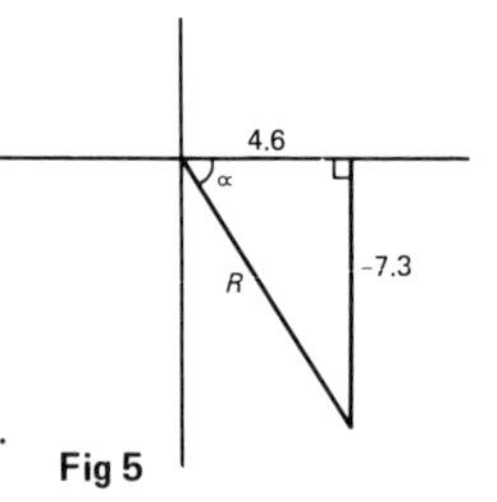

Fig 5

Problem 13 Express $-2.7 \sin \omega t - 4.1 \cos \omega t$ in the form $R \sin(\omega t + \alpha)$.

Let $-2.7 \sin \omega t - 4.1 \cos \omega t = R \sin(\omega t + \alpha)$
$= R[\sin \omega t \cos \alpha + \cos \omega t \sin \alpha]$
$= (R \cos \alpha) \sin \omega t + (R \sin \alpha) \cos \omega t$

Equating coefficients gives: $-2.7 = R \cos \alpha$, from which, $\cos \alpha = \dfrac{-2.7}{R}$

and $-4.1 = R \sin \alpha$, from which, $\sin \alpha = \dfrac{-4.1}{R}$

There is only one quadrant in which both cosine and sine are negative, i.e. the third quadrant, as shown in *Fig 6*. From *Fig 6*, $R = \sqrt{[(-2.7)^2 + (-4.1)^2]} = 4.909$,

and $\theta = \arctan \dfrac{4.1}{2.7} = 56.63°$.

Hence $\alpha = 180° + 56.63° = 236.63°$ or 4.130 radians.

Thus $-2.7 \sin \omega t - 4.1 \cos \omega t$
$= 4.909 \sin(\omega t + 4.130)$.

An angle of 236.63° is the same as −123.37° or −2.153 radians.

Hence $-2.7 \sin \omega t - 4.1 \cos \omega t$ may be expressed also as **$4.909 \sin(\omega t - 2.153)$**, which is preferred since it is the principal value (i.e. $-\pi \leqslant \alpha \leqslant \pi$).

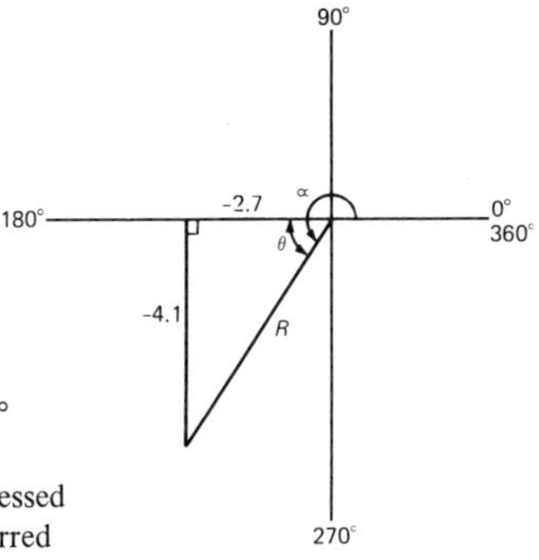

Fig 6

Problem 14 In *Problem 13*, state when the maximum value first occurs.

$4.909 \sin(\omega t - 2.153)$ has a maximum value of 4.909 which first occurs when $(\omega t - 2.153)$ is equal to 90° or $\pi/2$ rads (since a sine wave is at its maximum at 90°).
Hence $\omega t - 2.153 = \pi/2 = 1.571$, from which, $\omega t = 1.571 + 2.153 = 3.724$ rads or 213.37°.

Thus the maximum value of 4.909 occurs first when angle $\omega t = 213.37°$

Problem 15 Express $3 \sin \theta + 5 \cos \theta$ in the form $R \sin(\theta + \alpha)$, and hence solve the equation $3 \sin \theta + 5 \cos \theta = 4$, for values of θ between 0° and 360°.

Let $3 \sin \theta + 5 \cos \theta = R \sin(\theta + \alpha) = R [\sin \theta \cos \alpha + \cos \theta \sin \alpha]$
$= (R \cos \alpha) \sin \theta + (R \sin \alpha) \cos \theta$

Equating coefficients gives: $3 = R \cos \alpha$, from which, $\cos \alpha = \dfrac{3}{R}$

and $5 = R \sin \alpha$, from which, $\sin \alpha = \dfrac{5}{R}$

Since both $\sin \alpha$ and $\cos \alpha$ are positive, R lies in the first quadrant, as shown in *Fig 7*.

From *Fig 7*, $R = \sqrt{(3^2 + 5^2)} = 5.831$

and $\alpha = \arctan \dfrac{5}{3} = 59° \, 2'$.

Hence $3 \sin \theta + 5 \cos \theta = 5.831 \sin(\theta + 59° \, 2')$
However $3 \sin \theta + 5 \cos \theta = 4$
Hence $5.831 \sin(\theta + 59° \, 2') = 4$, from which,

$(\theta + 59° \, 2') = \arcsin\left(\dfrac{4}{5.831}\right)$

i.e. $\theta + 59° \, 2' = 43° \, 19'$ or $136° \, 41'$
Hence $\theta = 43° \, 19' - 59° \, 2' = -15° \, 43'$
or $\theta = 136° \, 41' - 59° \, 2' = 77° \, 39'$

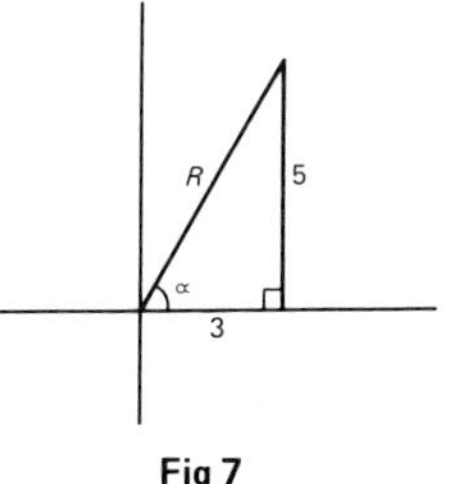

Fig 7

Since $-15° \, 43'$ is the same as $-15° \, 43' + 360°$, i.e. $344° \, 17'$, then the solutions are $\theta =$ **$77° \, 39'$ or $344° \, 17'$**, which may be checked by substituting into the original equation.

Problem 16 Solve the equation $3.5 \cos A - 5.8 \sin A = 6.5$ for $0° \leqslant A \leqslant 360°$.

Let $3.5 \cos A - 5.8 \sin A = R \sin(A+\alpha) = R\,[\sin A \cos\alpha + \cos A \sin\alpha]$
$= (R\cos\alpha)\sin A + (R\sin\alpha)\cos A$

Equating coefficients gives: $3.5 = R\sin\alpha$, from which, $\sin\alpha = \dfrac{3.5}{R}$

and $-5.8 = R\cos\alpha$, from which, $\cos\alpha = \dfrac{-5.8}{R}$

There is only one quadrant in which both sine is positive and cosine is negative, i.e. the second, as shown in *Fig 8*.

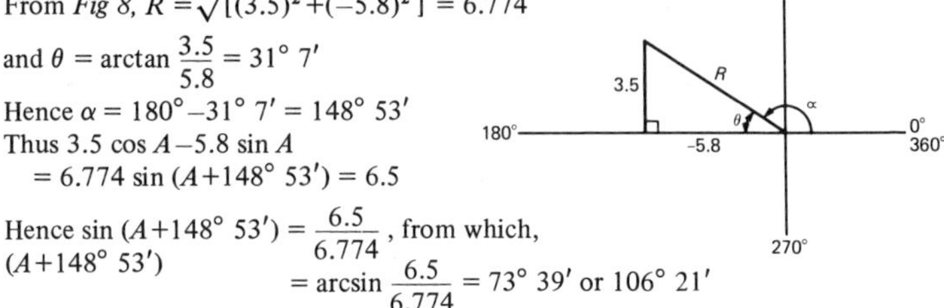

Fig 8

From *Fig 8*, $R = \sqrt{[(3.5)^2 + (-5.8)^2]} = 6.774$

and $\theta = \arctan \dfrac{3.5}{5.8} = 31°\,7'$

Hence $\alpha = 180° - 31°\,7' = 148°\,53'$

Thus $3.5 \cos A - 5.8 \sin A$
$= 6.774 \sin(A + 148°\,53') = 6.5$

Hence $\sin(A + 148°\,53') = \dfrac{6.5}{6.774}$, from which,

$(A + 148°\,53') = \arcsin \dfrac{6.5}{6.774} = 73°\,39'$ or $106°\,21'$

Thus $A = 73°\,39' - 148°\,53' = -75°\,14' \equiv (-75°\,14' + 360°) = 284°\,46'$

or $A = 106°\,21' - 148°\,53' = -42°\,32' \equiv (-42°\,32' + 360°) = 317°\,28'$

The solutions are thus $\boldsymbol{A = 284°\,46'}$ **or** $\boldsymbol{317°\,28'}$, which may be checked in the original equation.

Problem 17 If $\sin\theta = 4/5$, where θ is an acute angle, evaluate $\sin 2\theta$, $\cos 2\theta$ and $\tan 2\theta$, without evaluating angle θ.

$\sin\theta = \dfrac{4}{5} = \dfrac{\text{opposite side}}{\text{hypotenuse}}$. By Pythagoras' theorem, the adjacent side $= \sqrt{(5^2 - 4^2)}$
$= 3$

Hence $\cos\theta = \dfrac{3}{5}$ and $\tan\theta = \dfrac{4}{3}$

$$\sin 2\theta = 2\sin\theta\cos\theta = 2\left(\frac{4}{5}\right)\left(\frac{3}{5}\right) = \mathbf{\frac{24}{25}\ or\ 0.96}$$

$$\cos 2\theta = \cos^2\theta - \sin^2\theta = \left(\frac{3}{5}\right)^2 - \left(\frac{4}{5}\right)^2 = \mathbf{\frac{-7}{25}\ or\ -0.28}$$

$$\left(\text{or } \cos 2\theta = 1 - 2\sin^2\theta = 1 - 2\left(\frac{4}{5}\right)^2 = 1 - \frac{32}{25} = \frac{-7}{25}\right.$$

$$\left.\text{or } \cos 2\theta = 2\cos^2\theta - 1 = 2\left(\frac{3}{5}\right)^2 - 1 = \frac{18}{25} - 1 = \frac{-7}{25}\right)$$

$$\tan 2\theta = \frac{2\tan\theta}{1 - \tan^2\theta} = \frac{2\left(\frac{4}{3}\right)}{1 - \left(\frac{4}{3}\right)^2} = \frac{8/3}{-7/9} = \left(\frac{8}{3}\right)\left(-\frac{9}{7}\right) = \mathbf{\frac{-24}{7}\ or\ -3.4286}$$

Problem 18 $I_3 \sin 3\theta$ is the third harmonic of a waveform. Express the third harmonic in terms of the first harmonic $\sin\theta$, when $I_3 = 1$.

When $I_3 = 1, I_3 \sin 3\theta = \sin 3\theta = \sin(2\theta+\theta)$

$$= \sin 2\theta \cos\theta + \cos 2\theta \sin\theta, \text{ from the } \sin(A+B) \text{ formula}$$
$$= (2\sin\theta\cos\theta)\cos\theta + (1-2\sin^2\theta)\sin\theta, \text{from the double angle expansions}$$
$$= 2\sin\theta\cos^2\theta + \sin\theta - 2\sin^3\theta$$
$$= 2\sin\theta(1-\sin^2\theta) + \sin\theta - 2\sin^3\theta, \text{ (since } \cos^2\theta = 1-\sin^2\theta)$$
$$= 2\sin\theta - 2\sin^3\theta + \sin\theta - 2\sin^3\theta$$

i.e. $\mathbf{\sin 3\theta = 3\sin\theta - 4\sin^3\theta}$.

Problem 19 Prove that $\dfrac{1-\cos 2\theta}{\sin 2\theta} = \tan\theta$.

$$\text{LHS} = \frac{1-\cos 2\theta}{\sin 2\theta} = \frac{1-(1-2\sin^2\theta)}{2\sin\theta\cos\theta} = \frac{2\sin^2\theta}{2\sin\theta\cos\theta} = \frac{\sin\theta}{\cos\theta} = \tan\theta = \text{RHS}$$

Problem 20 Prove that $\cot 2x + \operatorname{cosec} 2x = \cot x$.

$$\text{LHS} = \cot 2x + \operatorname{cosec} 2x = \frac{\cos 2x}{\sin 2x} + \frac{1}{\sin 2x}$$
$$= \frac{\cos 2x + 1}{\sin 2x} = \frac{(2\cos^2 x - 1) + 1}{\sin 2x} = \frac{2\cos^2 x}{\sin 2x} = \frac{2\cos^2 x}{2\sin x\cos x}$$
$$= \frac{\cos x}{\sin x} = \cot x = \text{RHS}$$

Problem 21 Express $\sin 4x \cos 3x$ as a sum or difference of sines and cosines.

From equation (1), para. 7, $\sin 4x\cos 3x = \frac{1}{2}[\sin(4x+3x) + \sin(4x-3x)]$

$$= \mathbf{\frac{1}{2}(\sin 7x + \sin x)}$$

Problem 22 Express $2\cos 5\theta \sin 2\theta$ as a sum or difference of sines or cosines.

From equation (2), para. 7, $2\cos 5\theta\sin 2\theta = 2\left\{\frac{1}{2}[\sin(5\theta+2\theta) - \sin(5\theta-2\theta)]\right\}$

$$= \mathbf{\sin 7\theta - \sin 3\theta}$$

Problem 23 Express $3\cos 4t\cos t$ as a sum or difference of sines or cosines.

From equation (3), para. 7, $3\cos 4t\cos t = 3\left\{\frac{1}{2}[\cos(4t+t) + \cos(4t-t)]\right\}$

$$= \mathbf{\frac{3}{2}(\cos 5t + \cos 3t)}$$

Problem 24 Express $\sin\frac{\pi}{3}\sin\frac{\pi}{4}$ as a sum or difference of sines or cosines.

From equation (4), para. 7, $\sin\frac{\pi}{3}\sin\frac{\pi}{4} = -\frac{1}{2}\left[\cos\left(\frac{\pi}{3}+\frac{\pi}{4}\right) - \cos\left(\frac{\pi}{3}-\frac{\pi}{4}\right)\right]$

$$= -\frac{1}{2}\left[\cos\frac{7\pi}{12} - \cos\frac{\pi}{12}\right] = \mathbf{\frac{1}{2}\left[\cos\frac{\pi}{12} - \cos\frac{7\pi}{12}\right]}$$

Problem 25 In an alternating current circuit, voltage $v = 5 \sin \omega t$ and current $i = 10 \sin (\omega t - \pi/6)$. Find an expression for the instantaneous power p at time t given that $p = vi$, expressing the answer as a sum or difference of sines and cosines.

$$p = vi = (5 \sin \omega t)\left[10 \sin \left(\omega t - \frac{\pi}{6}\right)\right] = 50 \sin \omega t \sin \left(\omega t - \frac{\pi}{6}\right)$$

From equation (4), para. 7,

$$50 \sin \omega t \sin \left(\omega t - \frac{\pi}{6}\right) = 50\left\{-\frac{1}{2} \cos \left(\omega t + \omega t - \frac{\pi}{6}\right) - \cos \left[\omega t - \left(\omega t - \frac{\pi}{6}\right)\right]\right\}$$

$$= -25\left\{\cos \left(2\omega t - \frac{\pi}{6}\right) - \cos \frac{\pi}{6}\right\}$$

i.e., **instantaneous power,** $\quad \boldsymbol{p = 25\left[\cos \frac{\pi}{6} - \cos \left(2\omega t - \frac{\pi}{6}\right)\right]}$

Problem 26 Express $\sin 5\theta + \sin 3\theta$ as a product.

From equation (5), para. 8, $\sin 5\theta + \sin 3\theta = 2 \sin \left(\frac{5\theta + 3\theta}{2}\right) \cos \left(\frac{5\theta - 3\theta}{2}\right)$

$$= \mathbf{2 \sin 4\theta \cos \theta}$$

Problem 27 Express $\sin 7x - \sin x$ as a product.

From equation (6), para. 8, $\sin 7x - \sin x = 2 \cos \left(\frac{7x + x}{2}\right) \sin \left(\frac{7x - x}{2}\right)$

$$= \mathbf{2 \cos 4x \sin 3x}$$

Problem 28 Express $\cos 8x + \cos 4x$ as a product.

From equation (7), para. 8, $\cos 8x + \cos 4x = 2 \cos \left(\frac{8x + 4x}{2}\right) \cos \left(\frac{8x - 4x}{2}\right)$

$$= \mathbf{2 \cos 6x \cos 2x}$$

Problem 29 Express $\cos 2t - \cos 5t$ as a product.

From equation (8), para. 8, $\cos 2t - \cos 5t = -2 \sin \left(\frac{2t + 5t}{2}\right) \sin \left(\frac{2t - 5t}{2}\right)$

$$= -2 \sin \frac{7}{2}t \sin \left(-\frac{3}{2}t\right) = \mathbf{2 \sin \frac{7}{2}t \sin \frac{3}{2}t} \quad \left[\text{since } \sin \left(-\frac{3t}{2}\right) = -\sin \frac{3}{2}t\right]$$

Problem 30 Show that $\dfrac{\cos 6x + \cos 2x}{\sin 6x + \sin 2x} = \cot 4x$.

From equation (7), para. 8, $\cos 6x + \cos 2x = 2 \cos 4x \cos 2x$
From equation (5), para. 8, $\sin 6x + \sin 2x = 2 \sin 4x \cos 2x$

Hence $\dfrac{\cos 6x + \cos 2x}{\sin 6x + \sin 2x} = \dfrac{2 \cos 4x \cos 2x}{2 \sin 4x \cos 2x} = \dfrac{\cos 4x}{\sin 4x} = \cot 4x.$

C FURTHER PROBLEMS ON COMPOUND ANGLES

1 Show that the compound-angle formulae for $\sin(A+B)$ and $\cos(A+B)$ are true when $A = 32^\circ$ and $B = 59^\circ$.

2 Verify that the compound-angle formulae for $\sin(A-B)$ and $\tan(A+B)$ are true when $A = 115^\circ$ and $B = 51^\circ$.

3 Prove that the compound-angle formulae for $\tan(A-B)$ and $\cos(A-B)$ are true when $A = 278^\circ$ and $B = 141^\circ$.

4 Reduce the following to the sine of one angle:
(a) $\sin 37^\circ \cos 21^\circ + \cos 37^\circ \sin 21^\circ$;
(b) $\sin 2x \cos 5x + \cos 2x \sin 5x$;
(c) $\sin 7t \cos 3t - \cos 7t \cos 3t$. [(a) $\sin 58^\circ$; (b) $\sin 7x$; (c) $\sin 4t$]

5 Reduce the following to the cosine of one angle:
(a) $\cos 71^\circ \cos 33^\circ - \sin 71^\circ \sin 33^\circ$;
(b) $\cos 5\theta \cos 2\theta - \sin 5\theta \sin 2\theta$;
(c) $\cos\frac{\pi}{3}\cos\frac{\pi}{4} + \sin\frac{\pi}{3}\sin\frac{\pi}{4}$.
[(a) $\cos 104^\circ \equiv -\cos 76^\circ$; (b) $\cos 7\theta$; (c) $\cos\frac{\pi}{12}$]

6 Show that: (a) $\sin\left(x+\frac{\pi}{3}\right) + \sin\left(x+\frac{2\pi}{3}\right) = \sqrt{3}\cos x$

and (b) $-\sin\left(\frac{3\pi}{2} - \phi\right) = \cos\phi$

7 Prove that: (a) $\sin\left(\theta+\frac{\pi}{4}\right) - \sin\left(\theta-\frac{3\pi}{4}\right) = \sqrt{2}(\sin\theta + \cos\theta)$

and (b) $\dfrac{\cos(270^\circ+\theta)}{\cos(360^\circ-\theta)} = \tan\theta$

8 If $\sin A = \frac{5}{13}$ and $\cos B = \frac{9}{41}$ evaluate (a) $\sin(A+B)$; (b) $\cos(A+B)$; (c) $\tan(A-B)$, correct to 4 decimal places.
[(a) 0.9850; (b) –0.1726; (c) –1.4123]

9 Given $\cos A = 0.42$ and $\sin B = 0.73$ evaluate (a) $\sin(A-B)$; (b) $\cos(A-B)$; (c) $\tan(A+B)$, correct to 4 decimal places.
[(a) 0.3136; (b) 0.9495; (c) –2.4687]

10 Determine, in surd form, the value of $\cos 15^\circ$ without using trigonometric tables.
$\left[\dfrac{\sqrt{3}+1}{2\sqrt{2}}\right]$

11 Solve the following equations for values of θ between 0° and 360°:
(a) $3\sin(\theta+30^\circ) = 7\cos\theta$; (b) $4\sin(\theta-40^\circ) = 2\sin\theta$
[(a) 64° 43′ or 244° 43′; (b) 67° 31′ or 247° 31′]

In *Problems 12 to 16*, change the functions into the form $R\sin(\omega t \pm \alpha)$.

12 $5\sin\omega t + 8\cos\omega t$ [$9.434\sin(\omega t+1.012)$]

13 $9\sin\omega t + 5\cos\omega t$ [$10.30\sin(\omega t+0.507)$]

14 $4\sin\omega t - 3\cos\omega t$ [$5\sin(\omega t-0.644)$]

15 $-7\sin\omega t + 4\cos\omega t$ [$8.062\sin(\omega t+2.622)$]

16 $-3\sin\omega t - 6\cos\omega t$ [$6.708\sin(\omega t-2.034)$]

17 Solve the following equations for values of θ between 0° and 360°:
(a) $2\sin\theta + 4\cos\theta = 3$; (b) $12\sin\theta - 9\cos\theta = 7$
[(a) 74° 26′ or 338° 42′; (b) 64° 41′ or 189° 3′]

18 Solve the following equations for $0^\circ \leqslant A \leqslant 360^\circ$:
(a) $3\cos A + 2\sin A = 2.8$; (b) $12\cos A - 4\sin A = 11$.
[(a) 70° 44′ or 354° 38′; (b) 11° 9′ or 311° 59′]

19 The third harmonic of a wave motion is given by $4.3 \cos 3\theta - 6.9 \sin 3\theta$. Express this in the form $R \sin (3\theta \pm \alpha)$. [$8.13 \sin (3\theta+2.584)$]

20 The displacement x metres of a mass from a fixed point about which it is oscillating is given by $x = 2.4 \sin \omega t + 3.2 \cos \omega t$, where t is the time in seconds. Express x in the form $R \sin (\omega t+\alpha)$. [$x = 4.0 \sin (\omega t+0.927)$]

21 Alternating currents are given by $i_1 = 6 \sin \omega t$ amperes and $i_2 = 10 \cos \omega t$ amperes. Calculate the maximum value of the resultant (i_1+i_2) and its phase displacement relative to i_1. [11.66 A; 59° 2′ leading]

22 Two voltages, $v_1 = 5 \cos \omega t$ and $v_2 = -8 \sin \omega t$ are inputs to an analogue circuit. Determine an expression for the output voltage if this is given by (v_1+v_2). [$9.434 \sin (\omega t+2.583)$]

23 Given $\cos \theta = \dfrac{5}{13}$, where θ is an acute angle, evaluate (a) $\sin 2\theta$; (b) $\cos 2\theta$; (c) $\tan 2\theta$, without evaluating angle θ, each correct to 4 decimal places. [(a) 0.7101; (b) –0.7041; (c) –1.0084]

24 The power p in an electrical circuit is given by $p = \dfrac{v^2}{R}$. Determine the power in terms of V, R and $\cos 2t$ when $v = V \cos t$. $\left[\dfrac{V^2}{2R}(1+\cos 2t)\right]$

25 Prove the following identities: (a) $1 - \dfrac{\cos 2\phi}{\cos^2 \phi} = \tan^2 \phi$;

(b) $\dfrac{1+\cos 2t}{\sin^2 t} = 2 \cot^2 t$; (c) $\dfrac{(\tan 2x)(1+\tan x)}{\tan x} = \dfrac{2}{1-\tan x}$;

(d) $2 \operatorname{cosec} 2\theta \cos 2\theta = \cot \theta - \tan \theta$.

26 If the third harmonic of a waveform is given by $V_3 \cos 3\theta$, express the third harmonic in terms of the first harmonic $\cos \theta$, when $V_3 = 1$. [$\cos 3\theta = 4 \cos^3 \theta - 3 \cos \theta$]

In *Problems 27 to 32*, express as sums or differences:

27 $\sin 7t \cos 2t$ $\left[\dfrac{1}{2}(\sin 9t+\sin 5t)\right]$

28 $\cos 8x \sin 2x$ $\left[\dfrac{1}{2}(\sin 10x-\sin 6x)\right]$

29 $2 \sin 7t \sin 3t$ [$\cos 4t-\cos 10t$]

30 $4 \cos 3\theta \cos \theta$ [$2(\cos 4\theta+\cos 2\theta)$]

31 $3 \sin \dfrac{\pi}{3} \cos \dfrac{\pi}{6}$ $\left[\dfrac{3}{2}\left(\sin \dfrac{\pi}{2}+\sin \dfrac{\pi}{6}\right)\right]$

32 $\sin 23° \sin 68°$ $\left[\dfrac{1}{2}(\cos 45°-\cos 91°)\right]$

33 Solve the equation: $2 \sin 2\phi \sin \phi = \cos \phi$ in the range $\phi = 0$ to $\phi = 180°$. [30°, 90° or 150°]

In *Problems 34 to 39*, express as products:

34 $\sin 3x+\sin x$ [$2 \sin 2x \cos x$]

35 $\dfrac{1}{2}(\sin 9\theta-\sin 7\theta)$ [$\cos 8\theta \sin \theta$]

36 $\cos 5t+\cos 3t$ [$2 \cos 4t \cos t$]

37 $\dfrac{1}{8}(\cos 5t-\cos t)$ $\left[-\dfrac{1}{4}\sin 3t \sin 2t\right]$

38 $\dfrac{1}{2}\left(\cos \dfrac{\pi}{3}+\cos \dfrac{\pi}{4}\right)$ $\left[\cos \dfrac{7\pi}{12} \cos \dfrac{\pi}{12}\right]$

39 $3(\sin 7\phi+\sin 2\phi)$ $\left[6 \sin \dfrac{9}{2}\phi \cos \dfrac{5}{2}\phi\right]$

40 Show that: (a) $\dfrac{\sin 4x - \sin 2x}{\cos 4x + \cos 2x} = \tan x$

(b) $\dfrac{1}{2}\{\sin(5x-\alpha) - \sin(x+\alpha)\} = \cos 3x \sin(2x-\alpha)$.

14 Calculus (1) – Differentiation from first principles

A MAIN POINTS CONCERNED WITH DIFFERENTIATION FROM FIRST PRINCIPLES

1 **Calculus** is a branch of mathematics involving or leading to calculations dealing with continuously varying functions. Calculus is a subject which falls into two parts: (i) **differential calculus** (or **differentiation**) and (ii) **integral calculus** (or **integration**).

2 In an equation such as $y = 3x^2+2x-5$, y is said to be a function of x and may be written as $y = f(x)$. An equation written in the form $f(x) = 3x^2+2x-5$ is termed **functional notation.** The value of $f(x)$ when $x = 0$ is denoted by $f(0)$, and the value of $f(x)$ when $x = 2$ is denoted by $f(2)$, and so on.
Thus when $f(x) = 3x^2+2x-5$,
then $f(0) = 3(0)^2+2(0)-5 = -5$,
and $f(2) = 3(2)^2+2(2)-5 = 11$, and so on. (See *Problems 1 and 2.*)

3 If a tangent is drawn at a point P on a curve, then the gradient of this tangent is said to be the **gradient of the curve** at P. In *Fig 1*, the gradient of the curve at P is equal to the gradient of the tangent PQ.

Fig 1

4 For the curve shown in *Fig 2*, let the points A and B have co-ordinates (x_1, y_1) and (x_2, y_2) respectively. In functional notation $y_1 = f(x_1)$ and $y_2 = f(x_2)$ as shown.

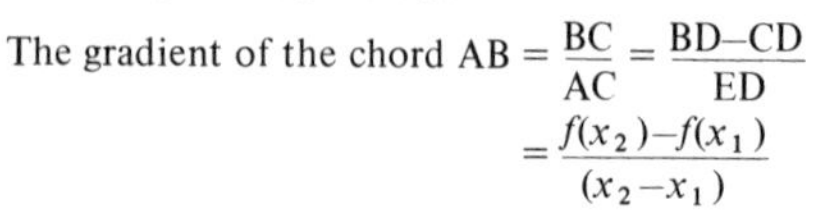

$$\text{The gradient of the chord AB} = \frac{BC}{AC} = \frac{BD-CD}{ED} = \frac{f(x_2)-f(x_1)}{(x_2-x_1)}$$

5 For the curve $f(x) = x^2$ shown in *Fig 3*:

(i) the gradient of chord AB $= \dfrac{f(3)-f(1)}{3-1} = \dfrac{9-1}{2} = 4,$

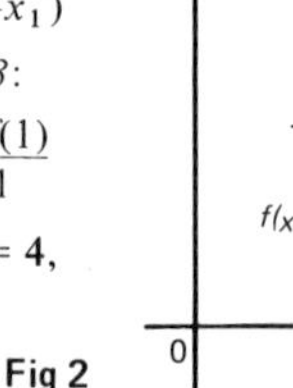

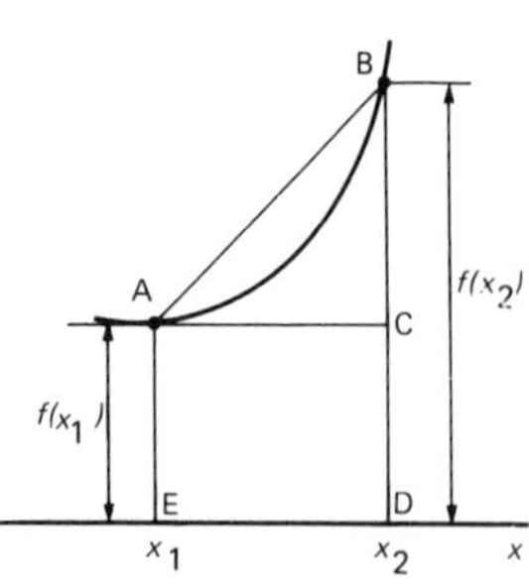

Fig 2

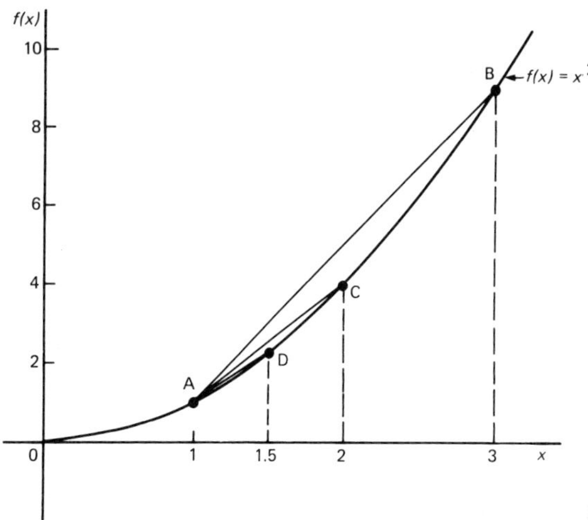

Fig 3

(ii) the gradient of chord AC $= \dfrac{f(2)-f(1)}{2-1} = \dfrac{4-1}{1} = 3,$

(iii) the gradient of chord AD $= \dfrac{f(1.5)-f(1)}{1.5-1} = \dfrac{2.25-1}{0.5} = 2.5,$

(iv) if E is the point on the curve $(1.1, f(1.1))$ then

the gradient of chord AE $= \dfrac{f(1.1)-f(1)}{1.1-1} = \dfrac{1.21-1}{0.1} = 2.1,$

(v) if F is the point on the curve $(1.01, f(1.01))$ then

the gradient of chord AF $= \dfrac{f(1.01)-f(1)}{1.01-1} = \dfrac{1.0201-1}{0.01} = 2.01$

Thus as point B moves closer and closer to point A the gradient of the chord approaches nearer and nearer to the value 2. This is called the **limiting value** of the gradient of the chord AB and when B coincides with A the chord becomes the tangent to the curve.

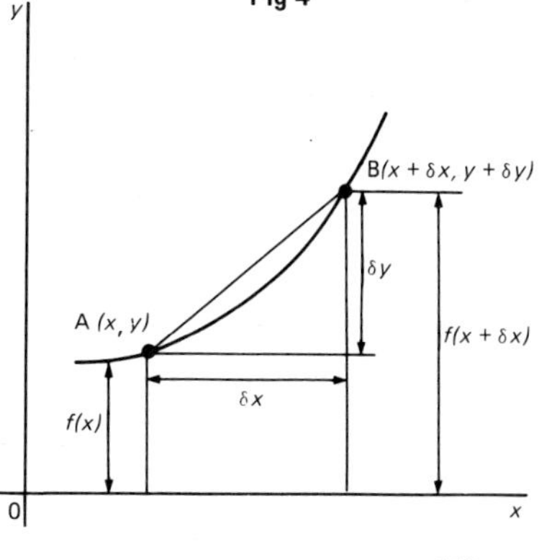

Fig 4

6 **Differentiation from first principles.**

(i) In *Fig. 4*, A and B are two points very close together on a curve, δx (delta x) and δy (delta y) representing small increments in the x and y directions respectively.

Gradient of chord AB $= \dfrac{\delta y}{\delta x}$

However $\delta y = f(x+\delta x)-f(x)$

Hence $\dfrac{\delta y}{\delta x} = \dfrac{f(x+\delta x)-f(x)}{\delta x}$

As δx approaches zero, $\frac{\delta y}{\delta x}$ approaches a limiting value and the gradient of the chord approaches the gradient of the tangent at A.

(ii) When determining the gradient of a tangent to a curve there are two notations used. The gradient of the curve at A in *Fig 4* can either be written as

$$\underset{\delta x \to 0}{\text{limit}} \frac{\delta y}{\delta x} \text{ or } \underset{\delta x \to 0}{\text{limit}} \left\{\frac{f(x+\delta x)-f(x)}{\delta x}\right\}$$

In Leibniz notation, $\frac{dy}{dx} = \underset{\delta x \to 0}{\textbf{limit}} \frac{\delta y}{\delta x}$

In functional notation, $f'(x) = \underset{\delta x \to 0}{\textbf{limit}} \left\{\frac{f(x+\delta x)-f(x)}{\delta x}\right\}$

(iii) $\frac{dy}{dx}$ is the same as $f'(x)$ and is called the **differential coefficient** or the **derivative.** The process of finding the differential coefficient is called **differentiation.** (See *Problems 3 to 7.*) Summarising,

the differential coefficient, $\frac{dy}{dx} = f'(x) = \underset{\delta x \to 0}{\text{limit}} \frac{\delta y}{\delta x} = \underset{\delta x \to 0}{\text{limit}} \left\{\frac{f(x+\delta x)-f(x)}{\delta x}\right\}$

7 When differentiating, results can be expressed in a number of ways. For example:

(i) if $y = 3x^2$ then $\frac{dy}{dx} = 6x$

(ii) if $f(x) = 3x^2$ then $f'(x) = 6x$,

(iii) the differential coefficient of $3x^2$ is $6x$,

(iv) the derivative of $3x^2$ is $6x$, and

(v) $\frac{d}{dx}(3x^2) = 6x$.

B WORKED PROBLEMS ON DIFFERENTIATION FROM FIRST PRINCIPLES

Problem 1 If $f(x) = 4x^2-3x+2$, find $f(0)$, $f(3)$, $f(-1)$ and $f(3)-f(-1)$.

$f(x) = 4x^2-3x+2$
$f(0) = 4(0)^2-3(0)+2 = 2$
$f(3) = 4(3)^2-3(3)+2 = 36-9+2 = 29$
$f(-1) = 4(-1)^2-3(-1)+2 = 4+3+2 = 9$
$f(3)-f(-1) = 29-9 = \mathbf{20}$

Problem 2 Given that $f(x) = 5x^2+x-7$, determine (i) $f(2) \div f(1)$; (ii) $f(3+a)$; (iii) $f(3+a)-f(3)$; (iv) $\frac{f(3+a)-f(3)}{a}$.

(i) $f(x) = 5x^2+x-7$.
$f(2) = 5(2)^2+2-7 = 15$
$f(1) = 5(1)+1-7 = -1$
$f(2) \div f(1) = \frac{15}{-1} = \mathbf{-15}$

(ii) $f(3+a) = 5(3+a)^2+(3+a)-7$
$= 5(9+6a+a^2)+(3+a)-7 = 45+30a+5a^2+3+a-7 = \mathbf{41+31a+5a^2}$

(iii) $f(3) = 5(3)^2+3-7 = 41$
$f(3+a)-f(3) = (41+31a+5a^2)-(41) = \mathbf{31a+5a^2}$

(iv) $\dfrac{f(3+a)-f(3)}{a} = \dfrac{31a+5a^2}{a} = \mathbf{31+5a}$

Problem 3 Differentiate from first principles $f(x) = x^2$ and determine the value of the gradient of the curve at $x = 2$.

To 'differentiate from first principles' means 'to find $f'(x)$ by using the expression

$$f'(x) = \lim_{\delta x \to 0} \left\{ \frac{f(x+\delta x)-f(x)}{\delta x} \right\}$$ (see para. 6)

$f(x) = x^2$.
Substituting $(x+\delta x)$ for x gives $f(x+\delta x) = (x+\delta x)^2 = x^2+2x\delta x+\delta x^2$

$$\text{Hence } f'(x) = \lim_{\delta x \to 0} \left\{ \frac{(x^2+2x\delta x+\delta x^2)-(x^2)}{\delta x} \right\}$$
$$= \lim_{\delta x \to 0} \left\{ \frac{2x\delta x+\delta x^2}{\delta x} \right\}$$
$$= \lim_{\delta x \to 0} \{2x+\delta x\}$$

As $\delta x \to 0$, $\{2x+\delta x\} \to \{2x+0\}$
Thus $\mathbf{f'(x) = 2x}$, i.e. the differential coefficient of x^2 is $2x$.
At $x = 2$, the gradient of the curve, $f'(x) = 2(2) = \mathbf{4}$

Problem 4 Find the differential coefficient of $y = 5x$.

By definition, $\dfrac{dy}{dx} = f'(x) = \lim_{\delta x \to 0} \left\{ \dfrac{f(x+\delta x)-f(x)}{\delta x} \right\}$

The function being differentiated is $y = f(x) = 5x$
Substituting $(x+\delta x)$ for x gives $f(x+\delta x) = 5(x+\delta x) = 5x+5\delta x$

$$\text{Hence } \frac{dy}{dx} = f'(x) = \lim_{\delta x \to 0} \left\{ \frac{(5x+5\delta x)-(5x)}{\delta x} \right\}$$
$$= \lim_{\delta x \to 0} \left\{ \frac{5\delta x}{\delta x} \right\} = \lim_{\delta x \to 0} \{5\}$$

Since the term δx does not appear in $\{5\}$ the limiting value as $\delta x \to 0$ of $\{5\}$ is 5.

Thus $\mathbf{\dfrac{dy}{dx} = 5}$, i.e. the differential coefficient of $5x$ is 5.
The equation $y = 5x$ represents a straight line of gradient 5.

The 'differential coefficient' (i.e. $\dfrac{dy}{dx}$ or $f'(x)$) means 'the gradient of the curve', and since the gradient of the line $y = 5x$ is 5 this result can be obtained by inspection.
Hence, in general, if $y = kx$, (where k is a constant), then the gradient of the line is k and $\dfrac{dy}{dx}$ or $f'(x) = k$.

Problem 5 Find the derivative of $y = 8$.

$y = f(x) = 8$.
Since there are no x-values in the original equation, substituting $(x+\delta x)$ for x still gives $f(x+\delta x) = 8$.

Hence $\frac{dy}{dx} = f'(x) = \lim_{\delta x \to 0} \left\{ \frac{f(x+\delta x) - f(x)}{\delta x} \right\} = \lim_{\delta x \to 0} \left\{ \frac{8-8}{\delta x} \right\} = 0$

Thus, when $y = 8$, $\frac{dy}{dx} = 0$

The equation $y = 8$ represents a straight horizontal line and the gradient of a horizontal line is zero, hence the result could have been determined by inspection. 'Finding the derivative' means 'finding the gradient', hence, in general, for any horizontal line if $y = k$ (where k is a constant) then $\frac{dy}{dx} = 0$.

Problem 6 Differentiate from first principles $f(x) = 2x^3$.

$f(x) = 2x^3$.
Substituting $(x+\delta x)$ for x gives

$$
\begin{aligned}
f(x+\delta x) &= 2(x+\delta x)^3 \\
&= 2(x+\delta x)(x^2+2x\delta x+\delta x^2) \\
&= 2(x^3+3x^2\delta x+3x\delta x^2+\delta x^3) \\
&= 2x^3+6x^2\delta x+6x\delta x^2+2\delta x^3)
\end{aligned}
$$

$$
\begin{aligned}
f'(x) &= \lim_{\delta x \to 0} \left\{ \frac{f(x+\delta x)-f(x)}{\delta x} \right\} \\
f'(x &= \lim_{\delta x \to 0} \left\{ \frac{(2x^3+6x^2\delta x+6x\delta x^2+2\delta x^3)-(2x^3)}{\delta x} \right\} \\
&= \lim_{\delta x \to 0} \left\{ \frac{6x^2\delta x+6x\delta x^2+2\delta x^3}{\delta x} \right\} \\
&= \lim_{\delta x \to 0} \{6x^2+6x\delta x+2\delta x^2\}
\end{aligned}
$$

Hence $\boldsymbol{f'(x) = 6x^2}$, i.e. the differential coefficient of $2x^3$ is $6x^2$.

Problem 7 Find the differential coefficient of $y = 4x^2+5x-3$ and determine the gradient of the curve at $x = -3$.

$$
\begin{aligned}
y = f(x) &= 4x^2+5x-3 \\
f(x+\delta x) &= 4(x+\delta x)^2+5(x+\delta x)-3 \\
&= 4(x^2+2x\delta x+\delta x^2)+5x+5\delta x-3 \\
&= 4x^2+8x\delta x+4\delta x^2+5x+5\delta x-3
\end{aligned}
$$

$$
\begin{aligned}
\frac{dy}{dx} = f'(x) &= \lim_{\delta x \to 0} \left\{ \frac{f(x+\delta x)-f(x)}{\delta x} \right\} \\
&= \lim_{\delta x \to 0} \left\{ \frac{(4x^2+8x\delta x+4\delta x^2+5x+5\delta x-3)-(4x^2+5x-3)}{\delta x} \right\}
\end{aligned}
$$

$$= \lim_{\delta x \to 0} \left\{ \frac{8x\delta x + 4\delta x^2 + 5\delta x}{\delta x} \right\}$$

$$= \lim_{\delta x \to 0} \{8x + 4\delta x + 5\}$$

i.e. $\dfrac{dy}{dx} = f'(x) = 8x+5$

At $x = -3$, the gradient of the curve $= \dfrac{dy}{dx} = f'(x) = 8(-3)+5 = \mathbf{-19}$

C FURTHER PROBLEMS ON DIFFERENTIATION FROM FIRST PRINCIPLES

1 If $f(x) = 6x^2-2x+1$, find $f(0), f(1), f(2), f(-1)$ and $f(-3)$. [1, 5, 21, 9, 61]

2 If $f(x) = 2x^2+5x-7$, find $f(1), f(2), f(-1)$ and $f(2)-f(-1)$. [0, 11, −10, 21]

3 Given $f(x) = 3x^3+2x^2-3x+2$, prove that $f(1) = \frac{1}{7}f(2)$.

4 If $f(x) = -x^2+3x+6$, find $f(2), f(2+a), f(2+a)-f(2)$ and $\dfrac{f(2+a)-f(2)}{a}$

[8; $-a^2-a+8$; $-a^2-a$; $-a-1$]

5 Plot the curve $f(x) = 4x^2-1$ for values of x from $x = -1$ to $x = +4$. Label the co-ordinates $(3, f(3))$ and $(1, f(1))$ as K and J respectively. Join points J and K to form the chord JK. Determine the gradient of chord JK. By moving J nearer and nearer to K, determine the gradient of the tangent of the curve at K. [16; 8]

In *Problems 6 to 17*, differentiate from first principles.

6 $y = x$ [1]

7 $y = 7x$ [7]

8 $y = 4x^2$ [$8x$]

9 $y = 5x^3$ [$15x^2$]

10 $y = -2x^2+3x-12$ [$-4x+3$]

11 $y = 23$ [0]

12 $f(x) = 9x$ [9]

13 $f(x) = \dfrac{2x}{3}$ $\left[\dfrac{2}{3}\right]$

14 $f(x) = 9x^2$ [$18x$]

15 $f(x) = -7x^3$ [$-21x^2$]

16 $f(x) = x^2+15x-4$ [$2x+15$]

17 $f(x) = 4$ [0]

18 Determine $\dfrac{d}{dx}(4x^3)$ from first principles. [$12x^2$]

19 Find $\dfrac{d}{dx}(3x^2+5)$ from first principles. [$6x$]

20 Differentiate from first principles $f(x) = 6x^2-3x+5$ and find the gradient of the curve at (a) $x = -1$ and (b) $x = 2$. [$12x-3$; (a) −15, (b) 21]

21 Find the differential coefficient of $y = 2x^3+3x^2-4x-1$ and determine the gradient of the curve at $x = 2$. [$6x^2+6x-4$; 32]

22 Determine the derivative of $y = -2x^3+4x+7$ and determine the gradient of the curve at $x = -1.5$ [$-6x^2+4$; -9.5]

23 The distance x metres moved by a body in a time t seconds is given by $x = 2t^2-3t+4$. Determine from first principles the velocity v, when time t is 3 seconds, if velocity is given by $v = dx/dt$. [9 m/s]

15 Calculus (2) – Methods of differentiation

A MAIN POINTS CONCERNED WITH METHODS OF DIFFERENTIATION

Differentiation of common functions

1 (i) From differentiation by first principles (see chapter 14), a general rule for differentiating $y = ax^n$ emerges, where a and n are constants.

The rule is: **if $y = ax^n$ then $\frac{dy}{dx} = anx^{n-1}$**,

(or, **if $f(x) = ax^n$ then $f'(x) = anx^{n-1}$**) and is true for all real values of a and n.

For example, if $y = 4x^3$ then $a = 4$ and $n = 3$,

and $\frac{dy}{dx} = anx^{n-1} = (4)(3)x^{3-1} = 12x^2$.

(ii) If $y = ax^n$ and $n = 0$ then $y = ax^0$ and $\frac{dy}{dx} = (a)(0)x^{0-1} = 0$,

i.e., **the differential coefficient of a constant is zero.**

2 *Fig 1(a)* shows a graph of $y = \sin x$. The gradient is continually changing as the curve moves from 0 to A to B to C to D. The gradient, given by dy/dx, may be plotted in a corresponding position below $y = \sin x$, as shown in *Fig 1(b).*

(i) At 0, the gradient is positive and is at its steepest. Hence 0′ is a maximum positive value.

(ii) Between 0 and A the gradient is positive but is decreasing in value until at A the gradient is zero, shown as A′.

(iii) Between A and B the gradient is negative but is increasing in value until at B the gradient is at its steepest negative value. Hence B′ is a maximum negative value.

(iv) If the gradient of $y = \sin x$ is further investigated between B and D then the resulting graph of dy/dx is seen to be a cosine wave.

Hence the rate of change of $\sin x$ is $\cos x$,

i.e., **if $y = \sin x$ then $\frac{dy}{dx} = \cos x$.**

By a similar construction to that shown in *Fig 1* it may be shown that:

if $y = \sin ax$ then $\frac{dy}{dx} = a \cos ax$.

3 If graphs of $y = \cos x$, $y = e^x$ and $y = \ln x$ are plotted and their gradients investigated, their differential coefficients may be determined in a similar

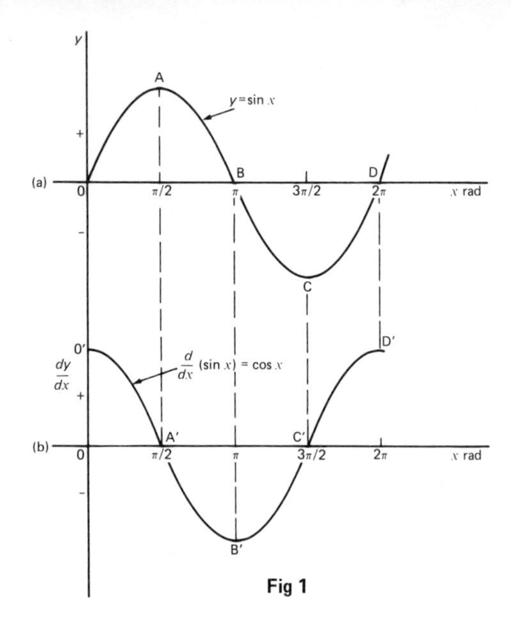

Fig 1

manner to that shown for $y = \sin x$. The rate of change of a function is a measure of the derivative. The **standard derivatives** summarised below may be proved theoretically and are true for all real values of x.

y or $f(x)$	$\frac{dy}{dx}$ or $f'(x)$
ax^n	anx^{n-1}
$\sin ax$	$a \cos ax$
$\cos ax$	$-a \sin ax$
e^{ax}	ae^{ax}
$\ln ax$	$\frac{1}{x}$

4 The **differential coefficient of a sum or difference** is the sum or difference of the differential coefficients of the separate terms.
Thus, if $f(x) = p(x)+q(x)-r(x)$, (where f, p, q and r are functions),
then $f'(x) = p'(x)+q'(x)-r'(x)$
(See *Problems 1 to 11.*)

5 **Differentiation of a product**
When $y = uv$, and u and v are both functions of x,
then $\frac{dy}{dx} = u\frac{dv}{dx} + v\frac{du}{dx}$

This is known as the **product rule**
(See *Problems 12 to 16.*)

6 **Differentiation of a quotient**

When $y = \frac{u}{v}$, and u and v are both functions of x

then $$\frac{dy}{dx} = \frac{v\frac{du}{dx} - u\frac{dv}{dx}}{v^2}$$

This is known as the **quotient rule.** (See *Problems 17 to 23.*)

7 **Function of a function**

It is often easier to make a substitution before differentiating.

If u is a function of x then $\frac{dy}{dx} = \frac{dy}{du} \times \frac{du}{dx}$

This is known as the **'function of a function'** rule (or sometimes the **chain rule**).
For example, if $y = (3x-1)^9$ then, by making the substitution $u = (3x-1)$
$y = u^9$, which is of the 'standard' form shown in para. 3.

Hence $\frac{dy}{du} = 9u^8$ and $\frac{du}{dx} = 3$.

Then $\frac{dy}{dx} = \frac{dy}{du} \times \frac{du}{dx} = (9u^8)(3) = 27u^8 = 27(3x-1)^8$.

Since y is a function of u, and u is a function of x, then y is a function of a function of x. (See *Problems 24 to 29.*)

8 **Successive differentiation**

When a function $y = f(x)$ is differentiated with respect to x the differential coefficient is written as $\frac{dy}{dx}$ or $f'(x)$. If the expression is differentiated again, the second differential coefficient is obtained and is written as $\frac{d^2y}{dx^2}$ (pronounced dee two y by dee x squared) or $f''(x)$ (pronounced f double-dash x).

By successive differentiation further higher derivatives such as $\frac{d^3y}{dx^3}$ and $\frac{d^4y}{dx^4}$ may be obtained.

Thus if $y = 5x^4$, $\frac{dy}{dx} = 20x^3$, $\frac{d^2y}{dx^2} = 60x^2$, $\frac{d^3y}{dx^3} = 120x$, $\frac{d^4y}{dx^4} = 120$ and $\frac{d^5y}{dx^5} = 0$.

B WORKED PROBLEMS ON METHODS OF DIFFERENTIATION

Problem 1 Find the differential coefficients of (a) $y = 12x^3$, (b) $y = \frac{12}{x^3}$.

If $y = ax^n$ then $\frac{dy}{dx} = anx^{n-1}$.

(a) Since $y = 12x^3$, $a = 12$ and $n = 3$ thus $\frac{dy}{dx} = (12)(3)x^{3-1} = \mathbf{36x^2}$.

(b) $y = \frac{12}{x^3}$ is rewritten in the standard ax^n form as $y = 12x^{-3}$ and in the general rule $a = 12$ and $n = -3$. Thus $\frac{dy}{dx} = (12)(-3)x^{-3-1} = -36x^{-4}$
$= \mathbf{\frac{-36}{x^4}}$.

Problem 2 Differentiate (a) $y = 6$; (b) $y = 6x$.

(a) $y = 6$ may be written as $y = 6x^0$, i.e., in the general rule $a = 6$ and $n = 0$.

Hence $\dfrac{dy}{dx} = (6)(0)x^{0-1} = \mathbf{0}$

In general, the differential coefficient of a constant is always zero.

(b) Since $y = 6x$, in the general rule $a = 6$ and $n = 1$.

Hence $\dfrac{dy}{dx} = (6)(1)x^{1-1} = 6x^0 = \mathbf{6}$

In general, the differential coefficient of kx, where k is a constant, is always k.

Problem 3 Find the derivatives of (a) $y = 3\sqrt{x}$; (b) $y = \dfrac{5}{\sqrt[3]{x^4}}$

(a) $y = 3\sqrt{x}$ is rewritten in the standard differential form as $y = 3x^{½}$.

In the general rule, $a = 3$ and $n = \dfrac{1}{2}$.

Thus $\dfrac{dy}{dx} = (3)\left(\dfrac{1}{2}\right)x^{½-1} = \dfrac{3}{2}x^{-½} = \dfrac{3}{2x^{½}} = \mathbf{\dfrac{3}{2\sqrt{x}}}$

(b) $y = \dfrac{5}{\sqrt[3]{x^4}} = \dfrac{5}{x^{\frac{4}{3}}} = 5x^{-\frac{4}{3}}$ in the standard differential form.

In the general rule, $a = 5$ and $n = -\dfrac{4}{3}$.

Thus $\dfrac{dy}{dx} = (5)\left(-\dfrac{4}{3}\right)x^{-\frac{4}{3}-1} = \dfrac{-20}{3}x^{-\frac{7}{3}} = \dfrac{-20}{3x^{7/3}} = \mathbf{\dfrac{-20}{3\sqrt[3]{x^7}}}$

Problem 4 Differentiate $y = 5x^4 + 4x - \dfrac{1}{2x^2} + \dfrac{1}{\sqrt{x}} - 3$ with respect to x.

$y = 5x^4 + 4x - \dfrac{1}{2x^2} + \dfrac{1}{\sqrt{x}} - 3$ is rewritten as:

$y = 5x^4 + 4x - \dfrac{1}{2}x^{-2} + x^{-\frac{1}{2}} - 3.$

When differentiating a sum, each term is differentiated in turn.

$$\text{Thus } \frac{dy}{dx} = (5)(4)x^{4-1} + (4)(1)x^{1-1} - \frac{1}{2}(-2)x^{-2-1} + (1)\left(-\frac{1}{2}\right)x^{-½-1} - 0$$

$$= 20x^3 + 4 + x^{-3} - \frac{1}{2}x^{-\frac{3}{2}},$$

$$\text{i.e. } \mathbf{\frac{dy}{dx} = 20x^3 + 4 - \frac{1}{x^3} - \frac{1}{2\sqrt{x^3}}}.$$

Problem 5 Differentiate (a) $s = (t+1)^2$; (b) $y = 3.5\theta^{1.2}$ with respect to the variable.

(a) $s = (t+1)^2 = t^2 + 2t + 1$. Hence $\dfrac{ds}{dt} = 2t + 2 = \mathbf{2(t+1)}$

(b) $y = 3.5\theta^{1.2}$. Hence $\dfrac{dy}{d\theta} = (3.5)(1.2)\theta^{1.2-1} = \mathbf{4.2\theta^{0.2}}$

b. could have been $y = 2\cos 3t$ if we use function notation $y = f(t)$ $\frac{dy}{dt} = f'(t)$ ✓

Problem 6 Find the differential coefficients of (a) $y = 3 \sin 4x$; (b) $f(t) = 2 \cos 3t$ with respect to the variable.

(a) When $y = 3 \sin 4x$ then $\frac{dy}{dx} = (3)(4 \cos 4x) = \mathbf{12 \cos 4x}$

(b) When $f(t) = 2 \cos 3t$ then $f'(t) = (2)(-3 \sin 3t) = \mathbf{-6 \sin 3t}$

Problem 7 Determine the derivatives of (a) $y = 3e^{5x}$; (b) $f(\theta) = \frac{2}{e^{3\theta}}$; (c) $y = 6 \ln 2x$.

(a) When $y = 3e^{5x}$ then $\frac{dy}{dx} = (3)(5)e^{5x} = \mathbf{15e^{5x}}$.

(b) $f(\theta) = \frac{2}{e^{3\theta}} = 2e^{-3\theta}$. Thus $f'(\theta) = (2)(-3)e^{-3\theta} = -6e^{-3\theta} = \mathbf{\frac{-6}{e^{3\theta}}}$

(c) When $y = 6 \ln 2x$ then $\frac{dy}{dx} = 6\left(\frac{1}{x}\right) = \mathbf{\frac{6}{x}}$.

Problem 8 Find the gradient of the curve $y = 3x^4 - 2x^2 + 5x - 2$ at the points $(0, -2)$ and $(1, 4)$.

The gradient of a curve at a given point is given by the corresponding value of the derivative. Thus, since $y = 3x^4 - 2x^2 + 5x - 2$

then the gradient $= \frac{dy}{dx} = 12x^3 - 4x + 5$

At the point $(0, -2)$, $x = 0$. Thus the gradient $= 12(0)^3 - 4(0) + 5 = \mathbf{5}$
At the point $(1, 4)$, $x = 1$. Thus the gradient $= 12(1)^3 - 4(1) + 5 = \mathbf{13}$

Problem 9 Determine the co-ordinates of the point on the graph $y = 3x^2 - 7x + 2$ where the gradient is -1.

The gradient of the curve is given by the derivative.

When $y = 3x^2 - 7x + 2$ then $\frac{dy}{dx} = 6x - 7$

Since the gradient is -1 then $6x - 7 = -1$, from which, $x = 1$
When $x = 1$, $y = 3(1)^2 - 7(1) + 2 = -2$
Hence the gradient is -1 at the point $(1, -2)$.

Problem 10 (a) Given $f(x) = \frac{2x^3 - 3\sqrt{x} + 4\sqrt[4]{x^3}}{x^2}$, determine $f'(x)$;
(b) Evaluate $f'(x)$ when $x = 1$.

(a) $f(x) = \frac{2x^3 - 3\sqrt{x} + 4\sqrt[4]{x^3}}{x^2} = \frac{2x^3 - 3x^{\frac{1}{2}} + 4x^{\frac{3}{4}}}{x^2} = \frac{2x^3}{x^2} - \frac{3x^{\frac{1}{2}}}{x^2} + \frac{4x^{\frac{3}{4}}}{x^2}$

$= 2x^{3-2} - 3x^{\frac{1}{2}-2} + 4x^{\frac{3}{4}-2}$, by the laws of indices

i.e. $f(x) = 2x - 3x^{-\frac{3}{2}} + 4x^{-\frac{5}{4}}$

Hence $f'(x) = 2-(3)\left(-\frac{3}{2}\right)x^{-\frac{5}{2}}+(4)\left(-\frac{5}{4}\right)x^{-\frac{9}{4}} = 2+\frac{9}{2}x^{-\frac{5}{2}}-5x^{-\frac{9}{4}}$

$$= 2+\frac{9}{2x^{\frac{5}{2}}} - \frac{5}{x^{\frac{9}{4}}} = 2+\frac{9}{2\sqrt{x^5}} - \frac{5}{\sqrt[4]{x^9}}$$

(b) When $x = 1, f'(x) = 2+\frac{9}{2\sqrt{1^5}} - \frac{5}{\sqrt[4]{1^9}} = 2+\frac{9}{2}-5 = 1\frac{1}{2}$.

Problem 11 (a) Find $\frac{dx}{d\theta}$ when $x = \frac{3}{\theta}-4(\sin 5\theta-2\cos 2\theta)+2\ln 3\theta-\frac{15}{2e^{4\theta}}$;

(b) Evaluate $\frac{dx}{d\theta}$ when $\theta = \frac{\pi}{2}$, correct to 4 significant figures.

(a) $x = \frac{3}{\theta}-4(\sin 5\theta-2\cos 2\theta)+2\ln 3\theta-\frac{15}{2e^{4\theta}}$

i.e. $x = 3\theta^{-1}-4\sin 5\theta+8\cos 2\theta+2\ln 3\theta-\frac{15}{2}e^{-4\theta}$

$$\frac{dx}{d\theta} = (3)(-1)\theta^{-2}-(4)(5\cos 5\theta)+(8)(-2\sin 2\theta)+\frac{2}{\theta}-\frac{15}{2}(-4)e^{-4\theta}$$

$$= -3\theta^{-2}-20\cos 5\theta-16\sin 2\theta+\frac{2}{\theta}+30e^{-4\theta}$$

i.e. $\mathbf{\frac{dx}{d\theta} = \frac{-3}{\theta^2}-20\cos 5\theta-16\sin 2\theta+\frac{2}{\theta}+\frac{30}{e^{4\theta}}}$

(b) When $\theta = \frac{\pi}{2}$, $\frac{dx}{d\theta} = \frac{-3}{\left(\frac{\pi}{2}\right)^2} - 20\cos 5\left(\frac{\pi}{2}\right)-16\sin 2\left(\frac{\pi}{2}\right)+\frac{2}{\left(\frac{\pi}{2}\right)}+\frac{30}{e^{4\left(\frac{\pi}{2}\right)}}$

$\cos\frac{5\pi}{2} = 0$ and $\sin\pi = 0$. Hence $\frac{dx}{d\theta} = \frac{-3}{\left(\frac{\pi}{2}\right)^2}+\frac{2}{\left(\frac{\pi}{2}\right)}+\frac{30}{e^{2\pi}}$

$$= -1.21585+1.27324+0.05602$$

Hence when $\theta = \frac{\pi}{2}$, $\mathbf{\frac{dx}{d\theta} = 0.1134}$, correct to 4 significant figures

Problem 12 Find the differential coefficient of $y = 3x^2 \sin 2x$.

$3x^2 \sin 2x$ is a product of two terms $3x^2$ and $\sin 2x$.
Let $u = 3x^2$ and $v = \sin 2x$.

Using the product rule: $\frac{dy}{dx} = u\frac{dv}{dx} + v\frac{du}{dx}$

gives $\frac{dy}{dx} = (3x^2)(2\cos 2x)+(\sin 2x)(6x)$

i.e. $\frac{dy}{dx} = 6x^2\cos 2x+6x\sin 2x = \mathbf{6x(x\cos 2x+\sin 2x)}$

Note that the differential coefficient of a product is **not** obtained by merely differentiating each term and multiplying the two answers together. The product rule formula **must** be used when differentiating products.

Problem 13 Find the rate of change of y with respect to x given $y = 3\sqrt{x}\ln 2x$.

The rate of change of y with respect to x is given by $\frac{dy}{dx}$.

$y = 3\sqrt{x} \ln 2x = 3x^{1/2} \ln 2x$, which is a product.
Let $u = 3x^{1/2}$ and $v = \ln 2x$.

$$\text{Then } \frac{dy}{dx} = u \frac{dv}{dx} + v \frac{du}{dx}$$

$$= (3x^{1/2})\left(\frac{1}{x}\right) + (\ln 2x)\left[3\left(\frac{1}{2}\right)x^{1/2-1}\right]$$

$$= 3x^{1/2-1} + (\ln 2x)\left(\frac{3}{2}x^{-1/2}\right) = 3x^{-1/2}\left(1+\frac{1}{2}\ln 2x\right)$$

i.e. $\boldsymbol{\frac{dy}{dx} = \frac{3}{\sqrt{x}}\left(1+\frac{1}{2}\ln 2x\right)}$

Problem 14 $x = 3e^{2\theta} \sin 5\theta$. Find $\frac{dx}{d\theta}$.

Let $u = 3e^{2\theta}$ and $v = \sin 5\theta$.

$$\text{Then } \frac{dx}{d\theta} = u \frac{dv}{d\theta} + v \frac{du}{d\theta}$$

$$= (3e^{2\theta})(5 \cos 5\theta) + (\sin 5\theta)(6e^{2\theta})$$
$$= 15e^{2\theta} \cos 5\theta + 6e^{2\theta} \sin 5\theta$$

Hence $\boldsymbol{\frac{dx}{d\theta} = 3e^{2\theta}(5 \cos 5\theta + 2 \sin 5\theta)}$

Problem 15 Differentiate $y = x^3 \cos 3x \ln x$.

Let $u = x^3 \cos 3x$ (i.e. a product) and $v = \ln x$

$$\text{Then } \frac{dy}{dx} = u \frac{dv}{dx} + v \frac{du}{dx}.$$

$$\frac{du}{dx} = (x^3)(-3 \sin 3x) + (\cos 3x)(3x^2) \text{ and } \frac{dv}{dx} = \frac{1}{x}$$

$$\text{Hence } \frac{dy}{dx} = (x^3 \cos 3x)\left(\frac{1}{x}\right) + (\ln x)[-3x^3 \sin 3x + 3x^2 \cos 3x]$$

$$= x^2 \cos 3x + 3x^2 \ln x (\cos 3x - x \sin 3x)$$

i.e. $\boldsymbol{\frac{dy}{dx} = x^2\{\cos 3x + 3 \ln x (\cos 3x - x \sin 3x)\}}$

Problem 16 Determine the rate of change of voltage, given $v = 5t \sin 2t$ volts when $t = 0.2$ s.

$$\text{Rate of change of voltage} = \frac{dv}{dt} = (5t)(2 \cos 2t) + (\sin 2t)(5)$$

$$= 10t \cos 2t + 5 \sin 2t$$

$$\text{When } t = 0.2, \frac{dv}{dt} = 10(0.2) \cos 2(0.2) + 5 \sin 2(0.2)$$

$= 2 \cos 0.4 + 5 \sin 0.4$ (where cos 0.4 means the cosine of 0.4 radians,

i.e. $\left(0.4 \times \frac{180}{\pi}\right)^\circ$ or 22.92°

Hence $\frac{dv}{dt} = 2 \cos 22.92^\circ + 5 \sin 22.92^\circ$

$= 1.8421 + 1.9471 = 3.7892$

i.e., **the rate of change of voltage when $t = 0.2$ s is 3.79 volts/s, correct to 3 significant figures.**

Problem 17 Find the differential coefficient of $y = \frac{4 \sin 5x}{5x^4}$.

$\frac{4 \sin 5x}{5x^4}$ is a quotient. Let $u = 4 \sin 5x$ and $v = 5x^4$

$$\frac{dy}{dx} = \frac{v\frac{du}{dx} - u\frac{dv}{dx}}{v^2} \quad \text{where } \frac{du}{dx} = (4)(5)\cos 5x = 20 \cos 5x$$

$$\text{and } \frac{dv}{dx} = (5)(4)x^3 = 20x^3$$

Hence $$\frac{dy}{dx} = \frac{(5x^4)(20 \cos 5x) - (4 \sin 5x)(20x^3)}{(5x^4)^2}$$

$$= \frac{100x^4 \cos 5x - 80x^3 \sin 5x}{25x^8} = \frac{20x^3[5x \cos 5x - 4 \sin 5x]}{25x^8}$$

i.e. $$\boldsymbol{\frac{dy}{dx} = \frac{4}{5x^5}(5x \cos 5x - 4 \sin 5x)}$$

The differential coefficient of a quotient is **not** obtained by merely differentiating each term in turn and then dividing the numerator by the denominator. The quotient formula **must** be used when differentiating quotients.

Problem 18 Determine the differential coefficient of $y = \tan ax$.

$y = \tan ax = \frac{\sin ax}{\cos ax}$. Differentiation of tan ax is thus treated as a quotient with $u = \sin ax$ and $v = \cos ax$.

$$\frac{dy}{dx} = \frac{v\frac{du}{dx} - u\frac{dv}{dx}}{v^2} = \frac{(\cos ax)(a \cos ax) - (\sin ax)(-a \sin ax)}{(\cos ax)^2}$$

$$= \frac{a \cos^2 ax + a \sin^2 ax}{(\cos ax)^2} = \frac{a(\cos^2 ax + \sin^2 ax)}{\cos^2 ax}$$

$$= \frac{a}{\cos^2 ax}, \text{ since } \cos^2 ax + \sin^2 ax = 1.$$

Hence $\boldsymbol{\frac{dy}{dx} = a \sec^2 ax.}$

Problem 19 Differentiate $y = \cot 2x$.

$y = \cot 2x = \frac{1}{\tan 2x} = \frac{\cos 2x}{\sin 2x}$. Let $u = \cos 2x$ and $v = \sin 2x$.

$$\frac{dy}{dx} = \frac{v\dfrac{du}{dx} - u\dfrac{dv}{dx}}{v^2} = \frac{(\sin 2x)(-2\sin 2x)-(\cos 2x)(2\cos 2x)}{(\sin 2x)^2}$$

$$= \frac{-2\sin^2 2x - 2\cos^2 2x}{\sin^2 2x} = \frac{-2(\sin^2 2x + \cos^2 2x)}{\sin^2 2x}$$

Hence $\dfrac{dy}{dx} = \dfrac{-2}{\sin^2 2x} = \mathbf{-2\,cosec^2\,2x}$.

Problem 20 Find the derivative of $y = \sec ax$.

$y = \sec ax = \dfrac{1}{\cos ax}$ (i.e., a quotient). Let $u = 1$ and $v = \cos ax$.

$$\frac{dy}{dx} = \frac{v\dfrac{du}{dx} - u\dfrac{dv}{dx}}{v^2} = \frac{(\cos ax)(0)-(1)(-a\sin ax)}{(\cos ax)^2} = \frac{a\sin ax}{\cos^2 ax}$$

$$= a\left(\frac{1}{\cos ax}\right)\left(\frac{\sin ax}{\cos ax}\right)$$

i.e. $\dfrac{dy}{dx} = \boldsymbol{a\sec ax\tan ax}$.

Problem 21 Show that $\dfrac{d}{d\theta}(\operatorname{cosec} a\theta) = -a\operatorname{cosec} a\theta\cot a\theta$.

Let $y = \operatorname{cosec} a\theta = \dfrac{1}{\sin a\theta}$ (i.e., a quotient). Let $u = 1$ and $v = \sin a\theta$.

$$\frac{dy}{d\theta} = \frac{v\dfrac{du}{d\theta} - u\dfrac{dv}{d\theta}}{v^2} = \frac{(\sin a\theta)(0)-(1)(a\cos a\theta)}{(\sin a\theta)^2} = \frac{-a\cos a\theta}{\sin^2 a\theta} = -a\left(\frac{1}{\sin a\theta}\right)\left(\frac{\cos a\theta}{\sin a\theta}\right)$$

i.e. $\boldsymbol{\dfrac{d}{d\theta}(\operatorname{cosec} a\theta) = -a\operatorname{cosec} a\theta\cot a\theta}$.

Problem 22 Differentiate $y = \dfrac{te^{2t}}{2\cos t}$.

The function $\dfrac{te^{2t}}{2\cos t}$ is a quotient, whose numerator is a product.

Let $u = te^{2t}$ and $v = 2\cos t$;

then $\dfrac{du}{dt} = (t)(2e^{2t})+(e^{2t})(1)$ and $\dfrac{dv}{dt} = -2\sin t$

Hence $$\frac{dy}{dt} = \frac{v\dfrac{du}{dt} - u\dfrac{dv}{dt}}{v^2} = \frac{(2\cos t)[2te^{2t}+e^{2t}]-(te^{2t})(-2\sin t)}{(2\cos t)^2}$$

$$= \frac{4te^{2t}\cos t + 2e^{2t}\cos t + 2te^{2t}\sin t}{4\cos^2 t}$$

$$= \frac{2e^{2t}[2t\cos t + \cos t + t\sin t]}{4\cos^2 t}$$

i.e. $\boldsymbol{\dfrac{dy}{dt} = \dfrac{e^{2t}}{2\cos^2 t}(2t\cos t + \cos t + t\sin t)}$

Problem 23 Determine the gradient of the curve $y = \frac{5x}{2x^2+4}$ at the point $\left(\sqrt{3}, \frac{\sqrt{3}}{2}\right)$.

Let $u = 5x$ and $v = 2x^2+4$

$$\frac{dy}{dx} = \frac{v\frac{du}{dx} - u\frac{dv}{dx}}{v^2} = \frac{(2x^2+4)(5)-(5x)(4x)}{(2x^2+4)^2} = \frac{10x^2+20-20x^2}{(2x^2+4)^2} = \frac{20-10x^2}{(2x^2+4)^2}$$

At the point $\left(\sqrt{3}, \frac{\sqrt{3}}{2}\right)$, $x = \sqrt{3}$. Hence the gradient $= \frac{dy}{dx} = \frac{20-10(\sqrt{3})^2}{[2(\sqrt{3})^2+4]^2}$

$$= \frac{20-30}{100} = -\frac{1}{10}$$

Problem 24 Differentiate $y = 3\cos(5x^2+2)$.

Let $u = 5x^2+2$ then $y = 3\cos u$.

Hence $\frac{du}{dx} = 10x$ and $\frac{dy}{du} = -3\sin u$.

Using the function of a function rule,

$$\frac{dy}{dx} = \frac{dy}{du} \times \frac{du}{dx} = (-3\sin u)(10x) = -30x\sin u.$$

Rewriting u as $5x^2+2$ gives: $\boldsymbol{\frac{dy}{dx} = -30x\sin(5x^2+2)}$.

Problem 25 Find the derivative of $y = (4t^3-3t)^6$.

Let $u = 4t^3-3t$, then $y = u^6$.

Hence $\frac{du}{dt} = 12t^2-3$ and $\frac{dy}{du} = 6u^5$.

Using the function of a function rule, $\frac{dy}{dt} = \frac{dy}{du} \times \frac{du}{dt} = (6u^5)(12t^2-3)$

Rewriting u as $4t^3-3t$ gives $\frac{dy}{dt} = 6(4t^3-3t)^5(12t^2-3)$

$$= \mathbf{18(4t^2-1)(4t^3-3t)^5}.$$

Problem 26 Determine the differential coefficient of $y = \sqrt{(3x^2+4x-1)}$.

$y = \sqrt{(3x^2+4x-1)} = (3x^2+4x-1)^{1/2}$

Let $u = 3x^2+4x-1$ then $y = u^{1/2}$

Hence $\frac{du}{dx} = 6x+4$ and $\frac{dy}{du} = \frac{1}{2}u^{-1/2} = \frac{1}{2\sqrt{u}}$

Using the function of a function rule,

$$\frac{dy}{dx} = \frac{dy}{du} \times \frac{du}{dx} = \left(\frac{1}{2\sqrt{u}}\right)(6x+4) = \frac{3x+2}{\sqrt{u}}$$

i.e. $\boldsymbol{\frac{dy}{dx} = \frac{3x+2}{\sqrt{(3x^2+4x-1)}}}$

Problem 27 Differentiate $y = 3\tan^4 3x$.

Let $u = \tan 3x$ then $y = 3u^4$.

Hence $\dfrac{du}{dx} = 3\sec^2 3x$, (from *Problem 18*), and $\dfrac{dy}{du} = 12u^3$.

Then $\dfrac{dy}{dx} = \dfrac{dy}{du} \times \dfrac{du}{dx} = (12u^3)(3\sec^2 3x) = 12(\tan 3x)^3(3\sec^2 3x)$

i.e. $\boldsymbol{\dfrac{dy}{dx} = 36\tan^3 3x \sec^2 3x}$.

Problem 28 Find the differential coefficient of $y = \dfrac{2}{(2t^3-5)^4}$

$y = \dfrac{2}{(2t^3-5)^4} = 2(2t^3-5)^{-4}$. Let $u = (2t^3-5)$, then $y = 2u^{-4}$.

Hence $\dfrac{du}{dt} = 6t^2$ and $\dfrac{dy}{du} = -8u^{-5} = \dfrac{-8}{u^5}$.

Then $\dfrac{dy}{dt} = \dfrac{dy}{du} \times \dfrac{du}{dt} = \left(\dfrac{-8}{u^5}\right)(6t^2) = \boldsymbol{\dfrac{-48t^2}{(2t^3-5)^5}}$

Problem 29 Differentiate $y = 4e^{(3\theta^2-2)}$

Let $u = 3\theta^2-2$ then $y = 4e^u$

Hence $\dfrac{du}{d\theta} = 6\theta$ and $\dfrac{dy}{du} = 4e^u$.

Then $\dfrac{dy}{d\theta} = \dfrac{dy}{du} \times \dfrac{du}{d\theta} = (4e^u)(6\theta) = 24\theta e^u = \boldsymbol{24\theta e^{(3\theta^2-2)}}$

Problem 30 If $f(x) = 4x^5-2x^3+x-3$, find $f''(x)$.

$f(x) = 4x^5-2x^3+x-3$
$f'(x) = 20x^4-6x^2+1$
$f''(x) = 80x^3-12x = \boldsymbol{4x(20x^2-3)}$.

Problem 31 Given $y = \dfrac{2}{3}x^3 - \dfrac{4}{x^2} + \dfrac{1}{2x} - \sqrt{x}$, determine $\dfrac{d^2y}{dx^2}$.

$y = \dfrac{2}{3}x^3 - \dfrac{4}{x^2} + \dfrac{1}{2x} - \sqrt{x} = \dfrac{2}{3}x^3 - 4x^{-2} + \dfrac{1}{2}x^{-1} - x^{½}$

$\dfrac{dy}{dx} = \left(\dfrac{2}{3}\right)(3)x^2 - 4(-2)x^{-3} + \left(\dfrac{1}{2}\right)(-1)x^{-2} - \left(\dfrac{1}{2}\right)x^{-½}$

i.e. $\dfrac{dy}{dx} = 2x^2 + 8x^{-3} - \dfrac{1}{2}x^{-2} - \dfrac{1}{2}x^{-½}$

$\dfrac{d^2y}{dx^2} = 4x - 24x^{-4} + x^{-3} + \dfrac{1}{4}x^{-\frac{3}{2}} = \boldsymbol{4x - \dfrac{24}{x^4} + \dfrac{1}{x^3} + \dfrac{1}{4\sqrt{x^3}}}$

Problem 32 If $y = \cos x - \sin x$, evaluate x, in the range $0 \leqslant x \leqslant \dfrac{\pi}{2}$, when $\dfrac{d^2y}{dx^2}$ is zero.

Since $y = \cos x - \sin x$, $\frac{dy}{dx} = -\sin x - \cos x$ and $\frac{d^2y}{dx^2} = -\cos x + \sin x$.

When $\frac{d^2y}{dx^2}$ is zero, $-\cos x + \sin x = 0$, i.e. $\sin x = \cos x$ or $\frac{\sin x}{\cos x} = 1$

Hence $\tan x = 1$ and $x = \arctan 1 =$ **45°** or $\frac{\pi}{4}$ **rads** in the range $0 \leqslant x \leqslant \frac{\pi}{2}$.

Problem 33 Given that $y = 2xe^{-3x}$ show that $\frac{d^2y}{dx^2} + 6\frac{dy}{dx} + 9y = 0$.

$y = 2xe^{-3x}$ (i.e. a product).

$$\text{Hence} \quad \frac{dy}{dx} = (2x)(-3e^{-3x}) + (e^{-3x})(2) = -6xe^{-3x} + 2e^{-3x}$$

$$\frac{d^2y}{dx^2} = [(-6x)(-3e^{-3x}) + (e^{-3x})(-6)] + (-6e^{-3x})$$
$$= 18xe^{-3x} - 6e^{-3x} - 6e^{-3x}$$

$$\text{i.e.} \quad \frac{d^2y}{dx^2} = 18xe^{-3x} - 12e^{-3x}$$

Substituting values into $\frac{d^2y}{dx^2} + 6\frac{dy}{dx} + 9y$ gives:

$$(18xe^{-3x} - 12e^{-3x}) + 6(-6xe^{-3x} + 2e^{-3x}) + 9(2xe^{-3x})$$
$$= (18xe^{-3x} - 12e^{-3x} - 36xe^{-3x} + 12e^{-3x} + 18xe^{-3x} = 0$$

Thus $\frac{d^2y}{dx^2} + 6\frac{dy}{dx} + 9y = 0$ when $y = 2xe^{-3x}$.

Problem 34 Evaluate $\frac{d^2y}{d\theta^2}$ when $\theta = 0$ given $y = 4 \sec 2\theta$.

Since $y = 4 \sec 2\theta$, then $\frac{dy}{d\theta} = (4)(2) \sec 2\theta \tan 2\theta$ (from *Problem 20*)
$= 8 \sec 2\theta \tan 2\theta$ (i.e., a product)

$$\frac{d^2y}{d\theta^2} = (8 \sec 2\theta)(2 \sec^2 2\theta) + (\tan 2\theta)[(8)(2) \sec 2\theta \tan 2\theta]$$
$$= 16 \sec^3 2\theta + 16 \sec 2\theta \tan^2 2\theta$$

$$\text{When } \theta = 0, \quad \frac{d^2y}{d\theta^2} = 16 \sec^3 0 + 16 \sec 0 \tan^2 0$$
$$= 16(1) + 16(1)(0) = \mathbf{16}$$

C FURTHER PROBLEMS ON METHODS OF DIFFERENTIATION

In *Problems 1 to 6* find the differential coefficients of the given functions with respect to the variable.

1 (a) $5x^5$; (b) $2.4x^{3.5}$; (c) $\frac{1}{x}$ $\left[\text{(a) } 25x^4;\ \text{(b) } 8.4x^{2.5};\ \text{(c) } -\frac{1}{x^2}\right]$

2 (a) $\frac{-4}{x^2}$; (b) 6; (c) $2x$ $\left[\text{(a) } \frac{8}{x^3};\ \text{(b) } 0;\ \text{(c) } 2\right]$

3 (a) $2\sqrt{x}$; (b) $3\sqrt[3]{x^5}$; (c) $\frac{4}{\sqrt{x}}$ $\left[\text{(a) } \frac{1}{\sqrt{x}};\ \text{(b) } 5\sqrt[3]{x^2};\ \text{(c) } -\frac{2}{\sqrt{x^3}}\right]$

4 (a) $\frac{-3}{\sqrt[3]{x}}$; (b) $(x-1)^2$; (c) $2 \sin 3x$ $\left[(a)\ \frac{1}{\sqrt[3]{x^4}}; (b)\ 2(x-1); (c)\ 6 \cos 3x\right]$

5 (a) $-4 \cos 2x$; (b) $2e^{6x}$; (c) $\frac{3}{e^{5x}}$ $\left[(a)\ 8 \sin 2x; (b)\ 12e^{6x}; (c)\ \frac{-15}{e^{5x}}\right]$

6 (a) $4 \ln 9x$; (b) $\frac{e^x - e^{-x}}{2}$; (c) $\frac{1-\sqrt{x}}{x}$

$\left[(a)\ \frac{4}{x}; (b)\ \frac{e^x + e^{-x}}{2}; (c)\ \frac{-1}{x^2} + \frac{1}{2\sqrt{x^3}}\right]$

7 Find the gradient of the curve $y = 2t^4 + 3t^3 - t + 4$ at the points (0,4) and (1,8) [−1; 16]

8 If $f(t) = 4 \ln t + 2$ evaluate $f'(t)$ when $t = 0.25$ [16]

9 Find the co-ordinates of the point on the graph $y = 5x^2 - 3x + 1$ where the gradient is 2. $\left[\left(\frac{1}{2}, \frac{3}{4}\right)\right]$

10 Given $f(x) = \frac{3x^4 - 4\sqrt{x} + 3\sqrt[4]{x^5}}{x^3}$ evaluate $f'(x)$ when $x = 1$. $[7\frac{3}{4}]$

11 (a) Differentiate $y = \frac{2}{\theta^2} + 2 \ln 2\theta - 2(\cos 5\theta + 3 \sin 2\theta) - \frac{2}{e^{3\theta}}$.

(b) Evaluate $\frac{dy}{d\theta}$ when $\theta = \frac{\pi}{2}$, correct to 4 significant figures.

$\left[(a)\ \frac{-4}{\theta^3} + \frac{2}{\theta} + 10 \sin 5\theta - 12 \cos 2\theta + \frac{6}{e^{3\theta}}; (b)\ 22.30\right]$

12 Evaluate $\frac{ds}{dt}$, correct to 3 significant figures, when $t = \frac{\pi}{6}$ given $s = 3 \sin t - 3 + \sqrt{t}$. [3.29]

In *Problems 13 to 18* differentiate the given products with respect to the variable.

13 $2x^3 \cos 3x$ $[6x^2(\cos 3x - x \sin 3x)]$

14 $\sqrt{x^3} \ln 3x$ $[\sqrt{x}(1 + \frac{3}{2} \ln 3x)]$

15 $e^{3t} \sin 4t$ $[e^{3t}(4 \cos 4t + 3 \sin 4t)]$

16 $\sqrt{x} \sin 2x$ $\left[2\sqrt{x} \cos 2x + \frac{\sin 2x}{2\sqrt{x}}\right]$

17 $e^{4\theta} \ln 3\theta$ $\left[e^{4\theta}\left(\frac{1}{\theta} + 4 \ln 3\theta\right)\right]$

18 $e^t \ln t \cos t$ $\left[e^t\left\{\left(\frac{1}{t} + \ln t\right) \cos t - \ln t \sin t\right\}\right]$

19 Evaluate $\frac{di}{dt}$, correct to 4 significant figures, when $t = 0.1$, and $i = 15t \sin 3t$. [8.732]

20 Evaluate $\frac{dz}{dt}$, correct to 4 significant figures, when $t = 0.5$, given that $z = 2e^{3t} \sin 2t$. [32.31]

In *Problems 21 to 26*, differentiate the quotients with respect to the variable.

21 $\frac{2 \cos 3x}{x^3}$ $\left[\frac{-6}{x^4}(x \sin 3x + \cos 3x)\right]$

22 $\frac{2x}{x^2+1}$ $\left[\frac{2(1-x^2)}{(x^2+1)^2}\right]$

23 $\frac{3\sqrt{\theta^3}}{2 \sin 2\theta}$ $\left[\frac{3\sqrt{\theta}(3 \sin 2\theta - 4\theta \cos 2\theta)}{4 \sin^2 2\theta}\right]$

24 $\dfrac{\ln 2t}{\sqrt{t}}$ $\left[\dfrac{1-\frac{1}{2}\ln 2t}{\sqrt{t^3}}\right]$

25 $\dfrac{3\tan p}{e^{3p}}$ $\left[\dfrac{3(\sec^2 p-3\tan p)}{e^{3p}}\right]$

26 $\dfrac{2xe^{4x}}{\sin x}$ $\left[\dfrac{2e^{4x}}{\sin^2 x}\{(1+4x)\sin x-x\cos x\}\right]$

27 Find the gradient of the curve $y=\dfrac{2x}{x^2-5}$ at the point (2, –4). [–18]

28 Evaluate $\dfrac{dy}{dx}$ at $x = 2.5$, correct to 3 significant figures, given $y=\dfrac{2x^2+3}{\ln 2x}$ [3.82]

29 Evaluate $f'(\frac{\pi}{3})$ when $f(t) = 2\tan 2t-\cot 4t$. $\left[21\frac{1}{3}\right]$

30 Show that $\dfrac{\frac{d}{dz}(2\sec 3z)}{\frac{d}{dz}(-\operatorname{cosec} 3z)} = 2\tan^3 3z$

In *Problems 31 to 40*, find the differential coefficients with respect to the variable.

31 $(2x^3-5x)^5$ $[5(6x^2-5)(2x^3-5x)^4]$

32 $2\sin(3\theta-2)$ $[6\cos(3\theta-2)]$

33 $\sqrt{(2t^3-4)}$ $\left[\dfrac{3t^2}{\sqrt{(2t^3-4)}}\right]$

34 $2\cos^5\alpha$ $[-10\cos^4\alpha\sin\alpha]$

35 $\dfrac{1}{(x^3-2x+1)^5}$ $\left[\dfrac{5(2-3x^2)}{(x^3-2x+1)^6}\right]$

36 $5e^{2t+1}$ $[10e^{2t+1}]$

37 $4\sec^3 x$ $[12\sec^3 x\tan x]$

38 $2\cot(5t^2+3)$ $[-20t\operatorname{cosec}^2(5t^2+3)]$

39 $6\tan(3y+1)$ $[18\sec^2(3y+1)]$

40 $2e^{\tan\theta}$ $[2\sec^2\theta\, e^{\tan\theta}]$

41 Differentiate $\theta\sin\left(\theta-\dfrac{\pi}{3}\right)$ with respect to θ, and evaluate, correct to 3 significant figures, when $\theta=\dfrac{\pi}{2}$. [1.86]

42 If $y = 3x^4+2x^3-3x+2$ find (a) $\dfrac{d^2y}{dx^2}$; (b) $\dfrac{d^3y}{dx^3}$ [(a) $36x^2+12x$; (b) $72x+12$]

43 (a) Given $f(t)=\dfrac{2}{5}t^2-\dfrac{1}{t^3}+\dfrac{3}{t}-\sqrt{t}+1$ determine $f''(t)$;
(b) Evaluate $f''(t)$ when $t = 1$. $\left[\text{(a) }\dfrac{4}{5}-\dfrac{12}{t^5}+\dfrac{6}{t^3}+\dfrac{1}{4\sqrt{t^3}}\text{; (b) }-4.95\right]$

In *Problems 44 and 45*, find the second differential coefficient with respect to the variable.

44 (a) $3\sin 2t+\cos t$; (b) $2\ln 4\theta$ $\left[\text{(a) }-(12\sin 2t+\cos t)\text{; (b) }\dfrac{-2}{\theta^2}\right]$

45 (a) $2\cos^2 x$; (b) $(2x-3)^4$ [(a) $4(\sin^2 x-\cos^2 x)$; (b) $48(2x-3)^2$]

46 If $y = Ae^{3t}+Be^{-2t}$ prove that $\dfrac{d^2y}{dt^2}-\dfrac{dy}{dt}-6y=0$.

47 Evaluate $f''(\theta)$ when $\theta = 0$ given $f(\theta) = 2 \sec 3\theta$. [18]

48 Show that the differential equation $\frac{d^2 y}{dx^2} - 4\frac{dy}{dx} + 4y = 0$ is satisfied when $y = xe^{2x}$

16 Calculus (3) – Applications of differentiation

A MAIN POINTS CONCERNED WITH APPLICATIONS OF DIFFERENTIATION

1 **Velocity and acceleration**
If a body moves a distance x metres in a time t seconds then:
(i) distance $x = f(t)$;
(ii) velocity $v = f'(t)$ or $\frac{dx}{dt}$, which is the gradient of the distance/time graph,
(iii) acceleration $a = \frac{dv}{dt} = f''(t)$ or $\frac{d^2x}{dt^2}$, which is the gradient of the velocity/time graph.
(See *Problems 1 to 6.*)

2 **Rates of change**
(i) If a quantity y depends on and varies with a quantity x then the rate of change of y with respect to x is dy/dx. Thus, for example, the rate of change of pressure p with height h is dp/dh.
(ii) A rate of change with respect to time is usually just called 'the rate of change', the 'with respect to time' being assumed. Thus, for example, a rate of change of voltage, v, is dv/dt and a rate of change of temperature, θ, is $d\theta/dt$, and so on.
(See *Problems 7 to 10.*)

3 **Turning points**
(i) In *Fig 1*, the gradient (or rate of change) of the curve changes from positive between 0 and P to negative between P and Q, and then positive again between Q and R. At point P, the gradient is zero and, as x increases, the gradient of the curve changes from positive just before P to negative just after. Such a point is called a **maximum point** and appears as the 'crest of a wave'. At point Q, the gradient is also zero and, as x increases, the gradient of the curve changes from negative just before Q to positive just after. Such a point is called a **minimum point**, and appears as the 'bottom of a valley'. Points such as P and Q are given the general name of **turning points**.

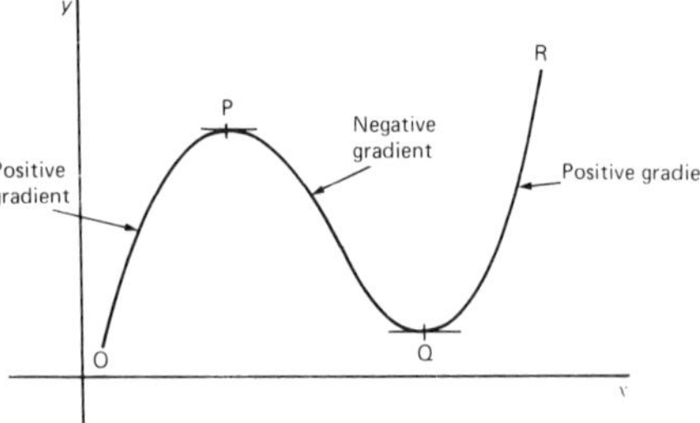

Fig 1

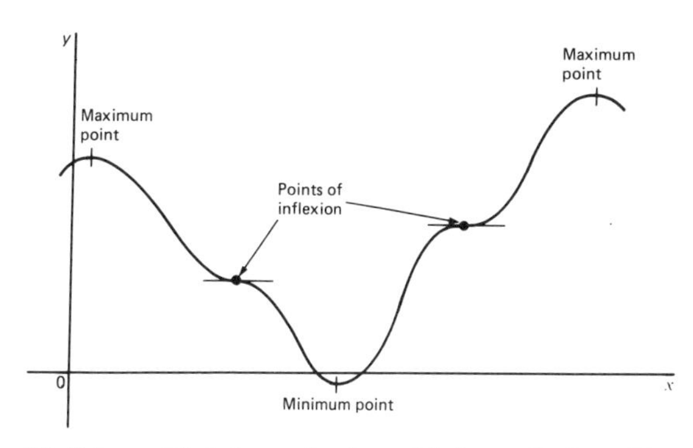

Fig 2

(ii) It is possible to have a turning point, the gradient on either side of which is the same. Such a point is given the special name of a **point of inflexion,** and examples are shown in *Fig 2.*

(iii) Maximum and minimum points and points of inflexion are given the general term of **stationary points.**

4 **Procedure for finding and distinguishing between stationary points**

(i) Given $y = f(x)$, determine $\frac{dy}{dx}$ (i.e., $f'(x)$).

(ii) Let $\frac{dy}{dx} = 0$ and solve for the values of x.

(iii) Substitute the values of x into the original equation, $y = f(x)$, to find the corresponding y-ordinate values. This establishes the co-ordinates of the stationary points.

To determine the nature of the stationary points:

Either (iv) Find $\frac{d^2y}{dx^2}$ and substitute into it the values of x found in (ii).

If the result is: (a) positive – the point is a minimum one,
(b) negative – the point is a maximum one,
(c) zero – the point is a point of inflexion.

or

(v) Determine the sign of the gradient of the curve just before and just after the stationary points. If the sign change for the gradient of the curve is:
(a) positive to negative – the point is a maximum one,
(b) negative to positive – the point is a minimum one,
(c) positive to positive, or negative to negative – the point is a point of inflexion.

(See *Problems 11 to 15.*)

5 There are many **practical problems** involving maximum and minimum values which occur in science and engineering. Usually, an equation has to be determined from given data, and rearranged where necessary, so that it contains only one variable. Some examples are demonstrated in *Problems 16 to 22.*

B WORKED PROBLEMS ON APPLICATIONS OF DIFFERENTIATION

Problem 1 The distance x metres moved by a car in a time t seconds is given by $x = 3t^3 - 2t^2 + 4t - 1$. Determine the velocity and acceleration where (a) $t = 0$, and (b) $t = 1.5$ s.

Distance $x = 3t^3 - 2t^2 + 4t - 1$ m. Velocity $v = \frac{dx}{dt} = 9t^2 - 4t + 4$ m/s

Acceleration $a = \frac{d^2x}{dt^2} = 18t - 4$ m/s^2

(a) When time $t = 0$, velocity $v = 9(0)^2 - 4(0) + 4 =$ **4 m/s**
and acceleration $a = 18(0) - 4 =$ **−4 m/s²** (i.e. a deceleration)

(b) When time $t = 1.5$ s, velocity $v = 9(1.5)^2 - 4(1.5) + 4 =$ **18.25 m/s**
and acceleration $a = 18(1.5) - 4 =$ **23 m/s²**

Problem 2 Supplies are dropped from a helicopter and the distance fallen in a time t seconds is given by $x = \frac{1}{2}gt^2$, where $g = 9.8$ m/s^2. Determine the velocity and acceleration of the supplies after it has fallen for 2 seconds.

Distance $x = \frac{1}{2}gt^2 = \frac{1}{2}(9.8)t^2 = 4.9t^2$ m.

Velocity $v = \frac{dv}{dt} = 9.8t$ m/s and acceleration, $a = \frac{d^2x}{dt^2} = 9.8$ m/s^2.

When time $t = 2$ s, velocity $v = (9.8)(2) =$ **19.6 m/s**
and acceleration $a =$ **9.8 m/s²**, which is the acceleration due to gravity.

Problem 3 The distance x metres travelled by a vehicle in time t seconds after the brakes are applied is given by $x = 20t - \frac{5}{3}t^2$. Determine (a) the speed of the vehicle (in km/h) at the instant the brakes are applied, and (b) the distance the car travels before it stops.

(a) Distance, $x = 20t - \frac{5}{3}t^2$. Hence velocity $v = \frac{dx}{dt} = 20 - \frac{10}{3}t$.
At the instance the brakes are applied, time $t = 0$.

Hence velocity $v = 20$ m/s $= \frac{20 \times 60 \times 60}{1000}$ km/h $=$ **72 km/h.**

(*Note*: changing from m/s to km/h merely involves multiplying by 3.6)

(b) When the car finally stops, the velocity is zero,
i.e. $v = 20 - \frac{10}{3}t = 0$, from which, $20 = \frac{10}{3}t$, giving $t = 6$ s
Hence the distance travelled before the car stops is given by:
$x = 20t - \frac{5}{3}t^2 = 20(6) - \frac{5}{3}(6)^2 = 120 - 60 =$ **60 m**

Problem 4 The angular displacement θ radians of a flywheel varies with time t seconds and follows the equation $\theta = 9t^2 - 2t^3$. Determine (a) the angular velocity and acceleration of the flywheel when time, $t = 1$ s, and (b) the time when the angular acceleration is zero.

(a) Angular displacement $\theta = 9t^2 - 2t^3$ rad.

Angular velocity $\omega = \frac{d\theta}{dt} = 18t - 6t^2$ rad/s.

When time $t = 1$ s, $\omega = 18(1) - 6(1)^2 =$ **12 rad/s.**

Angular acceleration $\alpha = \frac{d^2\theta}{dt^2} = 18 - 12t$ rad/s^2

When time $t = 1$ s, $\alpha = 18 - 12(1) =$ **6 rad/s²**.

(b) When the angular acceleration is zero, $18-12t = 0$, from which, $18 = 12t$, giving time, $\boldsymbol{t} = \mathbf{1.5}$ **s.**

Problem 5 The distance x metres moved by a body in t seconds is given by $x = 5t^3 - \frac{21}{2}t^2 + 6t - 4$ Determine (a) the initial velocity and the velocity after 3 s, (b) the values of t when the body came to rest, (c) its acceleration after 2 s, (d) the value of t when the acceleration is 24 m/s² and (e) the average velocity over the third second.

(a) Distance $x = 5t^3 - \frac{21}{2}t^2 + 6t - 4$

Velocity $v = \frac{dx}{dt} = 15t^2 - 21t + 6$

The initial velocity, i.e., when $t = 0$, is $\boldsymbol{v_0} = \mathbf{6}$ **m/s.**

Velocity after 3 s, $v_3 = 15(3)^2 - 21(3) + 6 = 135 - 63 + 6 =$ **78 m/s.**

(b) When the body comes to rest, velocity $v = 0$, i.e., $15t^2 - 21t + 6 = 0$

Rearranging and factorising gives: $3(5t^2 - 7t + 2) = 0$ and $3(5t-2)(t-1) = 0$

Hence $(5t-2) = 0$, from which, $\boldsymbol{t} = \frac{\mathbf{2}}{\mathbf{5}}$ **s**

or $(t-1) = 0$, from which, $\boldsymbol{t} = \mathbf{1}$ **s.**

(c) Acceleration $a = \frac{d^2x}{dt^2} = 30t - 21$

Acceleration after 2 s, $a_2 = 30(2) - 21 =$ **39 m/s².**

(d) When the acceleration is 24 m/s² then $30t - 21 = 24$, from which $t = \frac{45}{30} =$ **1.5 s**

(e) Distance travelled in the third second

= (distance travelled after 3 s) – (distance travelled after 2 s)

$= \left[5(3)^3 - \frac{21}{2}(3)^2 + 6(3) - 4\right] - \left[5(2)^3 - \frac{21}{2}(2)^2 + 6(2) - 4\right]$

$= 54.5 - 6 = 48.5$ m.

Average velocity over the third second $= \frac{\text{distance travelled}}{\text{time taken}} = \frac{48.5 \text{ m}}{1 \text{ s}}$

= **48.5 m/s**

Problem 6 The displacement x cm of the slide valve of an engine is given by $x = 2.2 \cos 5\pi t + 3.6 \sin 5\pi t$. Evaluate the velocity (in m/s) when time $t = 30$ ms.

Displacement $x = 2.2 \cos 5\pi t + 3.6 \sin 5\pi t$.

Velocity $v = \frac{dx}{dt} = (2.2)(-5\pi) \sin 5\pi t + (3.6)(5\pi) \cos 5\pi t$

$= -11\pi \sin 5\pi t + 18\pi \cos 5\pi t$ cm/s

When time $t = 30$ ms, velocity $= -11\pi \sin\left(5\pi \frac{30}{10^3}\right) + 18\pi \cos\left(5\pi \frac{30}{10^3}\right)$

$= -11\pi \sin 0.4712 + 18\pi \cos 0.4712$

$= -11\pi \sin 27° + 18\pi \cos 27° = -15.69 + 50.39$

= 34.7 cm/s = **0.347 m/s.**

Problem 7 The length l metres of a certain metal rod at temperature t°C is given by $l = 1 + 0.00003t + 0.0000003t^2$. Determine the rate of change of length, in mm/°C, when the temperature is (a) 100°C and (b) 250°C.

The rate of change of length means $\frac{dl}{dt}$

Since length $l = 1+0.00003t+0.0000003t^2$, then $\frac{dl}{dt} = 0.00003+0.0000006t$

(a) When $t = 100°C$, $\frac{dl}{dt} = 0.00003+(0.0000006)(100) = 0.00009$ m/°C
$= \mathbf{0.09}$ **mm/°C**

(b) When $t = 250°C$, $\frac{dl}{dt} = 0.00003+(0.0000006)(250) = 0.00018$ m/°C
$= \mathbf{0.18}$ **mm/°C.**

Problem 8 The luminous intensity I candelas of a lamp at varying voltage V is given by $I = 5 \times 10^{-4}\ V^2$. Determine the voltage at which the light is increasing at a rate of 0.4 candelas per volt.

The rate of change of light with respect to voltage is given by $\frac{dI}{dV}$.

Since $I = 5 \times 10^{-4}\ V^2$, $\frac{dI}{dV} = (5 \times 10^{-4})(2)V = 10 \times 10^{-4}\ V = 10^{-3}\ V$.
When the light is increasing at 0.4 candelas per volt then $+0.4 = 10^{-3}\ V$, from which

voltage $V = \frac{0.4}{10^{-3}} = 0.4 \times 10^{+3} =$ **400 volts.**

Problem 9 Newtons law of cooling is given by $\theta = \theta_0 e^{-kt}$, where the excess of temperature at zero time is θ_0°C and at time t seconds is θ°C. Determine the rate of change of temperature after 50 s, given that $\theta_0 = 15$°C and $k = -0.02$

The rate of change of temperature is $\frac{d\theta}{dt}$.

Since $\theta = \theta_0 e^{-kt}$ then $\frac{d\theta}{dt} = (\theta_0)(-k)e^{-kt} = -k\theta_0 e^{-kt}$

When $\theta_0 = 15$, $k = -0.02$ and $t = 50$ then $\frac{d\theta}{dt} = -(-0.02)(15)e^{-(-0.02)(50)}$
$= 0.3e^1 =$ **0.815°C/s.**

Problem 10 The pressure, p, of the atmosphere at height h above ground level is given by $p = p_0 e^{-h/c}$, where p_0 is the pressure at ground level and c is a constant. Determine the rate of change of pressure with height when $p_0 = 10^5$ pascals and $c = 6 \times 10^4$ at 1500 m.

The rate of change of pressure with height is $\frac{dp}{dh}$.

Since $p = p_0 e^{-h/c}$ then $\frac{dp}{dh} = (p_0)(-\frac{1}{c})\ e^{-h/c} = -\frac{p_0}{c}\ e^{-h/c}$

When $p_0 = 10^5$, $c = 6 \times 10^4$ and $h = 1500$ then $\frac{dp}{dh} = -\frac{10^5}{6 \times 10^4}\ e^{(-\frac{1500}{6 \times 10^4})}$
$= -\frac{5}{3}e^{-0.025} =$ **−1.63 Pa/m.**

Problem 11 Locate the turning point on the curve $y = 2x^2 - 4x$ and determine its nature by examining the sign of the gradient on either side.

Following the procedure of para. 4:

(i) Since $y = 2x^2 - 4x$, $\frac{dy}{dx} = 4x - 4$

(ii) At a turning point, $\frac{dy}{dx} = 0$. Hence $4x - 4 = 0$, from which, $x = 1$.

(iii) When $x = 1$, $y = 2(1)^2 - 4(1) = -2$.
Hence the co-ordinate of the turning point is (1, −2).

(iv) If x is slightly less than 1, say, 0.9, then $\frac{dy}{dx} = 4(0.9) - 4 = -0.4$, i.e. negative.
If x is slightly greater than 1, say, 1.1, then $\frac{dy}{dx} = 4(1.1) - 4 = 0.4$, i.e. positive.
Since the gradient of the curve is negative just before the turning point and positive just after (i.e. $-\vee+$), **(1, −2) is a minimum point.**

Problem 12 Find the maximum and minimum values of the curve $y = x^3 - 3x + 5$ by (a) examining the gradient on either side of the turning points, and (b) determining the sign of the second derivative.

Since $y = x^3 - 3x + 5$ then $\frac{dy}{dx} = 3x^2 - 3$

For a maximum or minimum value $\frac{dy}{dx} = 0$. Hence $3x^2 - 3 = 0$, from which, $3x^2 = 3$ and $x = \pm 1$.
When $x = +1$, $y = (1)^3 - 3(1) + 5 = 3$.
When $x = -1$, $y = (-1)^3 - 3(-1) + 5 = 7$.
Hence (1, 3) and (−1, 7) are the co-ordinates of the turning points.

(a) Considering the point (1, 3):

If x is slightly less than 1, say, 0.9, then $\frac{dy}{dx} = 3(0.9)^2 - 3$, which is negative.

If x is slightly more than 1, say 1.1, then $\frac{dy}{dx} = 3(1.1)^2 - 3$, which is positive.
Since the gradient changes from negative to positive, **the point (1, 3) is a minimum point.**
Considering the point (−1, 7):
If x is slightly less than −1, say, −1.1, then $\frac{dy}{dx} = 3(-1.1)^2 - 3$, which is positive.
If x is slightly more than −1, say, −0.9, then $\frac{dy}{dx} = 3(-0.9)^2 - 3$, which is negative.
Since the gradient changes from positive to negative, **the point (−1, 7) is a maximum point.**

(b) Since $\frac{dy}{dx} = 3x^2 - 3$, then $\frac{d^2y}{dx^2} = 6x$.

When $x = 1$, $\frac{d^2y}{dx^2}$ is positive, hence **(1, 3) is a minimum value.**

When $x = -1$, $\frac{d^2y}{dx^2}$ is negative, hence **(−1, 7) is a maximum value.**

Thus the maximum value is 7 and the minimum value is 3.
It can be seen that the second differential method of determining the nature of the turning points is, in this case, quicker than investigating the gradient.

Problem 13 Determine the co-ordinates of the maximum and minimum values of the graph $y = \frac{x^3}{3} - \frac{x^2}{2} - 6x + \frac{5}{3}$ and distinguish between them. Sketch the graph.

Following the procedure of para. 4:

(i) Since $y = \frac{x^3}{3} - \frac{x^2}{2} - 6x + \frac{5}{3}$ then $\frac{dy}{dx} = x^2 - x - 6$.

(ii) At a turning point, $\frac{dy}{dx} = 0$. Hence $x^2 - x - 6 = 0$,

i.e., $(x+2)(x-3) = 0$, from which $x = -2$ or $x = 3$.

(iii) When $x = -2$, $y = \frac{(-2)^3}{3} - \frac{(-2)^2}{2} - 6(-2) + \frac{5}{3} = 9$

When $x = 3$, $y = \frac{(3)^3}{3} - \frac{(3)^2}{2} = 6(3) + \frac{5}{3} = -11\frac{5}{6}$

Thus the co-ordinates of the turning points are $(-2, 9)$ and $(3, -11\frac{5}{6})$.

(iv) Since $\frac{dy}{dx} = x^2 - x - 6$ then $\frac{d^2y}{dx^2} = 2x - 1$.

When $x = -2$, $\frac{d^2y}{dx^2} = 2(-2) - 1 = -5$, which is negative.

Hence $(-2, 9)$ is a maximum point.

When $x = 3$, $\frac{d^2y}{dx^2} = 2(3) - 1 = 5$, which is positive.

Hence $(3, -11\frac{5}{6})$ is a minimum point.

Knowing $(-2, 9)$ is a maximum point (i.e. crest of a wave), and $(3, -11\frac{5}{6})$ is a minimum point (i.e., bottom of a valley) and that when $x = 0$, $y = \frac{5}{3}$, a sketch may be drawn as shown in *Fig 3*.

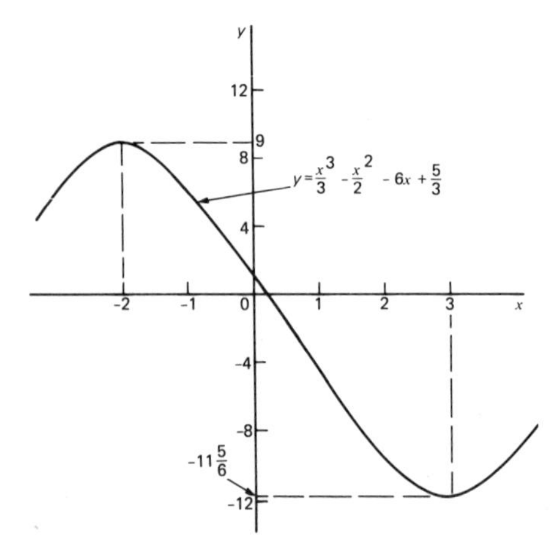

Fig 3

Problem 14 Locate the turning points on the following curves and determine whether they are maximum or minimum points: (a) $y = 4\theta + e^{-\theta}$, (b) $y = 3(\ln\theta - \theta)$.

(a) Since $y = 4\theta + e^{-\theta}$ then $\frac{dy}{d\theta} = 4 - e^{-\theta} = 0$ for a maximum or minimum value.

Hence $4 = e^{-\theta}$, $\frac{1}{4} = e^{\theta}$, giving $\theta = \ln\frac{1}{4} = -1.3863$ (see chapter 1).

When $\theta = 1.3863$, $y = 4(-1.3863) + e^{-(-1.3863)} = -5.5452 + 4.0000 = -1.5452$.

Thus $(-1.3863, -1.5452)$ are the co-ordinates of the turning point.

$\frac{d^2\theta}{d\theta^2} = e^{-\theta}$. When $\theta = -1.3863$, $\frac{d^2y}{d\theta^2} = e^{+1.3863} = 4.0$, which is positive.

Hence $(-1.3863, -1.5452)$ is a minimum point.

(b) Since $y = 3(\ln\theta - \theta) = 3\ln\theta - 3\theta$ then $\frac{dy}{d\theta} = \frac{3}{\theta} - 3 = 0$, for a maximum or minimum value.

Hence $\frac{3}{\theta} = 3$, $3 = 3\theta$ and $\theta = 1$.

When $\theta = 1$, $y = 3(\ln 1 - 1) = 3(0-1) = -3$.

Hence $(1, -3)$ are the co-ordinates of the turning point.

$\frac{d^2y}{d\theta^2} = -\frac{3}{\theta^2}$. When $\theta = 1$, $\frac{d^2y}{d\theta^2} = -3$, which is negative.

Hence $(1, -3)$ is a maximum value.

Problem 15 Determine the turning points on the curve $y = 4\sin x - 3\cos x$ in the range $x = 0$ to $x = 2\pi$ radians, and distinguish between them. Sketch the curve over one cycle.

Since $y = 4\sin x - 3\cos x$ then $\frac{dy}{dx} = 4\cos x + 3\sin x = 0$, for a turning point, from which, $4\cos x = -3\sin x$ and $\frac{-4}{3} = \frac{\sin x}{\cos x} = \tan x$.

Hence $x = \arctan\left(\frac{-4}{3}\right) = 126° \; 52'$ or $306° \; 52'$, since tangent is negative in the second and fourth quadrants.

When $x = 126° \; 52'$, $y = 4\sin 126° \; 52' - 3\cos 126° \; 52' = 5$.

When $x = 306° \; 52'$, $y = 4\sin 306° \; 52' - 3\cos 306° \; 52' = -5$.

$126° \; 52' = \left(126° \; 52' \times \frac{\pi}{180}\right)$ radians $= 2.214$ rad.

$306° \; 52' = \left(306° \; 52' \times \frac{\pi}{180}\right)$ radians $= 5.356$ rad.

Hence $(2.214, 5)$ and $(5.356, -5)$ are the co-ordinates of the turning points.

$\frac{d^2y}{dx^2} = 4\sin x + 3\cos x$.

When $x = 2.214$ rad, $\frac{d^2y}{dx^2} = -4\sin 2.214 + 3\cos 2.214$, which is negative.

Hence $(2.214, 5)$ is a maximum point

When $x = -5.356$ rad, $\frac{d^2y}{dx^2} = -4\sin 5.356 + 3\cos 5.356$, which is positive.

Hence $(5.356, -5)$ is a minimum point

A sketch of $y = 4\sin x - 3\cos x$ is shown in *Fig 4*.

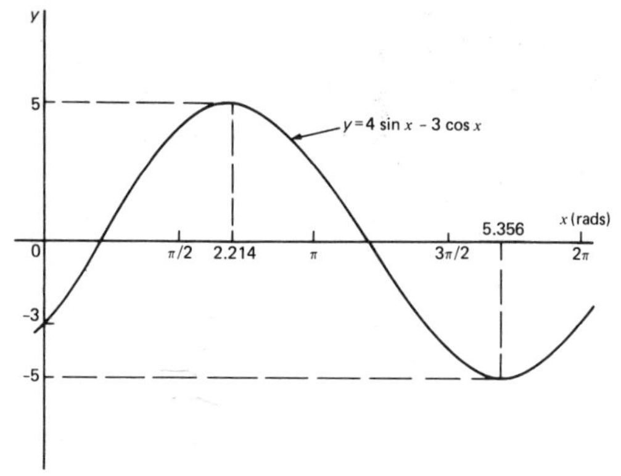

Fig 4

Problem 16 A rectangular area is formed having a perimeter of 40 cm. Determine the length and breadth of the rectangle if it is to enclose the maximum possible area.

Let the dimensions of the rectangle be x and y. Then the perimeter of the rectangle is $(2x+2y)$. Hence $2x+2y = 40$, or $x+y = 20$ (1)
Since the rectangle is to enclose the maximum possible area, a formula for area A must be obtained in terms of one variable only.
Area $A = xy$. From equation (1), $x = 20-y$.
Hence, area $A = (20-y)y = 20y-y^2$

$\frac{dA}{dy} = 20-2y = 0$ for a turning point, from which, $y = 10$ cm.

$\frac{d^2A}{dy^2} = -2$, which is negative, giving a maximum point.

When $y = 10$ cm, $x = 10$ cm, from equation (1).

Hence the length and breadth of the rectangle are each 10 cm, i.e., a square gives the maximum possible area. When the perimeter of a rectangle is 40 cm, the maximum possible area is $10 \times 10 =$ **100 cm²**.

Problem 17 A rectangular sheet of metal having dimensions 20 cm by 12 cm has squares removed from each of the four corners and the sides bent upwards to form an open box. Determine the maximum possible volume of the box.

The squares to be removed from each corner are shown shaded in *Fig 5*, having sides x cm. When the sides are bent upwards the dimensions of the box will be:
length $(20-2x)$ cm, breadth $(12-2x)$ cm and height, x cm.
Volume of box, $V = (20-2x)(12-2x)(x)$
$= 240x-64x^2+4x^3$

$\frac{dV}{dx} = 240-128x+12x^2 = 0$ for a turning point.

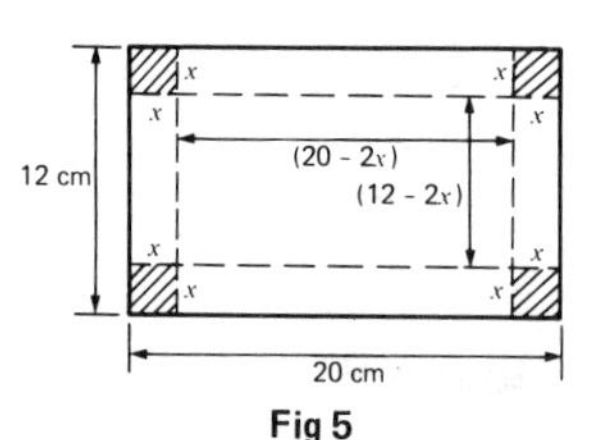

Fig 5

Hence $4(60-32x+3x^2) = 0$, i.e., $3x^2-32x+60 = 0$.

Using the quadratic formula, $x = \dfrac{32 \pm \sqrt{[(-32)^2-4(3)(60)]}}{(2)(3)} = 8.239$ cm or 2.427 cm. Since the breadth is $(12-2x)$ cm then $x = 8.239$ cm is not possible and is neglected. Hence $x = 2.427$ cm.

$\dfrac{d^2V}{dx^2} = -128+24x$. When $x = 2.427$, $\dfrac{d^2V}{dx^2}$ is negative, giving a maximum value.

The dimensions of the box are:

length $= 20-2(2.427) = 15.146$ cm,

breadth $= 12-2(2.427) = 7.146$ cm,

height $= 2.427$ cm.

Maximum volume $= (15.146)(7.146)(2.427) =$ **262.7 cm³**.

Problem 18 Determine the height and radius of a cylinder of volume 200 cm³ which has the least surface area.

Let the cylinder have radius r and perpendicular height h.

Volume of cylinder, $V = \pi r^2 h = 200$ (1)

Surface area of cylinder, $A = 2\pi rh+2\pi r^2$

Least surface area means minimum surface area and a formula for the surface area in terms of one variable only is required.

From equation (1), $h = \dfrac{200}{\pi r^2}$ (2)

Hence surface area, $A = 2\pi r\left(\dfrac{200}{\pi r^2}\right)+2\pi r^2 = \dfrac{400}{r}+2\pi r^2 = 400r^{-1}+2\pi r^2$

$\dfrac{dA}{dr} = \dfrac{-400}{r^2}+4\pi r = 0$, for a turning point.

Hence $4\pi r = \dfrac{400}{r^2}$ and $r^3 = \dfrac{400}{4\pi}$, from which, $r = \sqrt[3]{\left(\dfrac{100}{\pi}\right)} = 3.169$ cm.

$\dfrac{d^2A}{dr^2} = \dfrac{400}{r^3}+4\pi$. When $r = 3.169$ cm, $\dfrac{d^2A}{dr^2}$ is positive, giving a maximum value.

From equation (2), when $r = 3.169$ cm, $h = \dfrac{200}{\pi(3.169)^2} = 6.339$ cm.

Hence for the least surface area, a cylinder of volume 200 cm³ has a radius of 3.169 cm and height of 6.339 cm.

Problem 19 Determine the area of the largest piece of rectangular ground that can be enclosed by 100 m of fencing, if part of an existing straight wall is used as one side.

Let the dimensions of the rectangle be x and y as shown in *Fig 6*, where PQ represents the straight wall. From *Fig 6*,

$x+2y = 100$ (1)

Area of rectangle, $A = xy$ (2)

Since the maximum area is required, a formula for area A is needed in terms of one variable only. From equation (1), $x = 100-2y$.

Hence area $A = xy = (100-2y)y = 100y = 2y^2$.

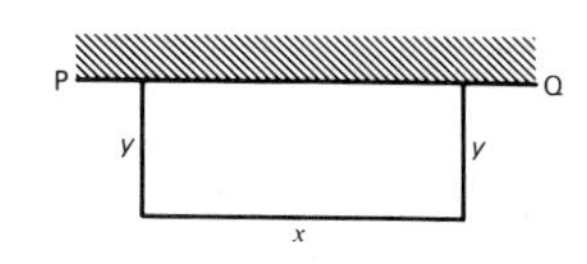

Fig 6

$\frac{dA}{dy} = 100-4y = 0$, for a turning point, from which, $y = 25$ m.

$\frac{d^2A}{dy^2} = -4$, which is negative, giving a maximum value.

When $y = 25$ m, $x = 50$ m from equation (1).

Hence the maximum possible area $= xy = (50)(25) =$ **1250 m²**.

Problem 20 Determine the height and radius of a cylinder of maximum volume which can be cut from a cone of height 30 cm and base radius 7.5 cm.

A cylinder of base radius r and height h is shown enclosed in a cone of height 30 cm and base radius 7.5 cm as shown in *Fig 7.*

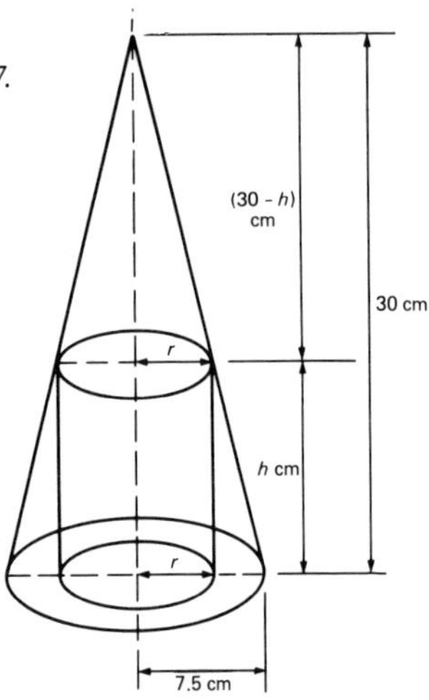

Fig 7

Volume of cylinder, $V = \pi r^2 h$ (1)

By similar triangles, $\frac{30-h}{r} = \frac{30}{7.5}$ (2)

Since the maximum volume is required a formula for the volume V must be obtained in terms of one variable only.

From equation (2), $\frac{30-h}{r} = 4$, from which,

$30-h = 4r$ and $h = 30-4r$ (3)

Substituting for h in equation (1) gives:

$V = \pi r^2 (30-4r) = 30\pi r^2 - 4\pi r^3$

$\frac{dV}{dr} = 60\pi r - 12\pi r^2 = 0$, for a maximum or minimum value.

Hence $12\pi r(5-r) = 0$, from which $r = 0$ or $r = 5$ cm.

$\frac{d^2V}{dr^2} = 60\pi - 24\pi r$.

When $r = 0$, $\frac{d^2V}{dr^2}$ is positive, giving a minimum value (obviously).

When $r = 5$, $\frac{d^2V}{dr^2}$ is negative, giving a maximum value.

From equation (3), height $h = 30-4r = 30-4(5) = 10$ cm.

Hence the cylinder of maximum volume which can be cut from a cone of height 30 cm and radius 7.5 cm, has a height of 10 cm and a radius of 5 cm.

Problem 21 An open rectangular box with square ends is fitted with an overlapping lid which covers the top and the front face. Determine the maximum volume of the box if 6 m² of metal are used in its construction.

A rectangular box having square ends of side x and length y is shown in *Fig 8.*

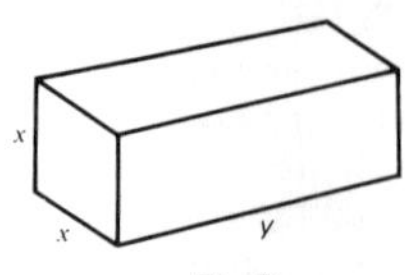

Fig 8

Surface area of box, A consists of two ends and five faces (since the lid also covers the front face).

Hence $A = 2x^2 + 5xy = 6$ (1)

Since it is the maximum volume required, a formula for the volumes in terms of one variable only is needed. Volume of box, $V = x^2 y$.

From equation (1), $y = \frac{6-2x^2}{5x} = \frac{6}{5x} - \frac{2x}{5}$ (2)

Hence volume $V = x^2 \left(\frac{6}{5x} - \frac{2x}{5}\right) = \frac{6x}{5} - \frac{2x^3}{5}$

$\frac{dV}{dx} = \frac{6}{5} - \frac{6x^2}{5} = 0$ for a maximum or minimum value.

Hence $6 = 6x^2$, giving $x = 1$ m ($x = -1$ is not possible, and is thus neglected).

$\frac{d^2 V}{dx^2} = \frac{-12x}{5}$. When $x = 1$, $\frac{d^2 V}{dx^2}$ is negative, giving a maximum value.

From equation (2), when $x = 1$, $y = \frac{6}{5(1)} - \frac{2(1)^3}{5} = \frac{4}{5}$

Hence the maximum volume of the box is given by $V = x^2 y = (1)^2 \left(\frac{4}{5}\right) = \mathbf{\frac{4}{5}\,m^3}$.

Problem 22 Find the diameter and height of a cylinder of maximum volume which can be cut from a sphere of radius 12 cm.

A cylinder of radius r and height h is shown enclosed in a sphere of radius $R = 12$ cm in *Fig 9*.

Volume of cylinder, $V = \pi r^2 h$ (1)

Using the right-angled triangle OPQ shown in *Fig 9*, $r^2 + \left(\frac{h}{2}\right)^2 = R^2$ by Pythagoras' theorem, i.e., $r^2 + \frac{h^2}{4} = 144$ (2)

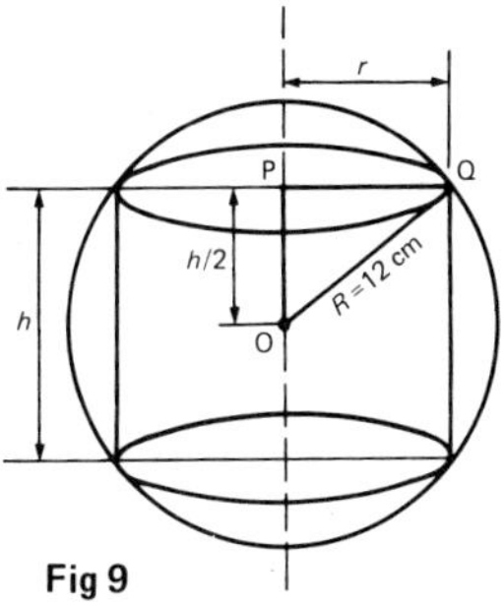

Fig 9

Since the maximum volume is required, a formula for the volume V is required in terms of one variable only. From equation (2), $r^2 = 144 - \frac{h^2}{4}$.

Substituting with equation (1) gives: $V = \pi\left(144 - \frac{h^2}{4}\right)h = 144\pi h - \frac{\pi h^3}{4}$

$\frac{dV}{dh} = 144\pi - \frac{3\pi h^2}{4} = 0$, for a maximum or minimum value.

Hence $144\pi = \frac{3\pi h^2}{4}$, from which, $h = \sqrt{\frac{(144)(4)}{3}} = 13.86$ cm.

$\frac{d^2 V}{dh^2} = \frac{-6\pi h}{4}$. When $h = 13.86$, $\frac{d^2 V}{dh^2}$ is negative, giving a maximum value.

From equation (2), $r^2 = 144 - \frac{h^2}{4} = 144 - \frac{(13.86)^2}{4}$, from which, radius $r = 9.80$ cm.

Diameter of cylinder $= 2r = 2\,(9.80) = 19.60$ cm.

Hence the cylinder having the maximum volume that can be cut from a sphere of radius 12 cm is one in which the diameter is 19.60 cm and the height is 13.86 cm.

C FURTHER PROBLEMS ON APPLICATIONS OF DIFFERENTIATION

1 The distance x metres moved by a body in a time t seconds is given by $x = 3t^3 - 4t^2 + 6t - 5$. Determine the velocity and acceleration when (a) $t = 0$ and (b) $t = 2.5$ s.

[(a) $v = 6$ m/s, $a = -8$ m/s^2; (b) $v = 42\frac{1}{4}$ m/s, $a = 37$ m/s^2]

2 The distance s metres travelled by a car in t seconds after the brakes are applied is given by $s = 25t - 2.5t^2$. Find (a) the speed of the car (in km/h) when the brakes are applied; (b) the distance the car travels before it stops.

[(a) 90 km/h; (b) 62.5 m]

3 The equation $\theta = 10\pi + 24t - 3t^2$ gives the angle θ, in radians, through which a wheel turns in t seconds. Determine (a) the time the wheel takes to come to rest; (b) the angle turned through in the last second of movement.

[(a) 4 s; (b) 3 rads]

4 At any time t seconds the distance x metres of a particle moving in a straight line from a fixed point is given by: $x = 4t + \ln(1-t)$. Determine (a) the initial velocity and acceleration; (b) the velocity and acceleration after 1.5 s; (c) the time when the velocity is zero.

[(a) 3 m/s; -1 m/s^2; (b) 6 m/s; -4 m/s^2; (c) $\frac{3}{4}$ s]

5 A missile fired from ground level rises x metres vertically upwards in t seconds and $x = 100t - \frac{25}{2}t^2$. Find (a) the initial velocity of the missile; (b) the time when the height of the missle is a maximum, (c) the maximum height reached; (d) the velocity with which the missile strikes the ground.

[(a) 100 m/s; (b) 4 s; (c) 200 m; (d) -100 m/s]

6 The distance, in metres, of a body moving in a straight line from a fixed point on the line is given by $x = t^4 - \frac{8t^3}{3} + \frac{3t^2}{2} - 6$, where t is the time in seconds. Determine the velocity and acceleration after 2 seconds. Find also the time when the body is momentarily at rest.

[6 m/s, 19 m/s^2; $t = 0$, $\frac{1}{2}$ or $1\frac{1}{2}$ s]

7 The angular displacement θ of a rotating disc is given by $\theta = 6 \sin \frac{t}{4}$, where t is the time in seconds. Determine (a) the angular velocity of the disc when t is 1.5 s, (b) the angular acceleration when t is 5.5 s, and (c) the first time when the angular velocity is zero.

[(a) $\omega = 1.40$ rad/s; (b) $\alpha = -0.37$ rad/s^2; (c) $t = 6.28$ s]

8 $x = \frac{20t^3}{3} - \frac{23t^2}{2} + 6t + 5$ represents the distance, x metres, moved by a body in t seconds. Determine (a) the velocity and acceleration at the start, (b) the velocity and acceleration when $t = 3$s, (c) the values of t when the body is at rest, (d) the value of t when the acceleration is 37 m/s^2 and (e) the distance travelled in the third second.

[(a) 6m/s; -23 m/s^2; (b) 117 m/s; 97 m/s^2; (c) $\frac{3}{4}$ s or $\frac{2}{5}$ s; (d) $1\frac{1}{2}$ s; (e) $75\frac{1}{6}$ m.]

9 The displacement s cm of the end of a stiff spring at time t seconds is given by $s = ae^{-kt} \sin 2\pi ft$. Find the velocity of the end of the spring after 2 s, if $a = 3$, $k = 0.8$ and $f = 10$. [38.06 cm/s]

10 The length, l metres, of a rod of metal at temperature θ°C is given by $l = 1 + 2 \times 10^{-4}\theta + 4 \times 10^{-6}\theta^2$. Determine the rate of change of l, in cm/°C, when the temperature is (a) 100°C; (b) 400°C.

[(a) 0.1 cm/°C; (b) 0.34 cm/°C]

11 An alternating current, i amperes, is given by $i = 100 \sin 2\pi f t$, where f is the frequency in hertz and t the time in seconds. Determine the rate of change of current when $t = 12$ ms, given that $f = 50$ Hz. [−25 420 A/s]

12 The luminous intensity, I candelas, of a lamp is given by $I = 8 \times 10^{-4} V^2$, where V is the voltage. Find (a) the rate of change of luminous intensity with voltage when $V = 100$ volts, and (b) the voltage at which the light is increasing at a rate of 0.5 candelas per volt. [(a) 0.16 cd/V; (b) 312.5 V]

13 The voltage across the plates of a capacitor at any time t seconds is given by $v = Ve^{-t/CR}$, where V, C and R are constants. Given $V = 200$ volts, $C = 0.1 \times 10^{-6}$ farads and $R = 2 \times 10^6$ ohms find (a) the initial rate of change of voltage, and (b) the rate of change of voltage after 0.2 s.
[(a) −1000 V/s; (b) 367.9 V/s]

14 Newton's law of cooling is given by $\theta = \theta_0 e^{-kt}$, where the excess of temperature at zero time is θ_0°C and at time t seconds is θ°C. Given $\theta_0 = 15$°C and $k = -0.02$, find the time when the rate of change of temperature is 1°C/s.
[60.2 s]

15 (a) Define a turning point of a graph.
(b) Find the co-ordinates of the turning points on the curve $y = 4x^3+3x^2-60x-12$. Distinguish between the turning points by examining the sign of the gradient on either side of the points.
[Minimum (2, −88); Maximum $(-2\frac{1}{2}, 94\frac{1}{4})$]

In *Problems 16 to 21*, find the turning points and distinguish between them.

16 $y = 3x^2-4x+2$ [Minimum at $(\frac{2}{3}, \frac{2}{3})$]

17 $x = \theta(6-\theta)$ [Maximum at (3, 9)]

18 $y = 5x-2 \ln x$ [Minimum at (0.4000, 3.8326)]

19 $y = 2x-e^x$ [Maximum at (0.6931, −0.6137)]

20 $y = t^3-\frac{t^2}{2}-2t+4$ [Minimum at $(1, 2\frac{1}{2})$; Maximum at $(-\frac{2}{3}, 4\frac{22}{27})$]

21 $x = 8t+\frac{1}{2t^2}$ [Minimum at $(\frac{1}{2}, 6)$]

22 Determine the maximum and minimum values on the graph $y = 12 \cos\theta-5 \sin\theta$ in the range $x = 0$ to $x = 360°$. Sketch the graph over one cycle showing relevant points [Maximum of 13 at 337° 23′; Minimum of −13 at 157° 23′]

23 Show that the curve $y = \frac{2}{3}(t-1)^3+2t(t-2)$ has a maximum value of $\frac{2}{3}$ and a minimum value of −2. What is a point of inflexion?

24 The speed, v, of a car (in m/s) is related to time t s by the equation $v = 3+12t-3t^2$. Determine the maximum speed of the car in km/h.
[54 km/h]

25 Determine the maximum area of a rectangular piece of land that can be enclosed by 1200 m of fencing. [90 000 m^2]

26 A shell is fired vertically upwards and its vertical height, x metres, is given by $x = 24t-3t^2$, where t is the time in seconds. Determine the maximum height reached. [48 m]

27 A lidless box with square ends is to be made from a thin sheet of metal. Determine the least area of the metal for which the volume of the box is 3.5 m^3. [11.42 m^2]

28 A closed cylindrical container has a surface area of 400 cm^2. Determine the dimensions for maximum volume. [radius = 4.607 cm; height = 9.212 cm]

29 Calculate the height of a cylinder of maximum volume which can be cut from a cone of height 20 cm and base radius 80 cm. [6.67 cm]

30 The power developed in a resistor R by a battery of emf E and internal resistance r is given by $P = \dfrac{E^2 R}{(R+r)^2}$. Differentiate P with respect to R and show that the power is a maximum when $R = r$.

31 Find the height and radius of a closed cylinder of volume 125 cm^3 which has the least surface area. [height = 5.42 cm; radius = 2.71 cm]

32 Resistance to motion, F, of a moving vehicle is given by $F = \dfrac{5}{x} + 100x$. Determine the minimum value of resistance. [44.72]

33 A right cylinder of maximum volume is to be cut from a sphere of radius 15 cm. Determine the base diameter and height of the cylinder. [diameter = 24.49 cm; height = 17.32 cm]

34 An electrical voltage E is given by $E = (15 \sin 50\pi t + 40 \cos 50\pi t)$ volts, where t is the time in seconds. Determine the maximum value of voltage. [42.72 volts]

35 A rectangular box with a lid which covers the top and front has a volume of 120 cm^3 and the length of the box is to be one and a half times the height. Determine the dimensions of the box so that the surface area shall be a minimum. [6.376 cm by 5.313 cm by 3.542 cm]

17 Calculus (4) – Methods of integration

A MAIN POINTS CONCERNED WITH METHODS OF INTEGRATION

1 The process of integration reverses the process of differentiation. In differentiation, if $f(x) = 2x^2$ then $f'(x) = 4x$. Thus the integral of $4x$ is $2x^2$, i.e., integration is the process of moving from $f'(x)$ to $f(x)$. By similar reasoning, the integral of $2t$ is t^2.

2 Integration is also a process of summation or adding parts together and an elongated S, shown as $\int$, is used to replace to words 'the integral of'. Hence, from para. 1, $\int 4x = 2x^2$ and $\int 2t = t^2$.

3 In differentiation, the differential coefficient $\frac{dy}{dx}$ indicates that a function of x is being differentiated with respect to x, the dx indicating that it is 'with respect to x'. In integration the variable of integration is shown by adding d(the variable) after th function to be integrated. Thus $\int 4x\, dx$ means 'the integral of $4x$ with respect to x', and $\int 2t\, dt$ means 'the integral of $2t$ with respect to t'.

4 As stated in para. 1, the differential coefficient of $2x^2$ is $4x$, hence $\int 4x\, dx = 2x^2$. However, the differential coefficient of $(2x^2 + 7)$ is also $4x$. Hence $\int 4x\, dx$ is also equal to $(2x^2 + 7)$. To allow for the possible presence of a constant, whenever the process of integration is performed, a constant 'c' is added to the result. Thus $\int 4x\, dx = 2x^2 + c$ and $\int 2t\, dt = t^2 + c$. 'c' is called the **arbitrary constant of integration.**

5 The general solution of integrals of the form $\int ax^n\, dx$, where a and n are constants is given by:

$$\int ax^n\, dx = \frac{ax^{n+1}}{n+1} + c$$

This rule is true when n is fractional, zero, or a positive or negative integer, with the exception of $n = -1$.
Using this rule gives:

(i) $\int 3x^4\, dx = \frac{3x^{4+1}}{4+1} + c = \frac{3}{5}x^5 + c$,

(ii) $\int \frac{2}{x^2}\, dx = \int 2x^{-2}\, dx = \frac{2x^{-2+1}}{-2+1} + c = \frac{2x^{-1}}{-1} + c = \frac{-2}{x} + c$, and

(iii) $\int \sqrt{x}\,dx = \int x^{1/2}dx = \dfrac{x^{1/2+1}}{\frac{1}{2}+1} + c = \dfrac{x^{3/2}}{\frac{3}{2}} + c = \frac{2}{3}\sqrt{x^3} + c$

Each of these three results may be checked by differentiation.

6 (i) The integral of a constant k is $kx+c$. For example, $\int 8dx = 8x+c$.

(ii) When a sum of several terms is integrated the result is the sum of the integrals of the separate terms. For example,

$$\int(3x+2x^2-5)dx = \int 3x\,dx + \int 2x^2\,dx - \int 5dx = \frac{3x^2}{2} + \frac{2x^3}{3} - 5x + c$$

7 Since integration is the reverse process of differentiation the **standard integrals** listed in table 1 may be deduced and readily checked by differentiation. (See *Problems 1 to 14.*)

TABLE 1 Standard integrals

(i)	$\int ax^n\,dx$	=	$\dfrac{ax^{n+1}}{n+1} + c$ (except when $n = -1$)
(ii)	$\int \cos ax\,dx$	=	$\dfrac{1}{a}\sin ax + c$
(iii)	$\int \sin ax\,dx$	=	$-\dfrac{1}{a}\cos ax + c$
(iv)	$\int \sec^2 ax\,dx$	=	$\dfrac{1}{a}\tan ax + c$
(v)	$\int \text{cosec}^2\, ax\,dx$	=	$-\dfrac{1}{a}\cot ax + c$
(vi)	$\int \text{cosec}\, ax \cot ax\,dx$	=	$-\dfrac{1}{a}\text{cosec}\, ax + c$
(vii)	$\int \sec ax \tan ax\,dx$	=	$\dfrac{1}{a}\sec ax + c$
(viii)	$\int e^{ax}\,dx$	=	$\dfrac{1}{a}e^{ax} + c$
(ix)	$\int \dfrac{a}{x}\,dx$	=	$\ln ax + c$

8 Integrals containing an arbitrary constant c in their results are called **indefinite integrals** since their precise value cannot be determined without further information. **Definite integrals** are those in which limits are applied. If an expression is written as $[x]_a^b$, 'b' is called the upper limit and 'a' the lower limit. The operation of applying the limits is defined as $[x]_a^b = (b-a)$. The

increase in the value of the integral x^2 as x increases from 1 to 3 is written as $\int_1^3 x^2\,dx$. Applying the limits gives:

$$\int_1^3 x^2 dx = \left[\frac{x^3}{3}+c\right]_1^3 = \left(\frac{(3)^3}{3}+c\right) - \left(\frac{(1)^3}{3}+c\right) = (9+c)-(\tfrac{1}{3}+c) = 8\tfrac{2}{3}.$$

Note that the 'c' term always cancels out when limits are applied and the arbitrary constant need not be shown with definite integrals. (See *Problems 15 to 20.*)

9 Functions which require integrating are not always in the standard integral form shown in *Table 1.* In such cases, it may be possible to change the function into a form which can be readily integrated by using an **algebraic substitution.** The substitution usually made is to let u be equal to $f(x)$ such that $f(u)du$ is a standard integral. (See *Problems 21 to 32.*)

B WORKED PROBLEMS ON METHODS OF INTEGRATION

Problem 1 Determine (a) $\int 5x^2\,dx$; (b) $\int 2t^3\,dt$

The standard integral of $\int ax^n\,dx = \dfrac{ax^{n+1}}{n+1}+c.$

(a) When $a = 5$ and $n = 2$ then $\int 5x^2\,dx = \dfrac{5x^{2+1}}{2+1}+c = \mathbf{\dfrac{5x^3}{3}+c.}$

(b) When $a = 2$ and $n = 3$ then $\int 2t^3\,dt = \dfrac{2t^{3+1}}{3+1}+c = \dfrac{2t^4}{4}+c = \mathbf{\dfrac{1}{2}t^4+c.}$

Each of these results may be checked by differentiating them.

Problem 2 Find (a) $\int 3dx$; (b) $\int \frac{2}{5}x\,dx$

(a) $\int 3dx$ is the same as $\int 3x^0\,dx$, and using the standard integral $\int ax^n\,dx$ when $a = 3$ and $n = 0$ gives $\int 3x^0 dx = \dfrac{3x^{0+1}}{0+1}+c = \mathbf{3x+c.}$

In general, if k is a constant then $\int k\,dx = kx+c.$

(b) When $a = \frac{2}{5}$ and $n = 1$ then $\int \frac{2}{5}x\,dx = \left(\frac{2}{5}\right)\dfrac{x^{1+1}}{1+1}+c = \left(\frac{2}{5}\right)\dfrac{x^2}{2}+c = \mathbf{\dfrac{1}{5}x^2+c.}$

Problem 3 Determine $\int(4+\frac{3}{7}x-6x^2)dx.$

$\int(4+\frac{3}{7}x-6x^2)dx$ may be written as $\int 4dx+\int\frac{3}{7}x\,dx-\int 6x^2\,dx$, i.e. each term is integrated separately. (This splitting up of terms only applies, however, for addition and subtraction.)

Hence $\int(4+\frac{3}{7}x-6x^2)dx = 4x+\left(\frac{3}{7}\right)\dfrac{x^{1+1}}{1+1}-(6)\dfrac{x^{2+1}}{2+1}+c$

$$= 4x+\left(\frac{3}{7}\right)\frac{x^2}{2}-(6)\frac{x^3}{3}+c = \mathbf{4x+\frac{3}{14}x^2-2x^3+c.}$$

Note that when an integral contains more than one term there is no need to have an arbitrary constant for each; just a single constant at the end is sufficient.

Problem 4 Determine (a) $\int\left(\frac{2x^3-3x}{4x}\right)dx$; (b) $\int(1-t)^2\,dt$.

(a) Rearranging into standard integral form gives:

$$\int\left(\frac{2x^3-3x}{4x}\right)dx = \int\left(\frac{2x^3}{4x}-\frac{3x}{4x}\right)dx = \int\left(\frac{x^2}{2}-\frac{3}{4}\right)dx = \left(\frac{1}{2}\right)\frac{x^{2+1}}{2+1}-\frac{3}{4}x+c$$

$$= \frac{1}{2}\left(\frac{x^3}{3}\right)-\frac{3}{4}x+c = \frac{1}{6}x^3-\frac{3}{4}x+c$$

(b) Rearranging $\int(1-t)^2\,dt$ gives: $\int(1-2t+t^2)dt = t-\frac{2t^{1+1}}{1+1}+\frac{t^{2+1}}{2+1}+c$

$$= t-\frac{2t^2}{2}+\frac{t^3}{3}+c = t-t^2+\frac{t^3}{3}+c.$$

This problem shows that functions often have to be rearranged into the standard form of $\int ax^n\,dx$ before it is possible to integrate them.

Problem 5 Determine (a) $\int\frac{3}{x^2}dx$; (b) $-\int\frac{2}{5x^3}dx$

(a) $\int\frac{3}{x^2}dx = \int 3x^{-2}\,dx$. Using the standard integral $\int ax^n\,dx$ when $a = 3$ and $n = -2$ gives: $\int 3x^{-2}dx = \frac{3x^{-2+1}}{-2+1}+c = \frac{3x^{-1}}{-1}+c = -3x^{-1}+c = \frac{-3}{x}+c$

(b) $-\int\frac{2}{5x^3}dx = \int -\frac{2}{5}x^{-3}\,dx$. Using the standard integral $\int ax^n\,dx$ when $a = -\frac{2}{5}$ and $n = -3$ gives:

$$\int -\frac{2}{5}x^{-3}dx = \left(-\frac{2}{5}\right)\frac{x^{-3+1}}{-3+1}+c = \left(-\frac{2}{5}\right)\frac{x^{-2}}{-2}+c = \frac{1}{5}x^{-2}+c = \frac{1}{5x^2}+c.$$

Problem 6 Determine (a) $\int 3\sqrt{x}\,dx$; (b) $\int\frac{2}{7}\sqrt[3]{x^5}\,dx$.

For fractional powers it is necessary to appreciate $\sqrt[n]{a^m} = a^{\frac{m}{n}}$.

(a) $$\int 3\sqrt{x}\,dx = \int 3x^{\frac{1}{2}}\,dx = \frac{(3)x^{\frac{1}{2}+1}}{\frac{1}{2}+1}+c = \frac{(3)x^{\frac{3}{2}}}{\frac{3}{2}}+c = 2x^{\frac{3}{2}}+c = 2\sqrt{x^3}+c$$

(b) $$\int\left(\frac{2}{7}\right)\sqrt[3]{x^5}\,dx = \int\left(\frac{2}{7}\right)x^{\frac{5}{3}}\,dx = \left(\frac{2}{7}\right)\frac{x^{\frac{5}{3}+1}}{\frac{5}{3}+1}+c = \left(\frac{2}{7}\right)\frac{x^{\frac{8}{3}}}{\frac{8}{3}}+c$$

$$= \left(\frac{2}{7}\right)\left(\frac{3}{8}\right)x^{\frac{8}{3}}+c = \frac{3}{28}\sqrt[3]{x^8}+c.$$

Problem 7 Determine (a) $\int\frac{2}{\sqrt{x}}dx$; (b) $\int\frac{-5}{9\sqrt[4]{t^3}}dt$.

(a) $\int \frac{2}{\sqrt{x}}\,dx = \int \frac{2}{x^{\frac{1}{2}}}\,dx = \int 2x^{-\frac{1}{2}}dx = (2)\frac{x^{-\frac{1}{2}+1}}{-\frac{1}{2}+1}+c = \frac{2x^{\frac{1}{2}}}{\frac{1}{2}}+c = \mathbf{4\sqrt{x}+c}$

(b) $\int \frac{-5}{9\sqrt[4]{t^3}}\,dt = \int \frac{-5}{9t^{\frac{3}{4}}}\,dt = \int \frac{-5}{9}\,t^{-\frac{3}{4}}\,dt = \left(\frac{-5}{9}\right)\frac{t^{-\frac{3}{4}+1}}{\frac{-3}{4}+1}+c$

$= \left(\frac{-5}{9}\right)\frac{t^{\frac{1}{4}}}{\frac{1}{4}}+c = \left(\frac{-5}{9}\right)\left(\frac{4}{1}\right)t^{\frac{1}{4}}+c = \mathbf{\frac{-20}{9}\sqrt[4]{t}+c}$

Problem 8 Find (a) $\int 2\theta^{2.6}\,d\theta$; (b) $\int \frac{3}{x^{1.2}}\,dx$

(a) $\int 2\,\theta^{2.6}\,d\theta = \frac{(2)\theta^{2.6+1}}{2.6+1}+c = \frac{2\theta^{3.6}}{3.6}+c = \mathbf{\frac{1}{1.8}\,\theta^{3.6}+c}$

(b) $\int \frac{3}{x^{1.2}}\,dx = \int 3x^{-1.2}dx = \frac{(3)x^{-1.2+1}}{-1.2+1}+c = \frac{3x^{-0.2}}{-0.2}+c$

$= -15x^{-0.2}+c = \mathbf{\frac{-15}{x^{0.2}}+c}\left(= \frac{-15}{x^{\frac{1}{5}}}+c = \mathbf{\frac{-15}{\sqrt[5]{x}}+c}\right)$

Problem 9 Determine $\int \frac{(1+\theta)^2}{\sqrt{\theta}}\,d\theta$.

$\int \frac{(1+\theta)^2}{\sqrt{\theta}}\,d\theta = \int \frac{(1+2\theta+\theta^2)}{\sqrt{\theta}}\,d\theta = \int\left(\frac{1}{\theta^{\frac{1}{2}}}+\frac{2\theta}{\theta^{\frac{1}{2}}}+\frac{\theta^2}{\theta^{\frac{1}{2}}}\,d\theta\right)$

$= \int(\theta^{-\frac{1}{2}}+2\theta^{1-\frac{1}{2}}+\theta^{2-\frac{1}{2}})d\theta = \int(\theta^{-\frac{1}{2}}+2\theta^{\frac{1}{2}}+\theta^{\frac{3}{2}})d\theta$

$= \frac{\theta^{-\frac{1}{2}+1}}{-\frac{1}{2}+1}+\frac{(2)\theta^{\frac{1}{2}+1}}{\frac{1}{2}+1}+\frac{\theta^{\frac{3}{2}+1}}{\frac{3}{2}+1}+c = \frac{\theta^{\frac{1}{2}}}{\frac{1}{2}}+\frac{2\theta^{\frac{3}{2}}}{\frac{3}{2}}+\frac{\theta^{\frac{5}{2}}}{\frac{5}{2}}+c$

$= 2\theta^{\frac{1}{2}}+\frac{4}{3}\theta^{\frac{3}{2}}+\frac{2}{5}\theta^{\frac{5}{2}}+c = \mathbf{2\sqrt{\theta}+\frac{4}{3}\sqrt{\theta^3}+\frac{2}{5}\sqrt{\theta^5}+c}$

Problem 10 Determine (a) $\int 4\cos 3x\,dx$; (b) $\int 5\sin 2\theta\,d\theta$

(a) From table 1(ii), $\int 4\cos 3x\,dx = (4)\left(\frac{1}{3}\right)\sin 3x+c = \mathbf{\frac{4}{3}\sin 3x+c}$

(b) From table 1(iii), $\int 5\sin 2\theta\,d\theta = (5)\left(-\frac{1}{2}\right)\cos 2\theta+c = \mathbf{-\frac{5}{2}\cos 2\theta+c}$

Problem 11 Determine (a) $\int 7\sec^2 4t\,dt$; (b) $3\int \mathrm{cosec}^2\,2\theta\,d\theta$

(a) From *Table 1(iv),* $\int 7\sec^2 4t\,dt = (7)\left(\frac{1}{4}\right)\tan 4t+c = \mathbf{\frac{7}{4}\tan 4t+c}$

(b) From *Table 1(v)*, $3\int \text{cosec}^2\, 2\theta\, d\theta = (3)\left(-\frac{1}{2}\right)\cot 2\theta + c = -\frac{3}{2}\cot 2\theta + c$

Problem 12 Determine (a) $\int 4\,\text{cosec}\, 3x \cot 3x\, dx$; (b) $\int 7 \sec 5t \tan 5t\, dt$

(a) From *Table 1(vi)*, $\int 4\,\text{cosec}\, 3x \cot 3x\, dx = (4)\left(-\frac{1}{3}\right)\text{cosec}\, 3x + c$

$$= -\frac{4}{3}\,\textbf{cosec}\, \mathbf{3x + c}$$

(b) From *Table 1(vii)*, $\int 7 \sec 5t \tan 5t\, dt = (7)\left(\frac{1}{5}\right)\sec 5t + c = \frac{7}{5}\sec 5t + c.$

Problem 13 Determine (a) $\int 5e^{3x}\, dx$; (b) $\int \frac{2}{3e^{4t}}\, dt$

(a) From *Table 1(viii)*, $\int 5e^{3x}\, dx = (5)\left(\frac{1}{3}\right)e^{3x} + c = \frac{5}{3}e^{3x} + c.$

(b) $\int \frac{2}{3e^{4t}}\, dt = \int \frac{2}{3}e^{-4t}\, dt = \left(\frac{2}{3}\right)\left(-\frac{1}{4}\right)e^{-4t} + c = -\frac{1}{6}e^{-4t} + c = -\frac{1}{6e^{4t}} + c$

Problem 14 Determine (a) $\int \frac{3}{5x}\, dx$; (b) $\int \left(\frac{2m^2+1}{m}\right) dm$

(a) $\int \frac{3}{5x}\, dx = \int \left(\frac{3}{5}\right)\left(\frac{1}{x}\right) dx = \frac{3}{5}\ln x + c$ (from *Table 1(ix)*).

(b) $\int \left(\frac{2m^2+1}{m}\right) dm = \int \left(\frac{2m^2}{m} + \frac{1}{m}\right) dm = \int (2m + \frac{1}{m})\, dm = \frac{2m^2}{2} + \ln m + c$

$$= m^2 + \ln m + c$$

Problem 15 Evaluate (a) $\int_1^2 3x\, dx$; (b) $\int_{-2}^{3} (4 - x^2)\, dx.$

(a) $\int_1^2 3x\, dx = \left[\frac{3x^2}{2}\right]_1^2 = \left\{\frac{3}{2}(2)^2\right\} - \left\{\frac{3}{2}(1)^2\right\} = 6 - 1\frac{1}{2} = 4\frac{1}{2}$

(b) $\int_{-2}^{3} (4 - x^2)\, dx = \left[4x - \frac{x^3}{3}\right]_{-2}^{3} = \left\{4(3) - \frac{(3)^3}{3}\right\} - \left\{4(-2) - \frac{(-2)^3}{3}\right\}$

$$= \{12 - 9\} - \left\{-8 - (-\frac{8}{3})\right\} = (3) - (-5\frac{1}{3}) = 8\frac{1}{3}$$

Problem 16 Evaluate $\int_1^4 \left(\frac{\theta+2}{\sqrt{\theta}}\right) d\theta$, taking positive square roots only.

$$\int_1^4 \left(\frac{\theta+2}{\sqrt{\theta}}\right) d\theta = \int_1^4 \left(\frac{\theta}{\theta^{\frac{1}{2}}} + \frac{2}{\theta^{\frac{1}{2}}}\right) d\theta = \int_1^4 (\theta^{\frac{1}{2}} + 2\theta^{-\frac{1}{2}})\, d\theta$$

$$= \left[\frac{\theta^{\frac{1}{2}+1}}{\frac{1}{2}+1} + \frac{2\theta^{-\frac{1}{2}+1}}{-\frac{1}{2}+1}\right]_1^4 = \left[\frac{\theta^{\frac{3}{2}}}{\frac{3}{2}} + \frac{2\theta^{\frac{1}{2}}}{\frac{1}{2}}\right]_1^4 = \left[\frac{2}{3}\sqrt{\theta^3} + 4\sqrt{\theta}\right]_1^4$$

$$= \left\{\frac{2}{3}\sqrt{(4)^3}+4\sqrt{4}\right\} - \left\{\frac{2}{3}\sqrt{(1)^3}+4\sqrt{(1)}\right\} = \left\{\frac{16}{3}+8\right\} - \left\{\frac{2}{3}+4\right\}$$

$$= 5\frac{1}{3}+8-\frac{2}{3}-4 = \mathbf{8\frac{2}{3}}$$

Problem 17 Evaluate $\int_0^{\frac{\pi}{2}} 3 \sin 2x\, dx$

$$\int_0^{\frac{\pi}{2}} 3 \sin 2x\, dx = \left[(3)\left(-\frac{1}{2}\right) \cos 2x\right]_0^{\frac{\pi}{2}} = \left[-\frac{3}{2} \cos 2x\right]_0^{\frac{\pi}{2}}$$

$$= \left\{-\frac{3}{2} \cos 2\left(\frac{\pi}{2}\right)\right\} - \left\{-\frac{3}{2} \cos 2(0)\right\} = \left\{-\frac{3}{2} \cos \pi\right\} - \left\{-\frac{3}{2} \cos 0\right\}$$

$$= \left\{-\frac{3}{2}(-1)\right\} - \left\{-\frac{3}{2}(1)\right\} = \frac{3}{2}+\frac{3}{2} = \mathbf{3}$$

Problem 18 Evaluate $\int_1^2 4 \cos 3t\, dt.$

$$\int_1^2 4 \cos 3t\, dt = \left[(4)\left(\frac{1}{3}\right) \sin 3t\right]_1^2 = \left[\frac{4}{3} \sin 3t\right]_1^2 = \left\{\frac{4}{3} \sin 6\right\} - \left\{\frac{4}{3} \sin 3\right\}$$

Limits of trigonometric functions are expressed in radians. Thus sin 6 means the sine of 6 radians. 6 rads $= (6 \times \frac{180}{\pi})° = 343.77°$

Similarly, 3 rads $= (3 \times \frac{180}{\pi})° = 171.89°$

Hence $\int_1^2 4 \cos 3t\, dt = \left\{\frac{4}{3} \sin 343.77°\right\} - \left\{\frac{4}{3} \sin 171.89°\right\}$

$$= (-0.3727)-(0.1881) = \mathbf{-0.5608}$$

Problem 19 Show that $\int_{\frac{\pi}{6}}^{\frac{\pi}{3}} (2 \sec^2 \theta - 3 \cos 2\theta)\, d\theta = \frac{4}{\sqrt{3}}$

$$\int_{\frac{\pi}{6}}^{\frac{\pi}{3}} (2 \sec^2 \theta - 3 \cos 2\theta)\, d\theta = \left[2 \tan \theta - \frac{3}{2} \sin 2\theta\right]_{\frac{\pi}{6}}^{\frac{\pi}{3}}$$

$$= \left\{2 \tan \frac{\pi}{3} - \frac{3}{2} \sin 2\left(\frac{\pi}{3}\right)\right\} - \left\{2 \tan \frac{\pi}{6} - \frac{3}{2} \sin 2\left(\frac{\pi}{6}\right)\right\}$$

$$= \left\{2(\sqrt{3}) - \frac{3}{2}\left(\frac{\sqrt{3}}{2}\right)\right\} - \left\{2\left(\frac{1}{\sqrt{3}}\right) - \frac{3}{2}\left(\frac{\sqrt{3}}{2}\right)\right\}$$

$$= 2\sqrt{3} - \frac{3}{4}\sqrt{3} - 2\left(\frac{\sqrt{3}}{3}\right) + \frac{3}{4}\sqrt{3} \quad \left(\text{since } \frac{1}{\sqrt{3}} \equiv \frac{\sqrt{3}}{3}\right)$$

$$= \frac{4}{3}\sqrt{3} = \frac{4}{\sqrt{3}}$$

Problem 20 Evaluate (a) $\int_1^2 4e^{2x}\,dx$; (b) $\int_1^4 \frac{3}{4u}\,du$, each correct to 4 significant figures.

(a) $\int_1^2 4e^{2x}\,dx = \left[\frac{4}{2}e^{2x}\right]_1^2 = 2[e^{2x}]_1^2 = 2[e^4 - e^2]$

$= 2[54.5982 - 7.3891]$
$= 94.42$

(b) $\int_1^4 \frac{3}{4u}\,du = \left[\frac{3}{4}\ln u\right]_1^4 = \frac{3}{4}[\ln 4 - \ln 1] = \frac{3}{4}[1.3863 - 0]$
$= \mathbf{1.040}$

Problem 21. Determine $\int \sin(8x-3)\,dx$.

$\sin(8x-3)$ is not a 'standard integral' of the form shown in *Table 1.* Thus an algebraic substitution is made.

Let $u = (8x-3)$, then $\frac{du}{dx} = 8$, and rearranging gives $du = 8dx$ or $dx = \frac{du}{8}$.

Hence $\int \sin(8x-3)dx = \int (\sin u)\,\frac{du}{8} = \frac{1}{8}\int \sin u\,du$ (which is a 'standard integral')

$= \frac{1}{8}(-\cos u) + c$

Rewriting u as $(8x-3)$ gives: $\int \sin(8x-3)\,dx = -\frac{1}{8}\cos(8x-3)+c$, which may be checked by differentiation.

Problem 22 Find $\int (5t+2)^8\,dt$

$(5t+2)$ may be multiplied by itself eight times and then each term of the result integrated. However this would be a lengthy process. Instead an algebraic substitution is made.

Let $u = (5t+2)$, then $\frac{du}{dt} = 5$ and $dt = \frac{du}{5}$.

Hence $\int (5t+2)^8\,dt = \int (u)^8\,\frac{du}{5} = \frac{1}{5}\int u^8\,du$ (which is a 'standard integral')

$= \frac{1}{5}\left(\frac{u^9}{9}\right) + c = \frac{1}{45}u^9 + c$

Rewriting u as $(5t+2)$ gives: $\int (5t+2)^8\,dt = \frac{1}{45}(5t+2)^9 + c$.

Problem 23 Find $\int \frac{1}{(7y+4)}\,dy$

Let $u = (7y+4)$, then $\frac{du}{dy} = 7$ and $dy = \frac{du}{7}$.

Hence $\int \frac{1}{(7y+4)}dy = \int \left(\frac{1}{u}\right)\frac{du}{7} = \frac{1}{7}\int \frac{1}{u}du = \frac{1}{7}\ln u + c = \frac{1}{7}\ln(7y+4)+c$.

Problem 24 Determine $\int 6e^{5t-1}\,dt$

Let $u = (5t-1)$, then $\frac{du}{dt} = 5$ and $dt = \frac{du}{5}$

Hence $\int 6e^{5t-1}\,dt = \int 6e^{u}\,\frac{du}{5} = \frac{6}{5}\int e^{u}\,du = \frac{6}{5}e^{u} + c = \mathbf{\frac{6}{5}e^{5t-1} + c.}$

Problem 25 Determine $\int 3x(2x^2-5)^5\,dx.$

Let $u = (2x^2-5)$, then $\frac{du}{dx} = 4x$ and $dx = \frac{du}{4x}$

Hence $\int 3x\,(2x^2-5)^5\,dx = \int 3x(u)^5\,\frac{du}{4x} = \frac{3}{4}\int u^5\,du$, by cancelling.

The original variable, 'x', has been completely removed, and the integral is now only in terms of u and is a 'standard integral'.

Hence $\frac{3}{4}\int u^5\,du = \frac{3}{4}\left(\frac{u^6}{6}\right) + c = \frac{3}{24}(2x^2-5)^6 + c = \mathbf{\frac{1}{8}(2x^2-5)^6 + c.}$

Problem 26 Determine $\int 2\sin^3\theta\cos\theta\,d\theta.$

Let $u = \sin\theta$, then $\frac{du}{d\theta} = \cos\theta$ and $d\theta = \frac{du}{\cos\theta}$

Hence $\int 2\sin^3\theta\cos\theta\,d\theta = \int 2(u)^3\cos\theta\,\frac{du}{\cos\theta} = 2\int u^3\,du$, by cancelling,

$$= 2\left(\frac{u^4}{4}\right) + c = \frac{1}{2}\sin^4\theta + c$$

(Note that $\sin^3\theta \equiv (\sin\theta)^3$)

Problem 27 Determine $\int \frac{t}{1+t^2}\,dt.$

Let $u = 1+t^2$, then $\frac{du}{dt} = 2t$ and $dt = \frac{du}{2t}$

Hence $\int \frac{t}{1+t^2}\,dt = \int\left(\frac{t}{u}\right)\frac{du}{2t} = \frac{1}{2}\int\frac{1}{u}\,du = \frac{1}{2}\ln u + c = \mathbf{\frac{1}{2}\ln(1+t^2) + c.}$

Problem 28 Determine $\int \frac{3x}{\sqrt{(4x^2-3)}}\,dx.$

Let $u = (4x^2-3)$, then $\frac{du}{dx} = 8x$ and $dx = \frac{du}{8x}$.

Hence $\int \frac{3x}{\sqrt{(4x^2-3)}}\,dx = \int\frac{3x}{\sqrt{u}}\left(\frac{du}{8x}\right) = \frac{3}{8}\int\frac{1}{\sqrt{u}}\,du = \frac{3}{8}\int u^{-\frac{1}{2}}\,du$

$$= \left(\frac{3}{8}\right)\frac{u^{-\frac{1}{2}+1}}{-\frac{1}{2}+1} + c = \left(\frac{3}{8}\right)\frac{u^{\frac{1}{2}}}{\frac{1}{2}} + c = \frac{3}{4}\sqrt{u} + c = \mathbf{\frac{3}{4}\sqrt{(4x^2-3)} + c}$$

Problem 29 Show that $\int \tan x\, dx = \ln(\sec x)+c$.

$\int \tan x\, dx = \int \frac{\sin x}{\cos x} dx$. Let $u = \cos x$, then $\frac{du}{dx} = -\sin x$ and $dx = -\frac{du}{\sin x}$

Hence $\int \frac{\sin x}{\cos x} dx = \int \frac{\sin x}{u}\left(-\frac{du}{\sin x}\right) = -\int \frac{1}{u} du = -\ln u+c$

$= -\ln(\cos x)+c$

$= \ln(\cos x)^{-1}+c$, by the law of logarithms,

Hence $\int \tan x\, dx = \ln\left(\frac{1}{\cos x}\right)+c = \mathbf{\ln(\sec x)+c}$.

Problem 30 Evaluate $\int_1^2 4\sec^2(3x-1)dx$, correct to 3 decimal places.

Let $u = (3x-1)$, then $\frac{du}{dx} = 3$, and $dx = \frac{du}{3}$.

Hence $\int 4\sec^2(3x-1)dx = \int 4\sec^2 u \frac{du}{3} = \frac{4}{3}\int \sec^2 u\, du = \frac{4}{3}\tan u+c$

$= \frac{4}{3}\tan(3x-1)+c$

Hence $\int_1^2 4\sec^2(3x-1)dx = \frac{4}{3}\left[\tan(3x-1)\right]_1^2 = \frac{4}{3}[\tan 5-\tan 2]$

$= \frac{4}{3}[\tan 286.48°-\tan 114.59°]$, since the limits are given in radians

$= \frac{4}{3}[(-3.3803)-(-2.1852)] = \mathbf{-1.593}$

Problem 31 Evaluate $\int_0^2 4x\sqrt{(4x^2+9)}dx$, taking positive square roots only.

Let $u = (4x^2+9)$, then $\frac{du}{dx} = 8x$ and $dx = \frac{du}{8x}$.

Hence $\int 4x\sqrt{(4x^2+9)}dx = \int 4x\sqrt{u}\left(\frac{du}{8x}\right) = \frac{1}{2}\int \sqrt{u}\, du = \frac{1}{2}\int u^{\frac{1}{2}} du$

$= \frac{1}{2}\left(\frac{u^{\frac{3}{2}}}{\frac{3}{2}}\right)+c = \frac{1}{3}\sqrt{u^3}+c = \frac{1}{3}(4x^2+9)^3+c$

Hence $\int_0^2 4x\sqrt{(4x^2+9)}dx = \frac{1}{3}\left[\sqrt{(4x^2+9)^3}\right]_0^2 = \frac{1}{3}[\sqrt{25^3}-\sqrt{9^3}] = \frac{1}{3}(125-27)$

$= \mathbf{32\frac{2}{3}}$

Problem 32 Evaluate, correct to 4 significant figures $\int_1^3 \frac{2e^{2t}}{4+e^{2t}} dt$.

Let $u = 4+e^{2t}$, then $\frac{du}{dt} = 2e^{2t}$ and $dt = \frac{du}{2e^{2t}}$

Hence $\int \frac{2e^{2t}}{4+e^{2t}} dt = \int \frac{2e^{2t}}{u}\left(\frac{du}{2e^{2t}}\right) = \frac{1}{u} du = \ln u+c = \ln(4+e^{2t})+c$

Hence $\int_1^3 \frac{2e^{2t}}{4+e^{2t}}\,dt = [\ln(4+e^{2t})]_1^3 = [\ln(4+e^6) - \ln(4+e^2)]$

$$= \ln\left(\frac{4+e^6}{4+e^2}\right) = \ln\left(\frac{407.43}{11.389}\right) = 3.577$$

It is noted from the above problems that integrals of the form $k\int[f(x)]^n f'(x)\,dx$ or $k\int\frac{f'(x)}{[f(x)]^n}\,dx$ can both be integrated by substituting u for $f(x)$. (Note, k and n are constants.)

C FURTHER PROBLEMS ON METHODS OF INTEGRATION

In *Problems 1 to 14,* determine the indefinite integrals.

1 (a) $\int 4dx$; (b) $\int 7x\,dx$ $\left[\text{(a) } 4x+c; \text{(b) } \frac{7x^2}{2}+c\right]$

2 (a) $\int\frac{2}{5}x^2\,dx$; (b) $\int\frac{5}{6}x^3\,dx$ $\left[\text{(a) } \frac{2}{15}x^3+c; \text{(b) } \frac{5}{24}x^4+c\right]$

3 (a) $\int(2+3x-4x^2)\,dx$; (b) $2\int(x-4x^2)\,dx$. $\left[\text{(a) } 2x+\frac{3x^2}{2}-\frac{4x^3}{3}+c; \text{(b) } x^2-\frac{8x^3}{3}+c\right]$

4 (a) $\int\left(\frac{3x^2-5x}{x}\right)dx$; (b) $\int(2+\theta)^2\,d\theta$ $\left[\text{(a) } \frac{3x^2}{2}-5x+c; \text{(b) } 4\theta+2\theta^2+\frac{\theta^3}{3}+c\right]$

5 (a) $\int\frac{4}{3x^2}\,dx$; (b) $\int\frac{3}{4x^4}\,dx$ $\left[\text{(a) } \frac{-4}{3x}+c; \text{(b) } \frac{-1}{4x^3}+c\right]$

6 (a) $2\int\sqrt{x^3}\,dx$; (b) $\int\frac{1}{4}\sqrt[4]{x^5}\,dx$ $\left[\text{(a) } \frac{4}{5}\sqrt{x^5}+c; \text{(b) } \frac{1}{9}\sqrt[4]{x^9}+c\right]$

7 (a) $\int\frac{-5}{\sqrt{t^3}}\,dt$; (b) $\int\frac{3}{7\sqrt[5]{x^4}}\,dx$ $\left[\text{(a) } \frac{10}{\sqrt{t}}+c; \text{(b) } \frac{15}{7}\sqrt[5]{x}+c\right]$

8 (a) $\int\frac{5}{2}x^{1.8}\,dx$; (b) $\int\frac{3}{2t^{1.4}}\,dt$ $\left[\text{(a) } \frac{x^{2.8}}{1.12}+c; \text{(b) } \frac{-3.75}{t^{0.4}}+c\right]$

9 (a) $\int 3\cos 2x\,dx$; (b) $\int 7\sin 3\theta\,d\theta$ $\left[\text{(a) } \frac{3}{2}\sin 2x+c; \text{(b) } -\frac{7}{3}\cos 3\theta+c\right]$

10 (a) $\int\frac{3}{4}\sec^2 3x\,dx$; (b) $\int 2\,\text{cosec}^2\,4\theta\,d\theta$ $\left[\text{(a) } \frac{1}{4}\tan 3x+c; \text{(b) } -\frac{1}{2}\cot 4\theta+c\right]$

11 (a) $5\int\cot 2t\,\text{cosec}\,2t\,dt$; (b) $\int\frac{4}{3}\sec 4t\tan 4t\,dt$ $\left[\text{(a) } -\frac{5}{2}\text{cosec}\,2t+c; \text{(b) } \frac{1}{3}\sec 4t+c\right]$

12 (a) $\int\frac{3}{4}e^{2x}\,dx$; (b) $\frac{2}{3}\int\frac{dx}{e^{5x}}$ $\left[\text{(a) } \frac{3}{8}e^{2x}+c; \text{(b) } \frac{-2}{15e^{5x}}+c\right]$

13 (a) $\int\frac{2}{3x}\,dx$; (b) $\int\left(\frac{u^2-1}{u}\right)du$ $\left[\text{(a) } \frac{2}{3}\ln x+c; \text{(b) } \frac{u^2}{2}-\ln u+c\right]$

14 (a) $\int\frac{(2+3x)^2}{\sqrt{x}}\,dx$; (b) $\int\left(\frac{1}{t}+2t\right)^2dt$ $\left[\text{(a) } 8\sqrt{x}+8\sqrt{x^3}+\frac{18}{5}\sqrt{x^5}+c; \text{(b) } -\frac{1}{t}+4t+\frac{4t^3}{3}+c\right]$

In *Problems 15 to 23*, evaluate the definite integrals (where necessary, correct to 4 significant figures).

15 (a) $\int_2^3 2dx$; (b) $\int_{-1}^2 2x\,dx$ [(a) 2; (b) 3]

16 (a) $\int_1^4 5x^2\,dx$; (b) $\int_{-1}^1 -\frac{3}{4}t^2\,dt$ [(a) 105; (b) $-\frac{1}{2}$]

17 (a) $\int_{-1}^2 (3-x^2)dx$; (b) $\int_1^3 (x^2-4x+3)dx$ [(a) 6; (b) $-1\frac{1}{3}$]

18 (a) $\int_0^\pi \frac{3}{2}\cos\theta\,d\theta$; (b) $\int_0^{\frac{\pi}{2}} 4\cos\theta\,d\theta$ [(a) 0; (b) 4]

19 (a) $\int_{\frac{\pi}{6}}^{\frac{\pi}{3}} 2\sin 2\theta\,d\theta$; (b) $\int_0^2 3\sin t\,dt$ [(a) 1; (b) 4.248]

20 (a) $\int_0^1 5\cos 3x\,dx$; (b) $\int_0^{\frac{\pi}{6}} 3\sec^2 2x\,dx$ [(a) 0.2352; (b) 2.598]

21 (a) $\int_1^2 \operatorname{cosec}^2 4t\,dt$; (b) $\int_{\frac{\pi}{4}}^{\frac{\pi}{2}} (3\sin 2x-2\cos 3x)dx$ [(a) 0.2527; (b) 2.638]

22 (a) $\int_0^1 3e^{3t}\,dt$; (b) $\int_{-1}^2 \frac{2}{3e^{2x}}\,dx$ [(a) 19.09; (b) 2.457]

23 (a) $\int_2^3 \frac{2}{3x}\,dx$; (b) $\int_1^3 \frac{2x^2+1}{x}\,dx$ [(a) 0.2703; (b) 9.099]

24 Show that $\int_1^2 \frac{(2x-1)(3x+4)}{x}\,dx = 2(7-2\ln 2)$.

In *Problems 25 to 33*, determine the indefinite integrals.

25 (a) $\int 3\cos(6x+1)dx$; (b) $\int (2x+3)^7\,dx$

$\left[\text{(a) } \frac{1}{2}\sin(6x+1)+c\text{; (b) } \frac{1}{16}(2x+3)^8+c\right]$

26 (a) $\int \frac{2}{4t+3}\,dt$; (b) $\int 2e^{4x-1}\,dx$ $\left[\text{(a) } \frac{1}{2}\ln(4t+3)+c\text{; (b) } \frac{1}{2}e^{4x-1}+c\right]$

27 (a) $\int 3\sec^2(7\theta-4)\,d\theta$; (b) $\int \frac{1}{2}e^{2-3x}\,dx$

$\left[\text{(a) } \frac{3}{7}\tan(7\theta-4)+c\text{; (b) } -\frac{1}{6}e^{2-3x}+c\right]$

28 (a) $\int x(2x^2+7)^9\,dx$; (b) $\int 2\cos^2\alpha\sin\alpha\,d\alpha$

$\left[\text{(a) } \frac{1}{40}(2x^2+7)^{10}+c\text{; (b)} -\frac{2}{3}\cos^3\alpha+c\right]$

29 (a) $\int \sin^5\theta\cos\theta\,d\theta$; (b) $\int 4\tan 2t\sec^2 2t\,dt$

$\left[\text{(a) } \frac{1}{6}\sin^6\theta+c\text{; (b) } \tan^2 2t+c\right]$

30 (a) $\int \frac{2x}{2+3x^2}\,dx$; (b) $\int 2\tan 3x\,dx$ $\left[\text{(a) } \frac{1}{3}\ln(2+3x^2)+c\text{; (b) } \frac{2}{3}\ln(\sec 3x)+c\right]$

31 (a) $\int \frac{u}{\sqrt{(3u^2-1)}}\,du$; (b) $\int \frac{2\ln x}{x}\,dx$ $\left[\text{(a) } \frac{1}{3}\sqrt{(3u^2-1)}+c\text{; (b) } (\ln x)^2+c\right]$

32 (a) $\int (2x^2-1)\sqrt{(2x^3-3x+4)}\,dx$; (b) $\int \dfrac{(6y+1)}{(3y^2+y-4)^6}\,dy$

$\left[\text{(a) } \frac{2}{9}\sqrt{(2x^3-3x+4)^3}+c;\ \text{(b) } \dfrac{-1}{5(3y^2+y-4)^5}+c\right]$

33 (a) $\int 3x \sin(2x^2+5)\,dx$; (b) $\int 4te^{4t^2+2}\,dt$

$\left[\text{(a) } -\frac{3}{4}\cos(2x^2+5)+c;\ \text{(b) } \frac{1}{2}e^{4t^2+2}+c\right]$

In *Problems 34 to 39* evaluate the definite integrals (where necessary, correct to 4 significant figures).

34 (a) $\int_0^1 (3x-1)^4\,dx$; (b) $\int_0^{\frac{\pi}{2}} 2\sin^4 t \cos t\,dt$ $\left[\text{(a) } 2\frac{1}{5};\ \text{(b) } \frac{2}{5}\right]$

35 (a) $\int_0^{\frac{\pi}{4}} 2\sin(4\theta+\frac{\pi}{3})\,d\theta$; (b) $\int_1^2 \sec^2(2x-3)\,dx$ $\left[\text{(a) } \frac{1}{2};\ \text{(b) } 1.557\right]$

36 (a) $\int_0^1 2\cos(3t-1)\,dt$; (b) $\int_1^3 \dfrac{1}{(2u-1)}\,du$ [(a) 1.167; (b) 0.8047]

37 (a) $\int_1^2 \dfrac{3\ln x}{x}\,dx$; (b) $\int_0^1 x\sqrt{(x^2+3)}\,dx$ [(a) 0.7207; (b) 0.9346]

38 (a) $\int_1^2 \dfrac{1}{(2x-1)^4}\,dx$; (b) $\int_1^2 x\cos(3x^2-2)\,dx$

$\left[\text{(a) } \frac{13}{81} \text{ or } 0.1605;\ \text{(b) } -0.2309\right]$

39 (a) $\int_0^1 3te^{2t^2-1}\,dt$; (b) $\int_1^2 \dfrac{4e^{2\theta}}{(e^{2\theta}-3)}\,d\theta$ [(a) 1.763; (b) 4.929]

18 Calculus (5) – Areas under and between curves

A MAIN POINTS CONCERNED WITH AREAS UNDER AND BETWEEN CURVES

1 The area shown shaded in *Fig 1* may be determined using approximate methods (such as the trapezoidal rule, the mid-ordinate rule or Simpson's rule, see *Mathematics 2 Checkbook*) or, more precisely, by using integration.

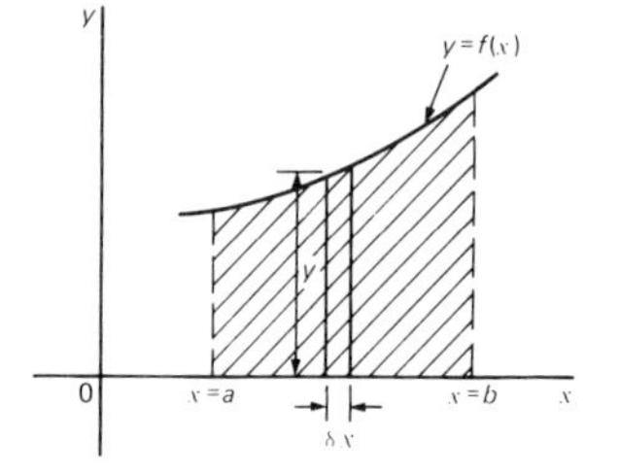

Fig 1

(i) Let A be the area shown shaded in *Fig 1* and let this area be divided into a number of strips each of width δx. One such strip is shown and let the area of this strip be δA. Then:

$$\delta A \simeq y\delta x \qquad (1)$$

The accuracy of statement (1) increases when the width of each strip is reduced, i.e., area A is divided into a greater number of strips.

(ii) Area A is equal to the sum of all the strips from $x = a$ to $x = b$,

$$\text{i.e.} \quad A = \lim_{\delta x \to 0} \sum_{x=a}^{x=b} y\delta x \qquad (2)$$

(iii) From statement (1), $\dfrac{\delta A}{\delta x} \simeq y$ (3)

In the limit, as δx approaches zero, $\dfrac{\delta A}{\delta x}$ becomes the differential coefficient $\dfrac{dA}{dx}$. Hence $\lim_{\delta x \to 0} \left(\dfrac{\delta A}{\delta x}\right) = \dfrac{dA}{dx} = y$, from statement (3).

By integration, $\int \dfrac{dA}{dx}\, dx = \int y\, dx$, i.e. $A = \int y\, dx$.

The ordinates $x = a$ and $x = b$ limit the area and such ordinate values are shown as limits. Hence

$$A = \int_a^b y\, dx \qquad (4)$$

(iv) Equating statements (2) and (4) gives:

$$\text{Area } A = \lim_{\delta x \to 0} \sum_{x=a}^{x=b} y\delta x = \int_a^b y\,dx = \int_a^b f(x)dx.$$

(v) If the area between a curve $x = f(y)$, the y-axis and ordinates $y = p$ and $y = q$ is required then area $= \int_p^q x\,dy$.

Thus determining the area under a curve by integration merely involves evaluating a definite integral.

2 There are several instances in engineering and science where the area beneath a curve needs to be accurately determined. For example, the areas between limits of a:
velocity/time graph gives distance travelled,
force/distance graph gives work done,
voltage/current graph gives power, and so on.

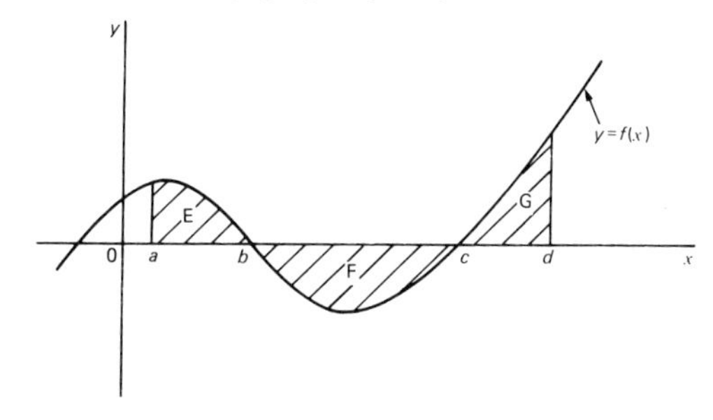

Fig 2

3 Should a curve drop below the x-axis, then y $(= f(x))$ becomes negative and $\int f(x)dx$ is negative. When determining such areas by integration, a negative sign is placed before the integral. For the curve shown in *Fig 2*, the total shaded area is given by (area E + area F + area G).

By integration, **total shaded area** $= \int_a^b f(x)dx - \int_b^c f(x)dx + \int_c^d f(x)dx$

(Note that this is **not** the same as $\int_a^d f(x)dx$.)

It is usually necessary to sketch a curve in order to check whether it crosses the x-axis.

4 The area enclosed between curves $y = f_1(x)$ and $y = f_2(x)$ (shown shaded in *Fig 3*) is given by:

$$\text{Shaded area} = \int_a^b f_2(x)dx - \int_a^b f_1(x)dx$$
$$= \int_a^b [f_2(x) - f_1(x)]\,dx$$

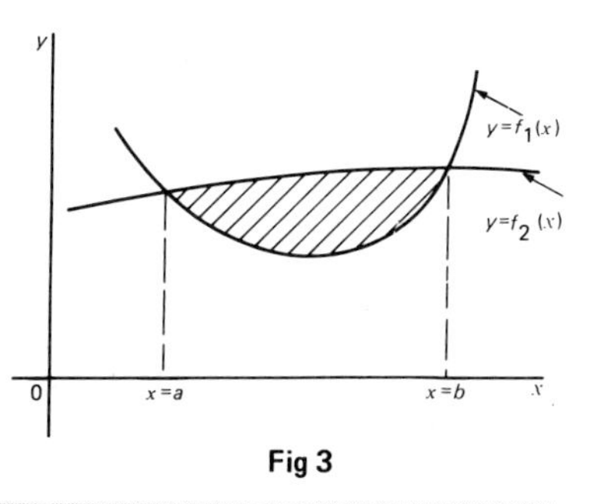

Fig 3

B WORKED PROBLEMS ON AREAS UNDER AND BETWEEN CURVES

Problem 1 Determine the area enclosed by $y = 2x+3$, the x-axis and ordinates $x = 1$ and $x = 4$.

$y = 2x+3$ is a straight line graph as shown in *Fig 4*, where the required area is shown shaded.

By integration, shaded area

$$= \int_1^4 y\,dx = \int_1^4 (2x+3)dx = \left[\frac{2x^2}{2}+3x\right]_1^4$$

$= [(16+12)-(1+3)] =$ **24 square units.**

[This answer may be checked since the shaded area is a trapezium.

Area of trapezium $= \frac{1}{2}$(sum of parallel sides)(perpendicular distance between parallel sides)

$= \frac{1}{2}(5+11)(3) =$ **24 square units.**]

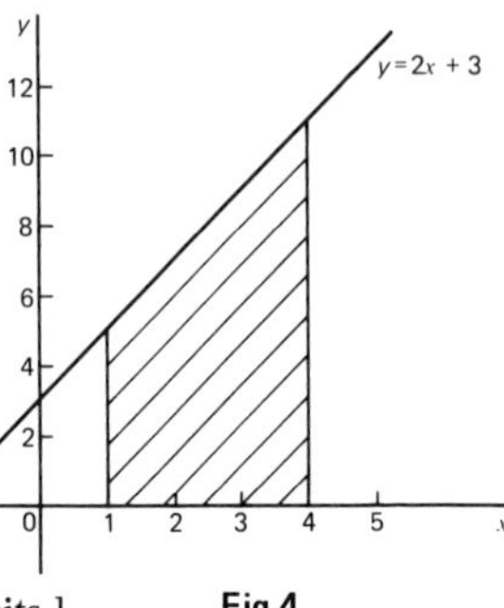

Fig 4

Problem 2 The velocity v of a body t seconds after a certain instant is $(2t^2+5)$m/s. Find by integration how far it moves in the interval from $t = 0$ to $t = 4$ s.

Since $2t^2+5$ is a quadratic expression, the curve $v = 2t^2+5$ is a parabola cutting the v-axis at $v = 5$, as shown in *Fig 5*. The distance travelled is given by the area under the v/t curve (shown shaded in *Fig 5*).

By integration, shaded area

$$= \int_0^4 v\,dt = \int_0^4 (2t^2+5)dt$$

$$= \left[\frac{2t^3}{3}+5t\right]_0^4$$

i.e., **distance travelled** $= \mathbf{62\frac{2}{3}}$ **m.**

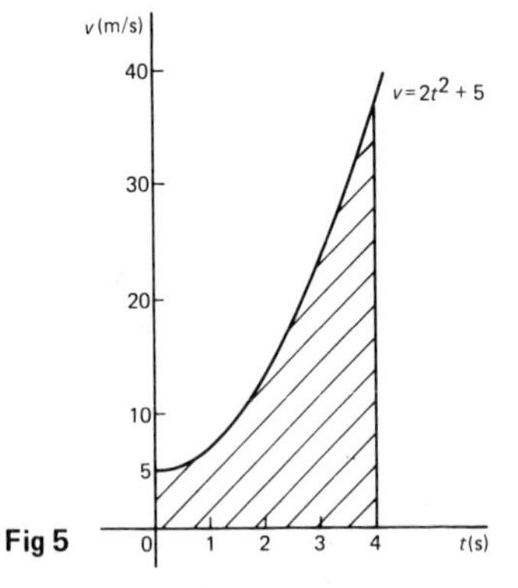

Fig 5

Problem 3 Sketch the graph $y = x^3+2x^2-5x-6$ between $x = -3$ and $x = 2$ and determine the area enclosed by the curve and the x-axis.

A table of values is produced and the graph sketched as shown in *Fig 6* where the area enclosed by the curve and the x-axis is shown shaded.

x	-3	-2	-1	0	1	2
x^3	-27	-8	-1	0	1	8
$2x^2$	18	8	2	0	2	8
$-5x$	15	10	5	0	-5	-10
-6	-6	-6	-6	-6	-6	-6
y	0	4	0	-6	-8	0

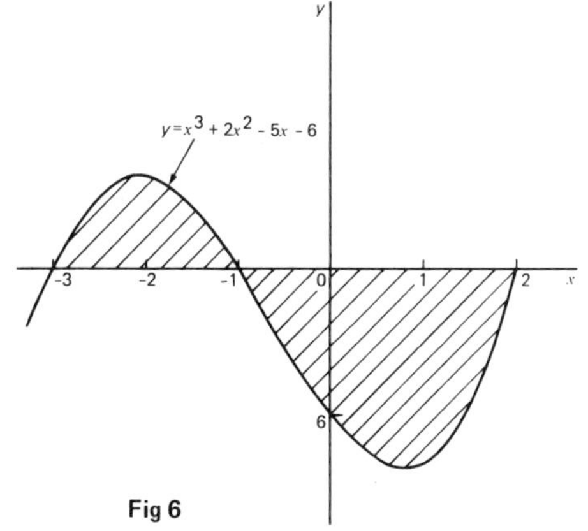

Fig 6

Shaded area $= \int_{-3}^{-1} y\,dx - \int_{-1}^{2} y\,dx$, the minus sign before the second integral being necessary since the enclosed area is below the x-axis.

Hence shaded area $= \int_{-3}^{-1} (x^2 + 2x^2 - 5x - 6)\,dx - \int_{1}^{2} (x^3 + 2x^2 - 5x - 6)\,dx.$

$$\int_{-3}^{-1} (x^3 + 2x^2 - 5x - 6)\,dx = \left[\frac{x^4}{4} + \frac{2x^3}{3} - \frac{5x^2}{2} - 6x\right]_{-3}^{-1}$$

$$= \left\{\frac{1}{4} - \frac{2}{3} - \frac{5}{2} + 6\right\} - \left\{\frac{81}{4} - 18 - \frac{45}{2} + 18\right\}$$

$$= \left\{3\frac{1}{12}\right\} - \left\{-2\frac{1}{4}\right\} = 5\frac{1}{3} \text{ square units}$$

$$\int_{-1}^{2} (x^3 + 2x^2 - 5x - 6)\,dx = \left[\frac{x^4}{4} + \frac{2x^3}{3} - \frac{5x^2}{2} - 6x\right]_{-1}^{2} = \left\{4 + \frac{16}{3} - 10 - 12\right\} - \left\{3\frac{1}{12}\right\}$$

$$= \left\{-12\frac{2}{3}\right\} - \left\{3\frac{1}{12}\right\}$$

$$= -15\frac{3}{4} \text{ square units}$$

Hence shaded area $= \left(5\frac{1}{3}\right) - \left(-15\frac{3}{4}\right) = $ **$21\frac{1}{12}$ square units.**

Problem 4 Determine the area enclosed by the curve $y = 3x^2 + 4$, the x-axis and ordinates $x = 1$ and $x = 4$ by (a) the trapezoidal rule; (b) the mid-ordinate rule, (e) Simpson's rule, and (d) integration.

The curve $y = 3x^2 + 4$ is shown plotted in *Fig 7.* (For approximate methods of determining areas under curves–see chapter 6 of *Mathematics 2 checkbook*).

(a) **By the trapezoidal rule**

$$\text{area} = (\text{width of interval})\left[\frac{1}{2}\begin{pmatrix}\text{first + last}\\ \text{ordinate}\end{pmatrix} + \text{sum of remaining ordinates}\right]$$

Selecting 6 intervals each of width 0.5 gives:

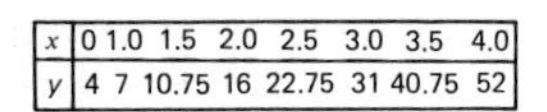

x	0	1.0	1.5	2.0	2.5	3.0	3.5	4.0
y	4	7	10.75	16	22.75	31	40.75	52

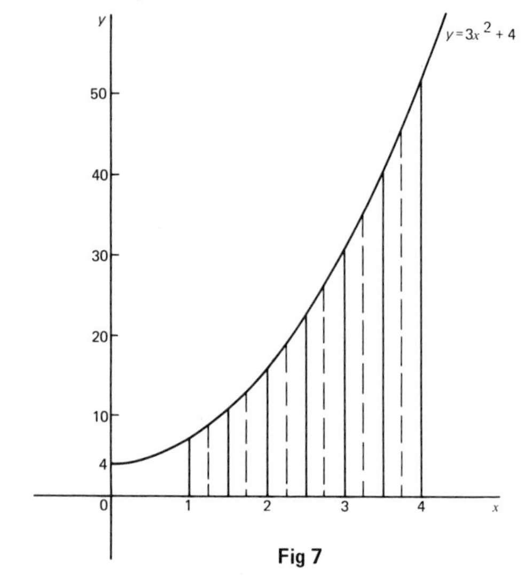

Fig 7

$$\text{Area} = (0.5)\left[\frac{1}{2}(7+52)+10.75+16+22.75+31+40.75\right]$$
$$= \mathbf{75.375\ square\ units.}$$

(b) **By the mid-ordinate rule**, area = (width of interval)(sum of mid-ordinates)
Selecting 6 intervals, each of width 0.5 gives the mid-ordinates as shown by the broken lines in *Fig 7*.
Thus, area = (0.5)(8.5+13+19+26.5+35.5+46) = **74.25 square units**

(c) **By Simpson's rule**, $\text{area} = \frac{1}{3}(\text{width of interval})\left[\begin{pmatrix}\text{first + last}\\ \text{ordinates}\end{pmatrix} + 4\begin{pmatrix}\text{sum of}\\ \text{even}\\ \text{ordinates}\end{pmatrix} + 2\begin{pmatrix}\text{sum of}\\ \text{remaining}\\ \text{odd ordinates}\end{pmatrix}\right]$

Selecting 6 intervals, each of width 0.5 gives:

$$\text{area} = \frac{1}{3}(0.5)\left[(7+52)+4(10.75+22.75+40.75)+2(16+31)\right]$$
$$= \mathbf{75\ square\ units.}$$

(d) **By integration,** shaded area $= \int_1^4 y\,dx = \int_1^4 (3x^2+4)dx$

$$= [x^3+4x]_1^4 = \mathbf{75\ square\ units.}$$

Integration gives the precise value for the area under a curve. In this case Simpson's rule is seen to be the most accurate of the three approximate methods.

Problem 5 Find the area enclosed by the curve $y = \sin 2x$, the x-axis and the ordinates $x = 0$ and $x = \frac{\pi}{3}$.

A sketch of $y = \sin 2x$ is shown in *Fig 8*. (Note that $y = \sin 2x$ has a period of $\frac{2\pi}{2}$, i.e. π radians).

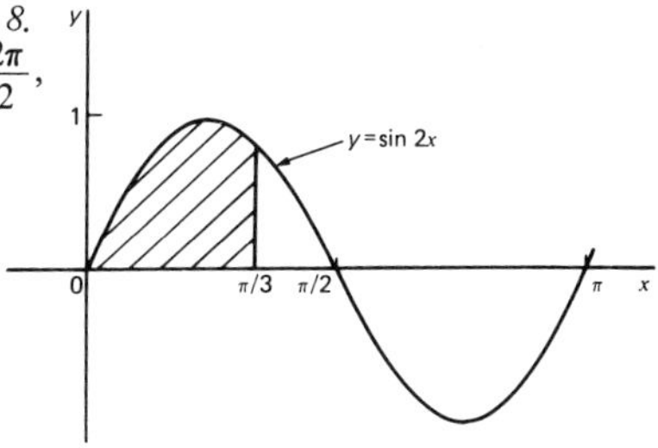

Shaded area $= \int_0^{\frac{\pi}{3}} y\,dx$

$= \int_0^{\frac{\pi}{3}} \sin 2x\,dx = \left[-\frac{1}{2}\cos 2x\right]_0^{\frac{\pi}{3}}$

$= \left\{-\frac{1}{2}\cos\frac{2\pi}{3}\right\} - \left\{-\frac{1}{2}\cos 0\right\}$

$= \left\{-\frac{1}{2}\left(-\frac{1}{2}\right)\right\} - \left\{-\frac{1}{2}(1)\right\} = \frac{1}{4} + \frac{1}{2} = \frac{3}{4}$ **square units.**

Fig 8

Problem 6 A gas expands according to the law pv = constant. When the volume is 3 m^3 the pressure is 150 kPa. Given that work done $= \int_{v_1}^{v_2} p\,dv$, determine the work done as the gas expands from 2 m^3 to a volume of 6 m^3.

pv = constant. When $v = 3$ m^3 and $p = 150$ kPa the constant is given by $(3 \times 150) = 450$ kPa m^3 or 450 kJ.

Hence $pv = 450$, or $p = \frac{450}{v}$

Work done $= \int_2^6 \frac{450}{v}\,dv = [450 \ln v]_2^6 = 450[\ln 6 - \ln 2]$

$= 450 \ln\left(\frac{6}{2}\right) = 450 \ln 3 =$ **494.4 kJ.**

Problem 7 Determine the area enclosed by the curve $y = 4\cos\frac{\theta}{2}$, the θ-axis and ordinates $\theta = 0$ and $\theta = \frac{\pi}{2}$.

Fig 9

The curve $y = 4\cos\frac{\theta}{2}$ is shown in *Fig 9*.

(Note that $y = 4\cos\frac{\theta}{2}$ has a maximum value of 4 and period $\frac{2\pi}{\frac{1}{2}}$, i.e. 4π rads).

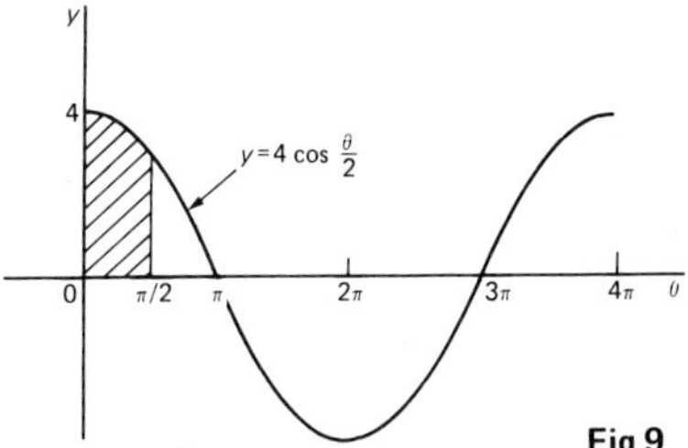

Shaded area $= \int_0^{\frac{\pi}{2}} y\,d\theta = \int_0^{\frac{\pi}{2}} 4\cos\frac{\theta}{2}\,d\theta = \left[4\left(\frac{1}{\frac{1}{2}}\right)\sin\frac{\theta}{2}\right]_0^{\frac{\pi}{2}}$

$= (8\sin\frac{\pi}{4}) - (8\sin 0) = \frac{8\sqrt{2}}{2} = 4\sqrt{2}$ **or 5.657 square units.**

Problem 8 Determine the area bounded by the curve $y = 3e^{\frac{t}{4}}$, the t-axis and ordinates $t = -1$ and $t = 4$, correct to 4 significant figures.

A table of values is produced as shown.

t	-1	0	1	2	3	4
$y = 3e^{\frac{t}{4}}$	2.34	3.0	3.85	4.95	6.35	8.15

Since all the values of y are positive the area required is wholly above the t-axis.

$$\text{Hence area} = \int_1^4 y\,dt = \int_1^4 3e^{\frac{t}{4}}dt = \left[\frac{3}{\frac{1}{4}}\ e^{\frac{t}{4}}\right]_{-1}^{4}$$

$$= 12\left[e^{\frac{t}{4}}\right]_{-1}^{4} = 12(e^1 - e^{-\frac{1}{4}}) = 12\,(2.7183 - 0.7788)$$

$$= 12(1.9395) = \mathbf{23.27\ square\ units.}$$

Problem 9. Sketch the curve $y = x^2 + 5$ between $x = -1$ and $x = 4$. Find the area enclosed by the curve, the x-axis and the ordinates $x = 0$ and $x = 3$. Determine also by integration the area enclosed by the curve and the y-axis, between the same limits.

A table of values is produced and the curve $y = x^2 + 5$ plotted as shown in *Fig 10.*

x	-1	0	1	2	3
y	6	5	6	9	14

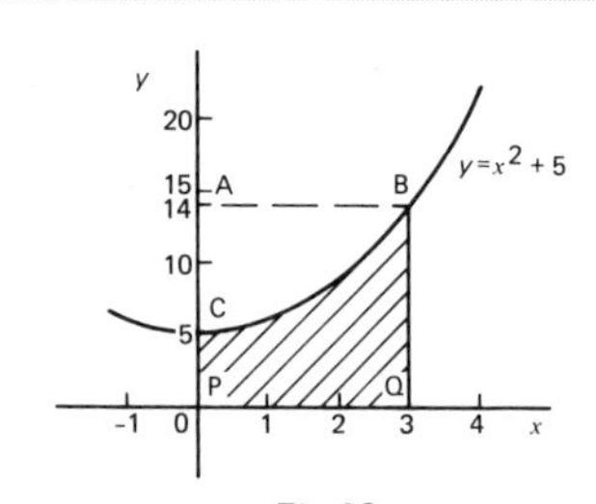

Fig 10

$$\text{Shaded area} = \int_0^3 y\,dx = \int_0^3 (x^2 + 5)dx$$

$$= \left[\frac{x^3}{3} + 5x\right]_0^3 = \mathbf{24\ square\ units.}$$

When $x = 3$, $y = 3^2 + 5 = 14$ and when $x = 0$, $y = 5$.
Since $y = x^2 + 5$ then $x^2 = y - 5$ and $x = \sqrt{(y-5)}$.
The area enclosed by the curve $y = x^2 + 5$ (i.e. $x = \sqrt{(y-5)}$), the y-axis and the ordinates $y = 5$ and $y = 14$ (i.e. area ABC of *Fig 10*) is given by:

$$\text{Area} = \int_{y=5}^{y=14} x\,dy = \int_5^{14} \sqrt{(y-5)}dy = \int_5^{14} (y-5)^{\frac{1}{2}}dy$$

Let $u = y - 5$, then $\frac{du}{dy} = 1$ and $dy = du$.

$$\text{Hence} \quad (y-5)^{\frac{1}{2}}\,dy = \int u^{\frac{1}{2}}du = \frac{2}{3}u^{\frac{3}{2}}$$

$$\text{Since } u = y - 5 \text{ then } \int_5^{14} \sqrt{(y-5)}dy = \frac{2}{3}\left[(y-5)^{\frac{3}{2}}\right]_5^{14} = \frac{2}{3}[\sqrt{9^3} - 0]$$

$$= \mathbf{18\ square\ units.}$$

(*Check*: From *Fig 10*, area BCPQ+area ABC = 24+18 = 42 square units, which is the area of rectangle ABQP).

Problem 10 Determine the area between the curve $y = x^3 - 2x^2 - 8x$ and the x-axis.

$y = x^3 - 2x^2 - 8x = x(x^2 - 2x - 8) = x(x+2)(x-4)$.
When $y = 0$, then $x = 0$ or $(x+2) = 0$ or $(x-4) = 0$,
i.e. when $y = 0$, $x = 0$ or -2 or 4, which means that the curve cuts the x-axis at 0, -2 and 4. Since the curve is a continuous function, only one other co-ordinate value needs to be calculated before a sketch of the curve can be produced. When $x = 1$, $y = -9$, showing that the part of the curve between $x = 0$ and $x = 4$ is negative. A sketch of $y = x^3 - 2x^2 - 8x$ is shown in *Fig 11*.

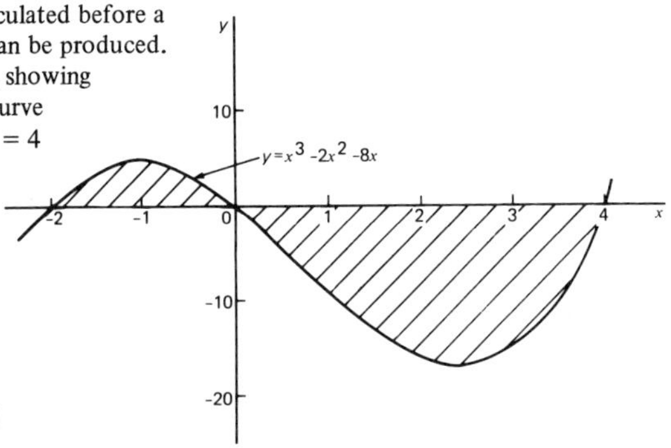

Fig 11

$$\text{Shaded area} = \int_{-2}^{0} (x^3 - 2x^2 - 8x)dx - \int_{0}^{4} (x^3 - 2x^2 - 8x)dx$$

$$= \left[\frac{x^4}{4} - \frac{2x^3}{3} - \frac{8x^2}{2}\right]_{-2}^{0} - \left[\frac{x^4}{4} - \frac{2x^3}{3} - \frac{8x^2}{2}\right]_{0}^{4} = (6\tfrac{2}{3}) - (-42\tfrac{2}{3})$$

$$= \mathbf{49\tfrac{1}{3}} \textbf{ square units.}$$

Problem 11 Determine the area enclosed between the curves $y = x^2 + 1$ and $y = 7 - x$.

At the points of intersection, the curves are equal. Thus, equating the y-values of each curve gives: $x^2 + 1 = 7 - x$, from which $x^2 + x - 6 = 0$. Factorising gives $(x-2)(x+3) = 0$, from which $x = 2$ and $x = -3$. By firstly determining the points of intersection the range of x-values have been found. Tables of values are produced as shown below.

x	-3	-2	-1	0	1	2
$y = x^2 + 1$	10	5	2	1	2	5

x	-3	0	2
$y = 7 - x$	10	7	5

$y = 7 - x$ is a straight line thus only two points are needed, plus one more as a check. A sketch of the two curves is shown in *Fig 12*.

$$\text{Shaded area} = \int_{-3}^{2} (7 - x)dx - \int_{-3}^{2} (x^2 + 1)dx$$

$$= \int_{-3}^{2} [(7 - x) - (x^2 + 1)]\, dx = \int_{-3}^{2} (6 - x - x^2)dx$$

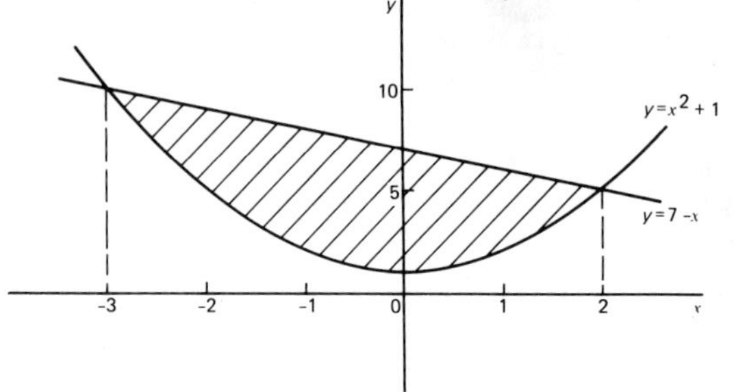

Fig 12

$$=\left[6x-\frac{x^2}{2}-\frac{x^3}{3}\right]_{-3}^{2}=(12-2-\tfrac{8}{3})-(-18-\tfrac{9}{2}+9)$$

$$=(7\tfrac{1}{3})-(-13\tfrac{1}{2})=\mathbf{20\tfrac{5}{6}\ square\ units.}$$

Problem 12 (a) Determine the co-ordinates of the points of intersection of the curves $y = x^2$ and $y^2 = 8x$; (b) sketch the curves $y = x^2$ and $y^2 = 8x$ on the same axes; (c) calculate the area enclosed by the two curves.

(a) At the points of intersection the co-ordinates of the curves are equal. When $y = x^2$ then $y^2 = x^4$.
Hence at the points of intersection $x^4 = 8x$, by equating the y^2 values.
Thus $x^4 - 8x = 0$, from which $x(x^3-8) = 0$
i.e. $x = 0$ or $(x^3-8) = 0$.
Hence at the points of intersection $x = 0$ or $x = 2$.
When $x = 0$, $y = 0$ and when $x = 2$, $y = 2^2 = 4$.
Hence the points of intersection of the curves $y = x^2$ and $y^2 = 8x$ are (0, 0) and (2, 4).

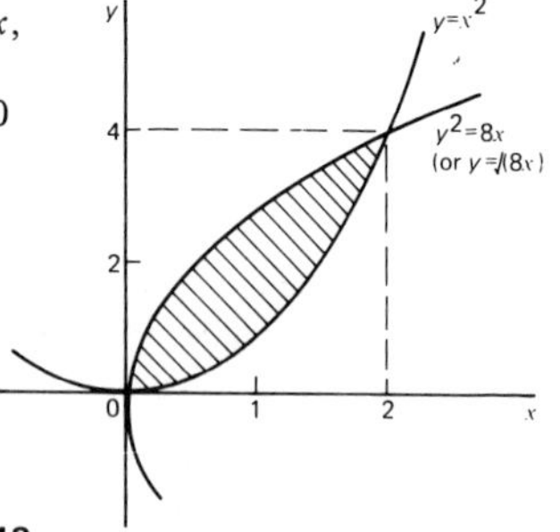

Fig 13

(b) A sketch of $y = x^2$ and $y^2 = 8x$ is shown in *Fig 13*.

(c) Shaded area $=\int_0^2 \{\sqrt{(8x)}-x^2\}\,dx=\int_0^2\{(\sqrt{8})x^{\frac{1}{2}}-x^2\}\,dx$

$$=\left[(\sqrt{8})\frac{x^{\frac{3}{2}}}{\frac{3}{2}}-\frac{x^3}{3}\right]_0^2=\left\{\frac{\sqrt{8}\sqrt{8}}{\frac{3}{2}}-\frac{8}{3}\right\}-\{0\}$$

$$=\frac{16}{3}-\frac{8}{3}=\frac{8}{3}=\mathbf{2\tfrac{2}{3}\ square\ units.}$$

Problem 13 Determine by integration the area bounded by the three straight lines $y = 4-x$, $y = 3x$ and $3y = x$.

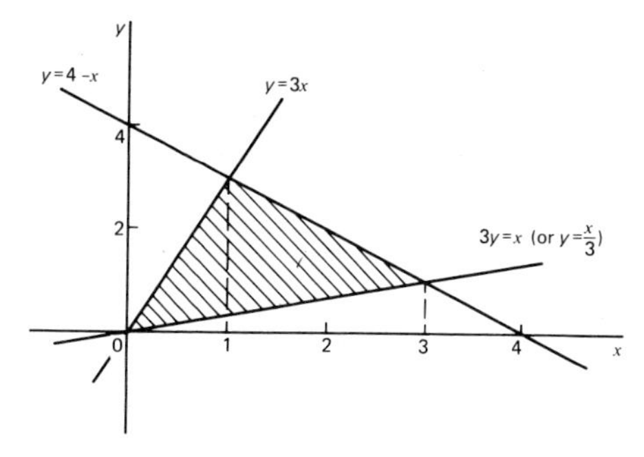

Fig 14

Each of the straight lines are shown sketched in *Fig 14*.

$$\text{Shaded area} = \int_0^1 (3x - \frac{x}{3})dx + \int_1^3 \left[(4-x) - \frac{x}{3}\right]dx$$

$$= \left[\frac{3x^2}{2} - \frac{x^2}{6}\right]_0^1 + \left[4x - \frac{x^2}{2} - \frac{x^2}{6}\right]_1^3 = (1\frac{1}{3}) + (6 - 3\frac{1}{3})$$

$$= \mathbf{4\ square\ units.}$$

C FURTHER PROBLEMS ON AREAS UNDER AND BETWEEN CURVES

Unless otherwise stated all answers are in square units.

1 Show by integration that the area of the triangle formed by the line $y = 2x$, the ordinates $x = 0$ and $x = 4$ and the x-axis is 16 square units.

2 Sketch the curve $y = 3x^2 + 1$ between $x = -2$ and $x = 4$. Determine by integration the area enclosed by the curve, the x-axis and ordinates $x = -1$ and $x = 3$. Use an approximate method to find the area and compare your result with that obtained by integration. [32]

In *Problems 3 to 10*, find the area enclosed between the given curves, the horizontal axis and the given ordinates.

3 $y = 5x;\quad x = 1, x = 4$ [$37\frac{1}{2}$]

4 $y = 2x^2 - x + 1;\quad x = -1, x = 2$ [$7\frac{1}{2}$]

5 $y = 2\sin 2\theta;\quad \theta = 0, \theta = \frac{\pi}{4}$ [1]

6 $\theta = t + e^t;\quad t = 0, t = 2$ [8.389]

7 $y = 5\cos 3t; t = 0, t = \frac{\pi}{6}$ [$1\frac{2}{3}$]

8 $y = (x-1)(x-3); x = 0, x = 3$ [$2\frac{2}{3}$]

9 $y = 2x^3;\ x = -2, x = 2$ [16]

10 $xy = 4;\ x = 1, x = 4$ [5.545]

11 The force F newtons acting on a body at a distance x metres from a fixed point is given by $F = 3x + 2x^2$. If work done $= \int_{x_1}^{x_2} F\,dx$, determine the work done when the body moves from the position where $x = 1$ m to that when $x = 3$ m. [$29\frac{1}{3}$ N m]

12 Find the area between the curve $y = 4x - x^2$ and the x-axis. [$10\frac{2}{3}$]

13 Sketch the curves $y = x^2 + 3$ and $y = 7 - 3x$ and determine the area enclosed by them. [$20\frac{5}{6}$]

14 Determine the area enclosed by the curves $y = \sin x$ and $y = \cos x$ and the y-axis. [0.4142]

15 The velocity v of a vehicle t seconds after a certain instant is given by $v = (3t^2 + 4)$ m/s. Determine how far it moves in the interval from $t = 1$ s to $t = 5$ s. [140 m]

16 Determine the co-ordinates of the points of intersection and the area enclosed between the parabolas $y^2 = 3x$ and $x^2 = 3y$. [(0, 0), (3, 3); 3]

17 Determine the area enclosed by the curve $y = 5x^2 + 2$, the x-axis and the ordinates $x = 0$ and $x = 3$. Find also the area enclosed by the curve and the y-axis between the same limits. [51; 90]

18 Calculate the area enclosed between $y = x^3 - 4x^2 - 5x$ and the x-axis using an approximate method and compare your result with the true area obtained by integration. [$73\frac{5}{6}$ or 73.83]

19 A gas expands according to the law pv = constant. When the volume is 2 m^3 the pressure is 250 kPa. Find the work done as the gas expands from 1 m^3 to a volume of 4 m^3 given that work done $= \int_{v_1}^{v_2} p\,dv$. [693.1 kJ]

20 Determine the area enclosed by the three straight lines $y = 3x$, $2y = x$ and $y + 2x = 5$. [$2\frac{1}{2}$]

19 Calculus (6) – Mean and root mean square values

A MAIN POINTS CONCERNED WITH MEAN AND ROOT MEAN SQUARE VALUES

1 **Mean or average values**

(i) The mean or average value of the curve shown in *Fig 1*, between $x = a$ and $x = b$, is given by:

$$\textbf{mean or average value, } \bar{y} = \frac{\textbf{area under curve}}{\textbf{length of base}}$$

(See *Mathematics 2 Checkbook,* chapter 6.)

(ii) When the area under a curve may be obtained by integration then:

$$\text{mean or average value,} \quad \bar{y} = \frac{\int_a^b y\, dx}{b-a}$$

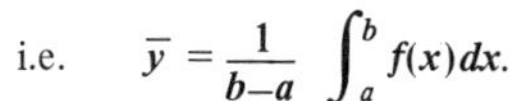

$$\text{i.e.} \quad \bar{y} = \frac{1}{b-a}\int_a^b f(x)\,dx.$$

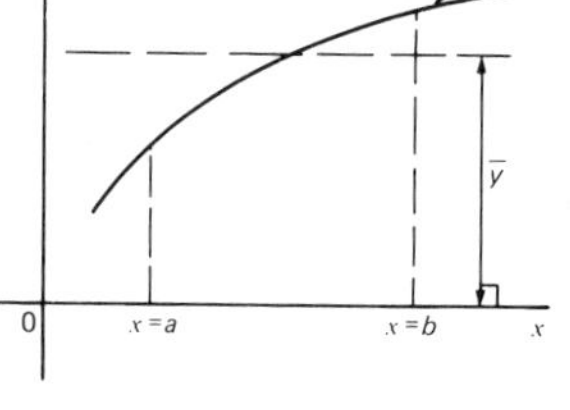

Fig 1

(iii) For a periodic function, such as a sine wave, the mean value is assumed to be 'the mean value over half a cycle', since the mean value over a complete cycle is zero.

2 The **root mean square value** of a quantity is 'the square root of the mean value of the squared values of the quantity' taken over an interval. With reference to *Fig 1*, the r.m.s. value of $y = f(x)$ over the range $x = a$ to $x = b$ is given by:

$$\textbf{r.m.s. value} = \sqrt{\left\{\frac{1}{b-a}\int_a^b y^2\, dx\right\}}$$

One of the principal applications of r.m.s. values is with alternating currents and voltages. The r.m.s. value of an alternating current is defined as that current which will give the same heating effect as the equivalent direct current.

B WORKED PROBLEMS ON MEAN AND ROOT MEAN SQUARE VALUES

Problem 1 Determine, using integration, the mean value of $y = 5x^2$ between $x = 1$ and $x = 4$.

$$\text{Mean value, } \bar{y} = \frac{1}{4-1}\int_1^4 y\,dx = \frac{1}{3}\int_1^4 5x^2\,dx = \frac{1}{3}\left[\frac{5x^3}{3}\right]_1^4$$

$$= \frac{5}{9}\left[x^3\right]_1^4 = \frac{5}{9}(64-1) = \mathbf{35}$$

Problem 2 A sinusoidal voltage is given by $v = 100 \sin \theta$ volts. Determine the mean value of the voltage over half a cycle using integration.

Half a cycle indicates that the limits are 0 and π rads.

$$\text{Hence mean value, } \bar{y} = \frac{1}{\pi-0}\int_0^\pi v\,d\theta = \frac{1}{\pi}\int_0^\pi 100 \sin\theta\,d\theta$$

$$= \frac{100}{\pi}\left[-\cos\theta\right]_0^\pi = \frac{100}{\pi}\{(-\cos\pi)-(-\cos 0)\}$$

$$= \frac{100}{\pi}\{(+1)-(-1)\} = \frac{200}{\pi}$$

$$= \mathbf{63.66\ volts.}$$

Note that for a sine wave, the **mean value** $= \frac{2}{\pi} \times$ **maximum value.** In this case, mean value $= 2/\pi \times 100 = 63.66$ V

Problem 3 Calculate the mean value of $y = 3x^2+2$ in the range $x = 0$ to $x = 3$ by (a) the mid-ordinate rule and (b) integration.

(a) A graph of $y = 3x^2+2$ over the required range is shown in *Fig 2* using the following table:

x	0	0.5	1.0	1.5	2.0	2.5	3.0
y	2.0	2.75	5.0	8.75	14.0	20.75	29.0

Using the mid-ordinate rule,

$$\text{mean value} = \frac{\text{area under curve}}{\text{length of base}}$$

$$= \frac{\text{sum of mid-ordinates}}{\text{number of mid-ordinates}}$$

Selecting 6 intervals, each of width 0.5, the mid-ordinates are erected as shown by the broken lines in *Fig 2*.

Mean value =

$$\frac{2.2+3.7+6.7+11.2+17.2+24.7}{6}$$

$$= \frac{65.7}{6} = \mathbf{10.95}$$

(b) By integration, mean value

$$= \frac{1}{3-0}\int_0^3 y\,dx = \frac{1}{3}\int_0^3 (3x^2+2)dx$$

$$= \frac{1}{3}\left[x^3+2x\right]_0^3 = \frac{1}{3}\{(27+6)-(0)\} = \mathbf{11}$$

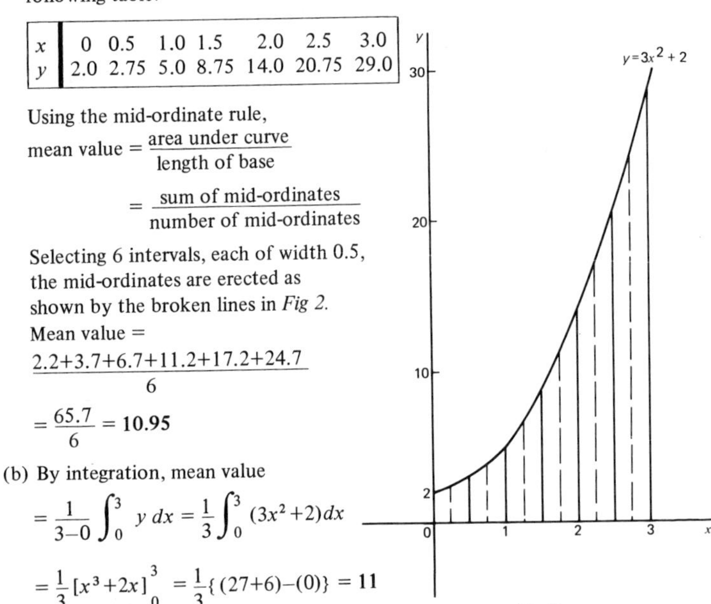

Fig 2

The answer obtained by integration is exact; greater accuracy may be obtained by the mid-ordinate rule if a larger number of intervals are selected.

Problem 4 The number of atoms, N, remaining in a mass of material during radioactive decay after time t seconds is given by $N = N_0 e^{-\lambda t}$, where N_0 and λ are constants. Determine the mean number of atoms in the mass of material for the time period $t = 0$ to $t = 1/\lambda$.

$$\text{Mean number of atoms} = \frac{1}{\frac{1}{\lambda} - 0} \int_0^{\frac{1}{\lambda}} N\,dt = \frac{1}{\frac{1}{\lambda}} \int_0^{\frac{1}{\lambda}} N_0 e^{-\lambda t}\,dt$$

$$= \lambda N_0 \int_0^{\frac{1}{\lambda}} e^{-\lambda t}\,dt = \lambda N_0 \left[\frac{e^{-\lambda t}}{-\lambda}\right]_0^{\frac{1}{\lambda}}$$

$$= -N_0\left[e^{-\lambda(\frac{1}{\lambda})} - e^0\right] = -N_0\,[e^{-1} - e^0] = +N_0\,[e^0 - e^{-1}]$$

$$= N_0\,[1 - e^{-1}] = \mathbf{0.632\,N_0}$$

Problem 5 Determine the r.m.s. value of $y = 2x^2$ between $x = 1$ and $x = 4$.

$$\text{r.m.s. value} = \sqrt{\left\{\frac{1}{4-1}\int_1^4 y^2\,dx\right\}} = \sqrt{\left\{\frac{1}{3}\int_1^4 (2x^2)^2\,dx\right\}} = \sqrt{\left\{\frac{1}{3}\int_1^4 4x^4\,dx\right\}}$$

$$= \sqrt{\left\{\frac{4}{3}\left[\frac{x^5}{5}\right]_1^4\right\}} = \sqrt{\left\{\frac{4}{15}(1024-1)\right\}} = \sqrt{(272.8)} = \mathbf{16.5}$$

Problem 6 A sinusoidal voltage has a maximum value of 10 V. Calculate its r.m.s. value.

A sinusoidal voltage v having a maximum value of 10 V may be written as $v = 10 \sin \theta$ V.

Over the range $\theta = 0$ to $\theta = \pi$,

$$\text{r.m.s. value} = \sqrt{\left\{\frac{1}{\pi - 0}\int_0^\pi v^2\,d\theta\right\}} = \sqrt{\left\{\frac{1}{\pi}\int_0^\pi (10 \sin \theta)^2\,d\theta\right\}}$$

$$= \sqrt{\left\{\frac{100}{\pi}\int_0^\pi \sin^2\theta\,d\theta\right\}}, \text{ which is not a 'standard' integral.}$$

In chapter 13, para. 6, it was shown that $\cos 2\theta = 1 - 2\sin^2\theta$ and this formula is used whenever $\sin^2\theta$ needs to be integrated.

Rearranging $\cos 2\theta = 1 - 2\sin^2\theta$ gives $\sin^2\theta = \frac{1}{2}(1 - \cos 2\theta)$.

$$\text{Hence} \sqrt{\left\{\frac{100}{\pi}\int_0^\pi \sin^2\theta\,d\theta\right\}} = \sqrt{\left\{\frac{100}{\pi}\int_0^\pi \frac{1}{2}(1 - \cos 2\theta)\,d\theta\right\}}$$

$$= \sqrt{\left\{\frac{100}{\pi}\left(\frac{1}{2}\right)\left[\theta - \frac{\sin 2\theta}{2}\right]_0^\pi\right\}}$$

$$= \sqrt{\left\{\frac{100}{\pi}\left(\frac{1}{2}\right)\left[\left(\pi - \frac{\sin 2\pi}{2}\right) - \left(0 - \frac{\sin 0}{2}\right)\right]\right\}}$$

$$=\sqrt{\left\{\frac{100}{\pi}\left(\frac{1}{2}\right)[\pi]\right\}}=\sqrt{\frac{100}{2}}=\frac{10}{\sqrt{2}}=\mathbf{7.071\ volts.}$$

Note that for a sine wave, the **r.m.s. value** $= \frac{1}{\sqrt{2}} \times$ **maximum value.** In this case,

r.m.s. value $= \frac{1}{\sqrt{2}} \times 10 = 7.071$ V.

Problem 7 In a frequency distribution the average distance from the mean, y, is related to the variable, x, by the equation $y = 2x^2 - 1$. Determine, correct to 3 significant figures, the r.m.s. deviation from the mean for values of x from -1 to $+4$.

$$\text{r.m.s. deviation} = \sqrt{\left\{\frac{1}{4--1}\int_{-1}^{4} y^2\,dx\right\}} = \sqrt{\left\{\frac{1}{5}\int_{-1}^{4}(2x^2-1)^2\,dx\right\}}$$

$$=\sqrt{\left\{\frac{1}{5}\int_{-1}^{4}(4x^4-4x^2+1)\,dx\right\}}=\sqrt{\left\{\frac{1}{5}\left[\frac{4x^5}{5}-\frac{4x^3}{3}+x\right]_{-1}^{4}\right\}}$$

$$=\sqrt{\left\{\frac{1}{5}\left[\left(\frac{4}{5}(4)^5-\frac{4}{3}(4)^3+4\right)-\left(\frac{4}{5}(-1)^5-\frac{4}{3}(-1)^3+(-1)\right)\right]\right\}}$$

$$=\sqrt{\left\{\frac{1}{5}[(737.87)-(-0.467)]\right\}}=\sqrt{\left\{\frac{1}{5}[738.34]\right\}}$$

$=\sqrt{(147.67)} = 12.152 =$ **12.2, correct to 3 significant figures.**

C FURTHER PROBLEMS ON MEAN AND ROOT MEAN SQUARE VALUES

1 Determine the mean value of (a) $y = 3\sqrt{x}$ from $x = 0$ to $x = 4$; (b) $y = \sin 2\theta$ from $\theta = 0$ to $\theta = \frac{\pi}{4}$; (c) $y = 4e^t$ from $t = 1$ to $t = 4$.
[(a) 4; (b) $\frac{2}{\pi}$ or 0.637; (c) 69.17]

2 Calculate the mean value of $y = 2x^2+5$ in the range $x = 1$ to $x = 4$ by (a) the mid-ordinate rule, and (b) integration. [19]

3 The speed v of a vehicle is given by $v = (4t+3)$ m/s, where t is the time in seconds. Determine the average value of the speed from $t = 0$ to $t = 3$ s.
[9 m/s]

4 Find the mean value of the curve $y = 6+x-x^2$ which lies above the x-axis by using an approximate method. Check the result using integration. [$4\frac{1}{6}$]

5 The vertical height h km of a missile varies with the horizontal distance d km, and is given by $h = 4d-d^2$. Determine the mean height of the missile from $d = 0$ to $d = 4$ km. [$2\frac{2}{3}$ km]

6 The velocity v of a piston moving with simple harmonic motion at any time t is given by $v = c \sin \omega t$, where c is a constant. Determine the mean velocity between $t = 0$ and $t = \pi/\omega$. $\left[\frac{2c}{\pi}\right]$

7 Determine the r.m.s. values of
(a) $y = 3x$ from $x = 0$ to $x = 4$.
(b) $y = t^2$ from $t = 1$ to $t = 3$.
(c) $y = 25 \sin\theta$ from $\theta = 0$ to $\theta = 2\pi$.
[(a) 6.928; (b) 4.919; (c) $\frac{25}{\sqrt{2}}$ or 17.68]

8 Calculate the r.m.s. values of

(a) $y = \sin 2\theta$ from $\theta = 0$ to $\theta = \frac{\pi}{4}$.

(b) $y = 1+\sin t$ from $t = 0$ to $t = 2\pi$.

(c) $y = 3 \cos 2x$ from $x = 0$ to $x = \pi$.

(Note that $\cos^2 t = \frac{1}{2}(1+\cos 2t)$, from para 6, chapter 13).

[(a) $\frac{1}{\sqrt{2}}$ or 0.707; (b) 1.225; (c) 2.121]

9 The distance, p, of points from the mean value of a frequency distribution are related to the variable, q, by the equation $p = (1/q)+q$. Determine the standard deviation (i.e. the r.m.s. value), correct to 3 significant figures, for values from $q = 1$ to $q = 3$. [2.58]

10 A current, $i = 30 \sin 100\,\pi t$ amperes is applied across an electric circuit. Determine its mean and r.m.s. values, each correct to 4 significant figures, over the range $t = 0$ to $t = 10$ ms. [19.10 A; 21.21 A]

11 A sinusoidal voltage has a peak value of 340 V. Calculate its mean and r.m.s. values, correct to 3 significant figures. [216 V; 240 V]

12 Determine the form factor, correct to 3 significant figures, of a sinusoidal voltage of maximum value 100 volts, given that

$$\text{form factor} = \frac{\text{r.m.s. value}}{\text{average value}}.$$

[1.11]

20 Calculus (7) – Volumes of solids of revolution

A MAIN POINTS CONCERNED WITH VOLUMES OF SOLIDS OF REVOLUTION

1 If the area under the curve $y = f(x)$, (shown in *Fig 1(a)*), between $x = a$ and $x = b$ is rotated 360° about the x-axis, then a volume known as a **solid of revolution** is produced as shown in *Fig 1(b)*. The volume of such a solid may be determined precisely using integration.

2 (i) Let the area shown in *Fig 1(a)* be divided into a number of strips each of width δx. One such strip is shown shaded.

(ii) When the area is rotated 360° about the x-axis, each strip produces a solid of revolution approximating to a circular disc of radius y and thickness δx. Volume of disc = (circular cross-sectional area)(thickness) = $(\pi y^2)(\delta x)$.

(iii) Total volume, V, between ordinate $x = a$ and $x = b$ is given by:

$$\textbf{Volume, } V = \lim_{\delta x \to 0} \sum_{x=a}^{x=b} \pi y^2\, \delta x = \int_a^b \pi y^2\, dx$$

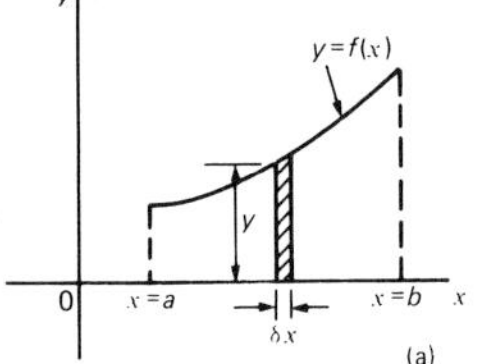

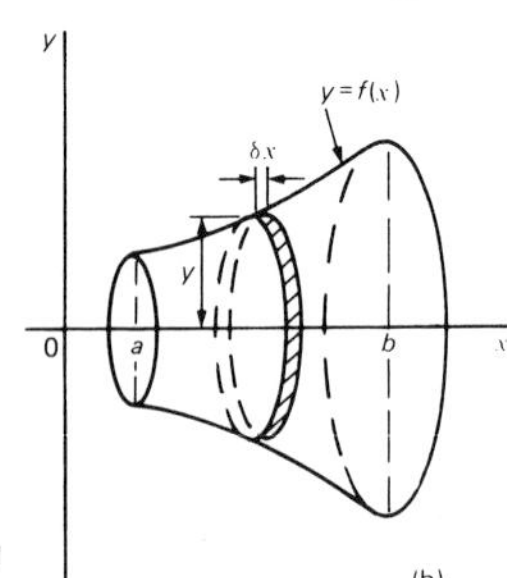

Fig 1

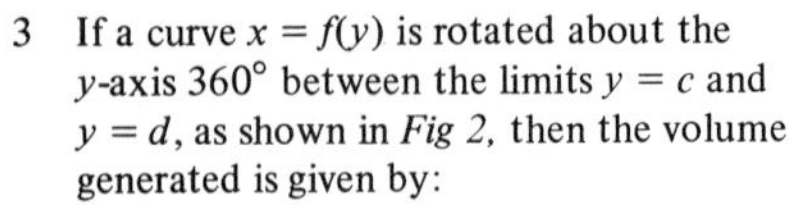

3 If a curve $x = f(y)$ is rotated about the y-axis 360° between the limits $y = c$ and $y = d$, as shown in *Fig 2*, then the volume generated is given by:

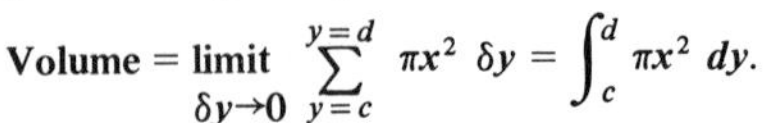

$$\textbf{Volume} = \lim_{\delta y \to 0} \sum_{y=c}^{y=d} \pi x^2\, \delta y = \int_c^d \pi x^2\, dy.$$

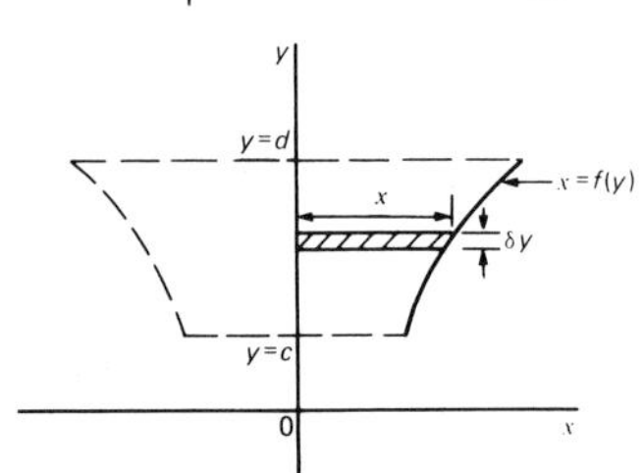

Fig 2

Problem 1 Determine the volume of the solid of revolution formed when the curve $y = 2$ is rotated 360° about the x-axis between the limits $x = 0$ to $x = 3$.

When $y = 2$ is rotated 360° about the x-axis between $x = 0$ and $x = 3$ (see *Fig 3*):

$$\text{volume generated} = \int_0^3 \pi y^2\, dx = \int_0^3 \pi(2)^2 dx = \int_0^3 4\pi\, dx$$

$$= 4\pi\,[x]_0^3 = \mathbf{12\pi \text{ cubic units.}}$$

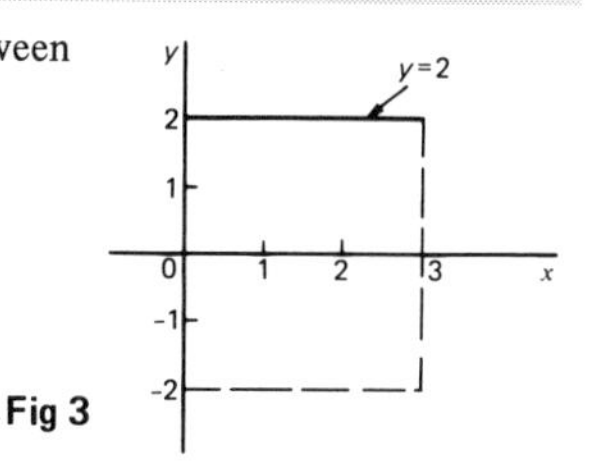

Fig 3

(*Check*: The volume generated is a cylinder of radius 2 and height 3.
Volume of cylinder $= \pi r^2 h = \pi(2)^2(3) = \mathbf{12\pi}$ **cubic units**).

Problem 2 Find the volume of the solid of revolution when the curve $y = 2x$ is rotated one revolution about the x-axis between the limits $x = 0$ and $x = 5$.

When $y = 2x$ is revolved one revolution about the x-axis between $x = 0$ and $x = 5$ (see *Fig 4*) then:

$$\text{volume generated} = \int_0^5 \pi y^2\, dx = \int_0^5 \pi(2x)^2\, dx = \int_0^5 4\pi x^2\, dx$$

$$= 4\pi\left[\frac{x^3}{3}\right]_0^5 = \frac{500\pi}{3} = \mathbf{166\tfrac{2}{3}\pi \text{ cubic units.}}$$

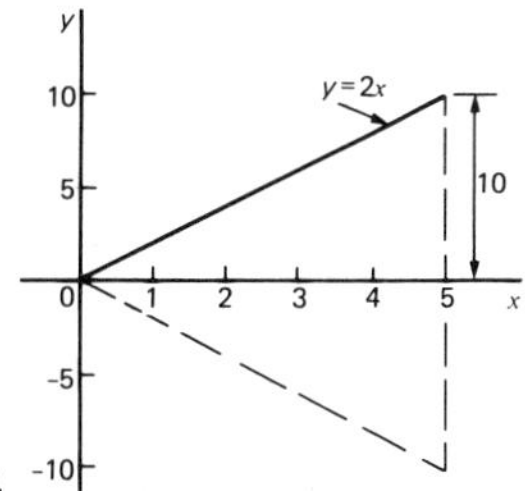

Fig 4

(*Check*: The volume generated is a cone of radius 10 and height 5.

Volume of cone $= \frac{1}{3}\pi r^2 h = \frac{1}{3}\pi(10)^2 5 = \frac{500\pi}{3} = \mathbf{166\tfrac{2}{3}\pi}$ **cubic units.**)

Problem 3 The curve $y = x^2 + 4$ is rotated one revolution about the x-axis between the limits $x = 1$ and $x = 4$. Determine the volume of the solid of revolution produced.

Revolving the shaded area shown in *Fig 5* 360° about the x-axis produces a solid of revolution given by:

$$\text{Volume} = \int_1^4 \pi y^2\, dx = \int_1^4 \pi(x^2+4)^2 dx$$

$$= \pi \int_1^4 (x^4 + 8x^2 + 16) dx$$

$$= \pi\left[\frac{x^5}{5} + \frac{8x^3}{3} + 16x\right]_1^4 = \pi\,[(204.8+170.67+64)-(0.2+2.67+16)]$$

$$= \mathbf{420.6\pi \text{ cubic units.}}$$

Problem 4 If the curve in *Problem 3* is revolved about the y-axis between the same limits, determine the volume of the solid of revolution produced.

The volume produced when the curve $y = x^2 + 4$ is rotated about the y-axis between $y = 5$ (when $x = 1$) and $y = 20$ (when $x = 4$)–i.e. rotating area ABCD of *Fig 5* about the y-axis is given by:

$$\text{volume} = \int_5^{20} \pi x^2 \, dy.$$

Since $y = x^2 + 4$, then $x^2 = y - 4$.

$$\text{Hence volume} = \int_5^{20} \pi (y-4) dy$$

$$= \pi \left[\frac{y^2}{2} - 4y \right]_5^{20} = \pi [(120) - (-7\tfrac{1}{2})]$$

$$= \mathbf{127\tfrac{1}{2}\pi \text{ cubic units.}}$$

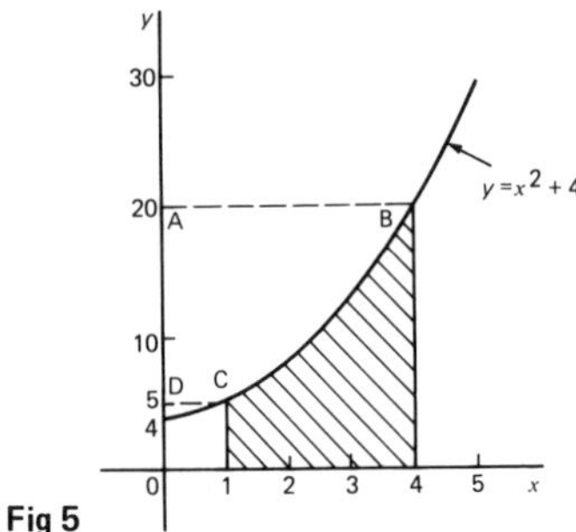

Fig 5

Problem 5 The area enclosed by the curve $y = 3e^{\frac{x}{3}}$, the x-axis and ordinates $x = -1$ and $x = 3$ is rotated 360° about the x-axis. Determine the volume generated.

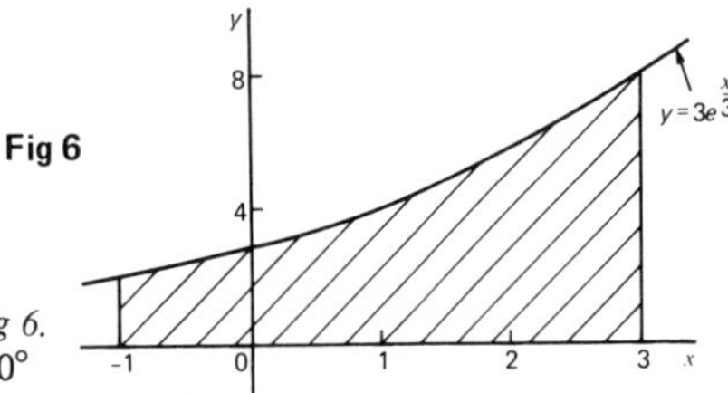

Fig 6

A sketch of $y = 3e^{\frac{x}{3}}$ is shown in *Fig 6*. When the shaded area is rotated 360° about the x-axis then;

$$\text{volume generated} = \int_{-1}^{3} \pi y^2 \, dx = \int_{-1}^{3} \pi (3e^{\frac{x}{3}})^2 \, dx = 9\pi \int_{-1}^{3} e^{\frac{2x}{3}} \, dx$$

$$= 9\pi \left[\frac{e^{\frac{2x}{3}}}{\frac{2}{3}} \right]_{-1}^{3} = \frac{27\pi}{2}(e^2 - e^{-\frac{2}{3}})$$

$$= \mathbf{92.82\pi \text{ cubic units.}}$$

Problem 6 Determine the volume generated when the area above the x-axis bounded by the curve $x^2 + y^2 = 9$ and the ordinates $x = 3$ and $x = -3$ is rotated one revolution about the x-axis.

Fig 7 shows the part of the curve $x^2 + y^2 = 9$ lying above the x-axis. Since, in general, $x^2 + y^2 = r^2$ represents a circle, centre 0 and radius r, then $x^2 + y^2 = 9$ represents a circle, centre 0 and radius 3. When the semicircular area of *Fig 7* is rotated one revolution about the x-axis then:

volume generated $= \int_{-3}^{3} \pi y^2 \, dx$

$= \int_{-3}^{3} \pi(9-x^2)dx = \pi\left[9x - \frac{x^3}{3}\right]_{-3}^{3}$

$= \pi\,[(18)-(-18)] = \mathbf{36\pi}$ **cubic units.**

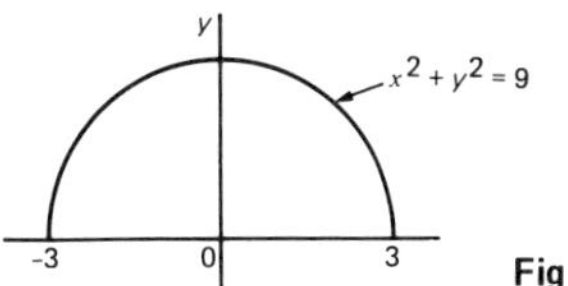

Fig 7

(*Check*: The volume generated is a sphere of radius 3

Volume of sphere $= \frac{4}{3}\pi r^3 = \frac{4}{3}\pi(3)^3 = \mathbf{36\pi}$ **cubic units.**)

Problem 7 Calculate the volume of a frustum of a sphere of radius 4 cm which lies between two parallel planes at 1 cm and 3 cm from the centre and on the same side of it.

The volume of a frustum of a sphere may be determined by integration by rotating the curve $x^2 + y^2 = 4^2$ (i.e. a circle, centre 0, radius 4) one revolution about the x-axis, between the limits $x = 1$ and $x = 3$ (i.e. rotating the shaded area of *Fig 8*.)

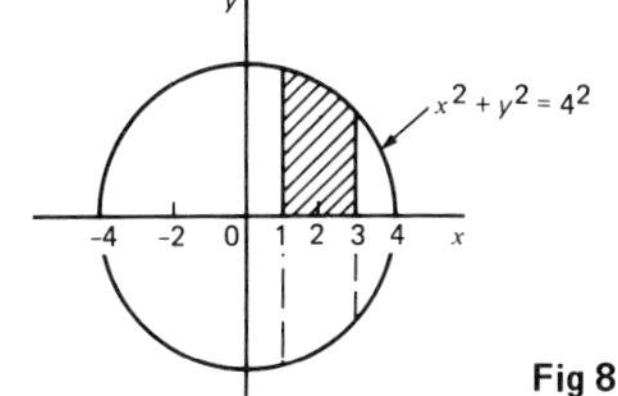

Fig 8

Volume of frustum $= \int_{1}^{3} \pi y^2 \, dx = \int_{1}^{3} \pi(4^2 - x^2)dx = \pi\left[16x - \frac{x^3}{3}\right]_{1}^{3}$

$= \pi\,[(39)-(15\tfrac{2}{3})] = \mathbf{23\tfrac{1}{3}\pi}$ **cubic units.**

Problem 8 The area enclosed between the two parabolas $y = x^2$ and $y^2 = 8x$ of *Problem 12*, chapter 18, page 200, is rotated 360° about the x-axis. Determine the volume of the solid produced.

The area enclosed by the two curves is shown in *Fig 13*, page 200. The volume produced by revolving the shaded area 360° about the x-axis is given by {(volume produced by revolving $y^2 = 8x$)–(volume produced by revolving $y = x^2$)}

i.e. volume $= \int_{0}^{2} \pi(8x)dx - \int_{0}^{2} \pi(x^4)dx$

$= \pi\int_{0}^{2} (8x - x^4)dx = \pi\left[\frac{8x^2}{2} - \frac{x^5}{5}\right]_{0}^{2} = \pi\left[(16 - \frac{32}{5}) - (0)\right]$

$= \mathbf{9.6\pi}$ **cubic units.**

C FURTHER PROBLEMS ON VOLUMES OF SOLIDS OF REVOLUTION

Answers to *Problems 1 to 17* are in cubic units and in terms of π. In *Problems 1 to 7*, determine the volume of the solid of revolution formed by revolving the areas enclosed by the given curves, the x-axis and the given ordinates through one revolution about the x-axis.

1 $y = 5x$; $x = 1,\ x = 4$ [525π]

2 $y = x^2$; $x = -2,\ x = 3$ [55π]

3 $y = 2x^2+3$; $x = 0,\ x = 2$ [75.6π]

4 $\frac{y^2}{4} = x$; $x = 1,\ x = 5$ [48π]

5 $xy = 3$; $x = 2,\ x = 3$ [1.5π]

6 $y = 4e^x$; $x = 0, x = 2$ [428.8π]

7 $y = \sec x$; $x = 0, x = \frac{\pi}{4}$ [π]

In *Problems 8 to 12*, determine the volume of the solid of revolution formed by revolving the areas enclosed by the given curves, the y-axis and the given ordinates through one revolution about the y-axis.

8 $y = x^2$; $y = 1, y = 3$ [4π]

9 $y = 3x^2-1$; $y = 2, y = 4$ [$2\frac{2}{3}\pi$]

10 $y = \frac{2}{x}$; $y = 1, y = 3$ [$2\frac{2}{3}\pi$]

11 $x^2+y^2 = 16$; $y = 0, y = 4$ [$42\frac{2}{3}\pi$]

12 $x\sqrt{y} = 2$; $y = 2, y = 3$ [1.622π]

13 The curve $y = 2x^2+3$ is rotated about (a) the x-axis, between the limits $x = 0$ and $x = 3$, and (b) the y-axis, between the same limits. Determine the volume generated in each case. [(a) 329.4π; (b) 81π]

14 Determine the volume of a plug formed by the frustum of a sphere of radius 6 cm which lies between two parallel planes at 2 cm and 4 cm from the centre and on the same side of it. (The equation of a circle, centre 0, radius r is $x^2+y^2 = r^2$). [$53\frac{1}{3}\pi$]

15 The area enclosed between the two curves $x^2 = 3y$ and $y^2 = 3x$ is rotated about the x- axis. Determine the volume of the solid formed. [8.1π]

16 The portion of the curve $y = x^2+\frac{1}{x}$ lying between $x = 1$ and $x = 3$ is revolved 360° about the x-axis. Determine the volume of the solid formed. [$57\frac{1}{15}\pi$]

17 Calculate the volume of the frustum of a sphere of radius 5 cm which lies between two parallel planes at 3 cm and 2 cm from the centre and on opposite sides of it. [$113\frac{1}{3}\pi$]

18 Sketch the curves $y = x^2+2$ and $y-12 = 3x$ from $x = -3$ to $x = 6$. Determine (a) the co-ordinates of the points of intersection of the two curves; and (b) the area enclosed by the two curves; (c) if the enclosed area is rotated 360° about the x-axis, calculate the volume of the solid produced.

[(a) (−2, 6) and (5, 27); (b) $57\frac{1}{6}$ square units; (c) 1326π cubic units]

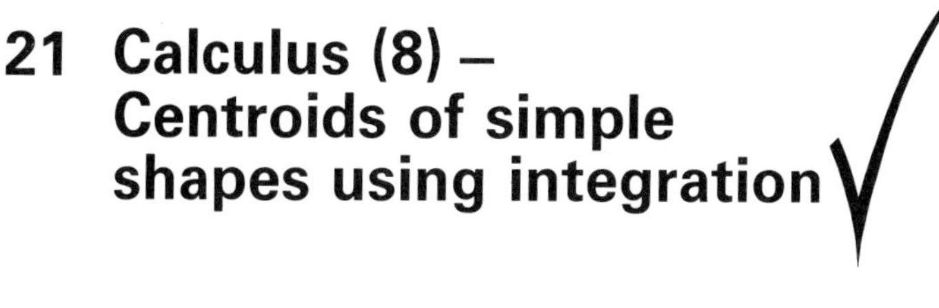

21 Calculus (8) – Centroids of simple shapes using integration

A MAIN POINTS CONCERNED WITH CENTROIDS OF SIMPLE SHAPES USING INTEGRATION

1 A **lamina** is a think flat sheet having uniform thickness. The **centre of gravity** of a lamina is the point where it balances perfectly, i.e. the lamina's **centre of mass.** When dealing with an area (i.e. a lamina of negligible thickness and mass) the term **centre of area** or **centroid** is used for the point where the centre of gravity of a lamina of that shape would lie.

2 The **first moment of area** is defined as the product of the area and the perpendicular distance of its centroid from a given axis in the plane of the area. In *Fig 1*, the first moment of area A about axis XX is given by (Ay) cubic units.

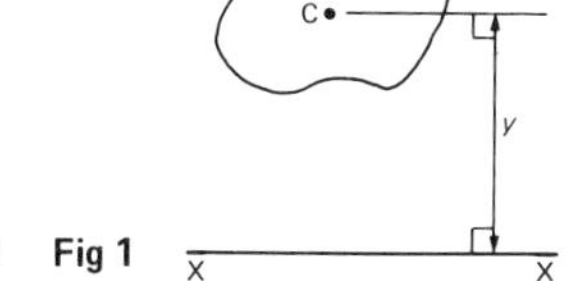

Fig 1

3 **Centroid of area between a curve and the x-axis**

(i) *Fig 2* shows an area PQRS bounded by the curve $y = f(x)$, the x-axis and ordinates $x = a$ and $x = b$. Let this area be divided into a large number of strips, each of width δx. A typical strip is shown shaded drawn at point (x, y) on $f(x)$. The area of the strip is approximately rectangular and is given by $y\delta x$.

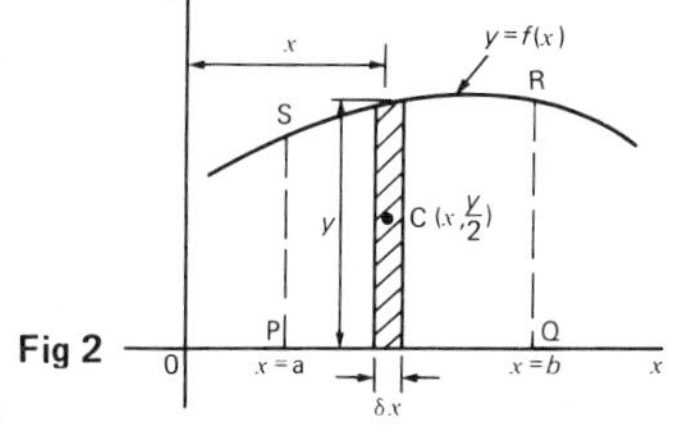

Fig 2

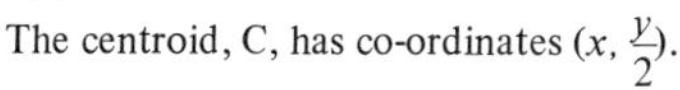

The centroid, C, has co-ordinates $(x, \frac{y}{2})$.

(ii) First moment of area of shaded strip about axis OY $= (y\delta x)(x) = xy\delta x$.

Total first moment of area PQRS about axis OY $= \lim_{\delta x \to 0} \sum_{x=a}^{x=b} xy\delta x = \int_a^b xy\,dx$.

(iii) First moment of area of shaded strip about axis OX $= (y\delta x)(\frac{y}{2}) = \frac{1}{2}y^2\,\delta x$.

Total first moment of area PQRS about axis OX $= \lim_{\delta x \to 0} \sum_{x=a}^{x=b} \frac{1}{2}y^2\,\delta x = \frac{1}{2}\int_a^b y^2\,dx$.

(iv) Area of PQRS, $A = \int_a^b y\,dx$ (from chapter 18).

(v) Let $\bar{x}$ and $\bar{y}$ be the distances of the centroid of area A about OY and OX respectively then:

$(\bar{x})(A)$ = total first moment of area A about axis OY $= \int_a^b xy\,dx$,

from which, $\bar{x} = \dfrac{\int_a^b xy\,dx}{\int_a^b y\,dx}$,

and $(\bar{y})(A)$ = total first moment of area A about axis OX $= \frac{1}{2}\int_a^b y^2\,dx$,

from which, $\bar{y} = \dfrac{\frac{1}{2}\int_a^b y^2\,dx}{\int_a^b y\,dx}$.

4 **Centroid of area between a curve and the y-axis**
If $\bar{x}$ and $\bar{y}$ are the distances of the centroid of area EFGH in *Fig 3* from OY and OX respectively, then, by similar reasoning as in para. 3:

$(\bar{x})(\text{total area}) = \lim_{\delta y \to 0} \sum_{y=c}^{y=d} (x\delta y)\left(\frac{x}{2}\right) = \frac{1}{2}\int_c^d x^2\,dy$,

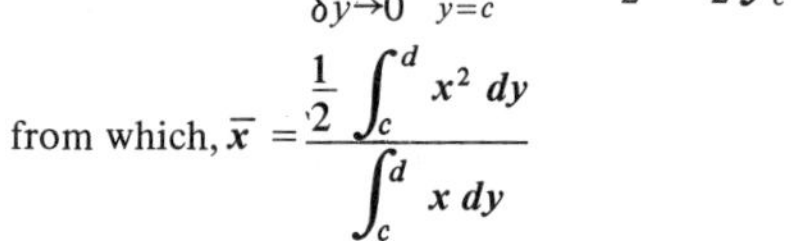

from which, $\bar{x} = \dfrac{\frac{1}{2}\int_c^d x^2\,dy}{\int_c^d x\,dy}$

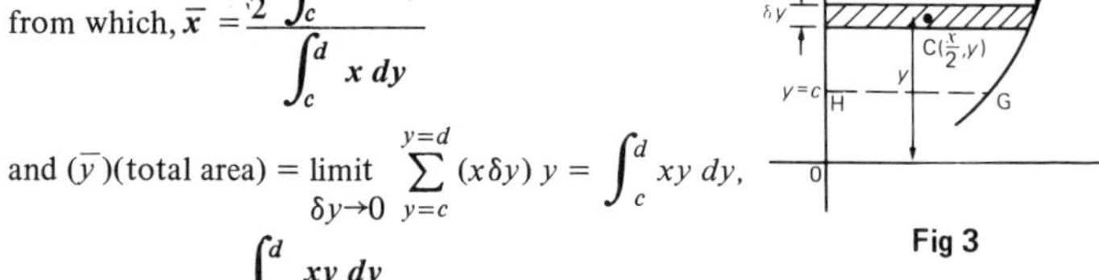

and $(\bar{y})(\text{total area}) = \lim_{\delta y \to 0} \sum_{y=c}^{y=d} (x\delta y)\,y = \int_c^d xy\,dy$,

Fig 3

from which, $\bar{y} = \dfrac{\int_c^d xy\,dy}{\int_c^d x\,dy}$

5 **A theorem of Pappus** states:
'If a plane area is rotated about an axis in its own plane but not intersecting it, the volume of the solid formed is given by the product of the area and the distance moved by the centroid of the area'.
With reference to *Fig 4*, when the curve $y = f(x)$ is rotated one revolution about the x-axis between the limits $x = a$ and $x = b$, the volume V generated is given by:

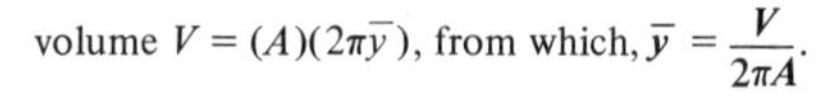

volume $V = (A)(2\pi\bar{y})$, from which, $\bar{y} = \dfrac{V}{2\pi A}$.

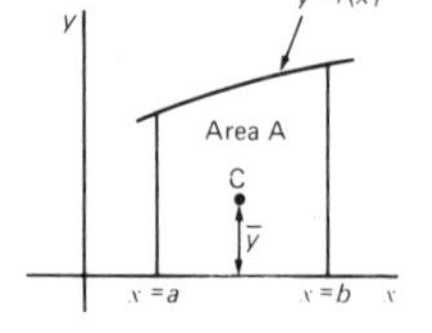

Fig 4

B WORKED PROBLEMS ON CENTROIDS OF SIMPLE SHAPES BY INTEGRATION

Problem 1 Show, by integration, that the centroid of a rectangle lies at the intersection of the diagonals.

Let a rectangle be formed by the line $y = b$, the x-axis and ordinates $x = 0$ and $x = l$ as shown in *Fig 5*. Let the co-ordinates of the centroid C of this area be $(\bar{x}, \bar{y})$.

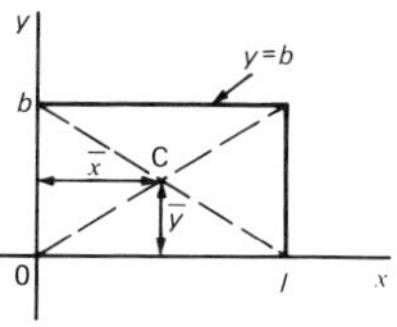

Fig 5

By integration, (from para. 3), $$\bar{x} = \frac{\int_0^l xy\,dx}{\int_0^l y\,dx} = \frac{\int_0^l (x)(b)dx}{\int_0^l b\,dx} = \frac{\left[b\frac{x^2}{2}\right]_0^l}{[bx]_0^l} = \frac{\frac{bl^2}{2}}{bl} = \frac{l}{2}$$

and $$\bar{y} = \frac{\frac{1}{2}\int_0^l y^2\,dx}{\int_0^l y\,dx} = \frac{\frac{1}{2}\int_0^l b^2\,dx}{bl} = \frac{\frac{1}{2}[b^2x]_0^l}{bl} = \frac{\frac{1}{2}b^2l}{bl} = \frac{b}{2}$$

i.e. **the centroid lies at $(\frac{l}{2}, \frac{b}{2})$ which is at the intersection of the diagonals.**

Problem 2 Find the position of the centroid of the area bounded by the curve $y = 3x^2$, the x-axis and the ordinates $x = 0$ and $x = 2$.

If $(\bar{x}, \bar{y})$ are the co-ordinates of the centroid of the given area then, from para. 3,

$$\bar{x} = \frac{\int_0^2 xy\,dx}{\int_0^2 y\,dx} = \frac{\int_0^2 x(3x^2)dx}{\int_0^2 3x^2\,dx} = \frac{\int_0^2 3x^3\,dx}{\int_0^2 3x^2\,dx} = \frac{\left[\frac{3x^4}{4}\right]_0^2}{[x^3]_0^2} = \frac{12}{8} = 1.5$$

$$\bar{y} = \frac{\frac{1}{2}\int_0^2 y^2\,dx}{\int_0^2 y\,dx} = \frac{\frac{1}{2}\int_0^2 (3x^2)^2dx}{8} = \frac{\frac{1}{2}\int_0^2 9x^4\,dx}{8} = \frac{\frac{9}{2}\left[\frac{x^5}{5}\right]_0^2}{8} = \frac{\frac{9}{2}\left(\frac{32}{5}\right)}{8}$$
$$= \frac{18}{5} = 3.6$$

Hence the centroid lies at (1.5, 3.6).

Problem 3 Determine by integration the position of the centroid of the area enclosed by the line $y = 4x$, the x-axis and ordinates $x = 0$ and $x = 3$.

Let the co-ordinates of the area be $(\bar{x}, \bar{y})$ as shown in *Fig 6*.

From para. 3, $$\bar{x} = \frac{\int_0^3 xy\,dx}{\int_0^3 y\,dx} = \frac{\int_0^3 (x)(4x)dx}{\int_0^3 4x\,dx}$$
$$= \frac{\int_0^3 4x^2\,dx}{\int_0^3 4x\,dx} = \frac{\left[\frac{4x^3}{3}\right]_0^3}{[2x^2]_0^3} = \frac{36}{18} = 2$$

$$\bar{y} = \frac{\frac{1}{2}\int_0^3 y^2\,dx}{\int_0^3 y\,dx} = \frac{\frac{1}{2}\int_0^3 (4x)^2\,dx}{18} = \frac{\frac{1}{2}\int_0^3 16x^2\,dx}{18} = \frac{\frac{1}{2}\left[16\frac{x^3}{3}\right]_0^3}{18}$$

$$= \frac{72}{18} = 4$$

Hence the centroid lies at (2, 4).
In *Fig 6*, ABD is a right-angled triangle. The centroid lies 4 units from AB and 1 unit from BD showing that the centroid of a triangle lies at one-third of the perpendicular height above any side as base.

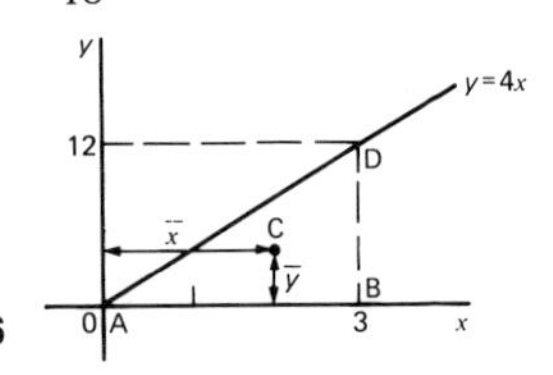

Fig 6

Problem 4 Determine the co-ordinates of the centroid of the area lying between the curve $y = 5x - x^2$ and the x-axis.

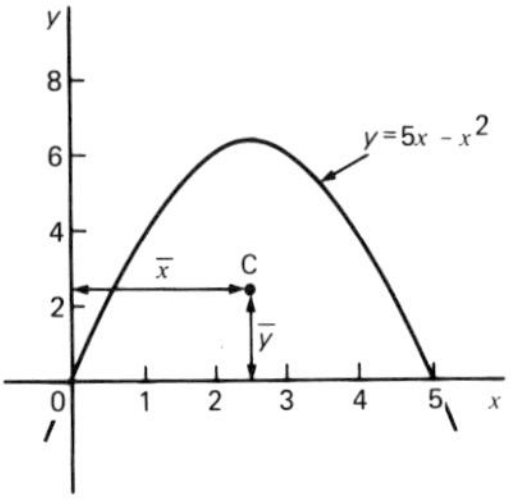

$y = 5x - x^2 = x(5 - x)$. When $y = 0$, $x = 0$ or $x = 5$. Hence the curve cuts the x-axis at 0 and 5 as shown in *Fig 7*. Let the co-ordinates of the centroid be $(\bar{x}, \bar{y})$ then, by integration, (from para. 3):

Fig 7

$$\bar{x} = \frac{\int_0^5 xy\,dx}{\int_0^5 y\,dx} = \frac{\int_0^5 x(5x - x^2)dx}{\int_0^5 (5x - x^2)dx} = \frac{\int_0^5 (5x^2 - x^3)\,dx}{\int_0^5 (5x - x^2)dx}$$

$$= \frac{\left[5\frac{x^3}{3} - \frac{x^4}{4}\right]_0^5}{\left[\frac{5x^2}{2} - \frac{x^3}{3}\right]_0^5} = \frac{\frac{625}{3} - \frac{625}{4}}{\frac{125}{2} - \frac{125}{3}} = \frac{\frac{625}{12}}{\frac{125}{6}} = \left(\frac{625}{12}\right)\left(\frac{6}{125}\right)$$

$$= \frac{5}{2} = 2.5$$

$$\bar{y} = \frac{\frac{1}{2}\int_0^5 y^2\,dx}{\int_0^5 y\,dx} = \frac{\frac{1}{2}\int_0^5 (5x - x^2)^2\,dx}{\int_0^5 (5x - x^2)\,dx} = \frac{\frac{1}{2}\int_0^5 (25x^2 - 10x^3 + x^4)\,dx}{\frac{125}{6}}$$

$$= \frac{\frac{1}{2}\left[\frac{25x^3}{3} - \frac{10x^4}{4} + \frac{x^5}{5}\right]_0^5}{\frac{125}{6}} = \frac{\frac{1}{2}\left(\frac{25(125)}{3} - \frac{6250}{4} + 625\right)}{\frac{125}{6}}$$

$$= 2.5$$

Hence the centroid of the area lies at (2.5, 2.5).
(Note from *Fig 7* that the curve is symmetrical about $x = 2.5$ and thus $\bar{x}$ could have been determined 'on sight'.)

Problem 5 Locate the centroid of the area enclosed by the curve $y = 2x^2$, the y-axis and ordinates $y = 1$ and $y = 4$, correct to 3 decimal places.

$$\text{From para. 4, } \bar{x} = \frac{\frac{1}{2}\int_1^4 x^2\,dy}{\int_1^4 x\,dy} = \frac{\frac{1}{2}\int_1^4 \frac{y}{2}\,dy}{\int_1^4 \sqrt{\left(\frac{y}{2}\right)}dy} = \frac{\frac{1}{2}\left[\frac{y^2}{4}\right]_1^4}{\left[\frac{2y^{3/2}}{3\sqrt{2}}\right]_1^4} = \frac{\frac{15}{8}}{\frac{14}{3\sqrt{2}}} = 0.568$$

$$\bar{y} = \frac{\int_1^4 xy\,dy}{\int_1^4 x\,dy} = \frac{\int_1^4 \sqrt{\left(\frac{y}{2}\right)}(y)\,dy}{\frac{14}{3\sqrt{2}}} = \frac{\int_1^4 \frac{y^{3/2}}{\sqrt{2}}\,dy}{\frac{14}{3\sqrt{2}}}$$

$$= \frac{\frac{1}{\sqrt{2}}\left[\frac{y^{5/2}}{\frac{5}{2}}\right]_1^4}{\frac{14}{3\sqrt{2}}} = \frac{\frac{2}{5\sqrt{2}}(31)}{\frac{14}{3\sqrt{2}}} = 2.657$$

Hence the position of the centroid is at (0.568, 2.657).

Problem 6 Determine the position of the centroid of a semicircle of radius r by using the theorem of Pappus. Check the answer by using integration (given that the equation of a circle, centre 0, radius r is $x^2 + y^2 = r^2$).

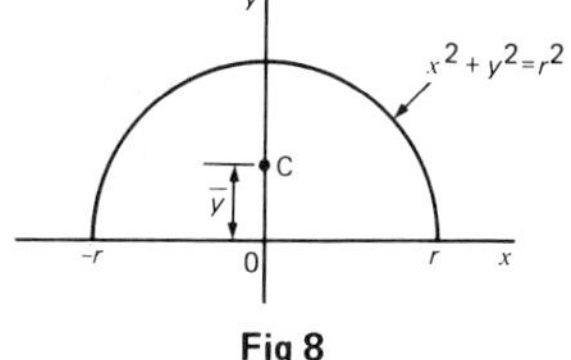

Fig 8

A semicircle is shown in *Fig 8* with its diameter lying on the x-axis and its centre at the origin. Area of semicircle $= \frac{\pi r^2}{2}$

When the area is rotated about the x-axis one revolution a sphere is generated of volume $\frac{4}{3}\pi r^3$.

Let centroid C be at a distance $\bar{y}$ from the origin as shown in *Fig 8.*

From the theorem of Pappus, volume generated = area × distance moved through by centroid

$$\text{i.e. } \frac{4}{3}\pi r^3 = \left(\frac{\pi r^2}{2}\right)(2\pi\bar{y})$$

$$\text{Hence } \bar{y} = \frac{\frac{4}{3}\pi r^3}{\pi^2 r^2} = \frac{4r}{3\pi}$$

$$\text{By integration, } \bar{y} = \frac{\frac{1}{2}\int_{-r}^{r} y^2\,dx}{\text{area}} = \frac{\frac{1}{2}\int_{-r}^{r}(r^2-x^2)\,dx}{\frac{\pi r^2}{2}} = \frac{\frac{1}{2}\left[r^2x - \frac{x^3}{3}\right]_{-r}^{r}}{\frac{\pi r^2}{2}}$$

$$= \frac{\frac{1}{2}\left[\left(r^3 - \frac{r^3}{3}\right) - \left(-r^3 + \frac{r^3}{3}\right)\right]}{\frac{\pi r^2}{2}} = \frac{4r}{3\pi}$$

Hence the centroid of a semicircle lies on the axis of symmetry, distance $\frac{4r}{3\pi}$ (or 0.424 r) from its diameter.

Problem 7 (a) Calculate the area bounded by the curve $y = 2x^2$, the x-axis and ordinates $x = 0$ and $x = 3$. (b) If this area is revolved (i) about the x-axis and (ii) about the y-axis, find the volumes of the solids produced. (c) Locate the area of the centroid using (i) integration, and (ii) the theorem of Pappus.

(a) The required area is shown shaded in *Fig 9*.

$$\text{Area} = \int_0^3 y\,dx = \int_0^3 2x^2\,dx = \left[\frac{2x^3}{3}\right]_0^3 = \mathbf{18\ square\ units.}$$

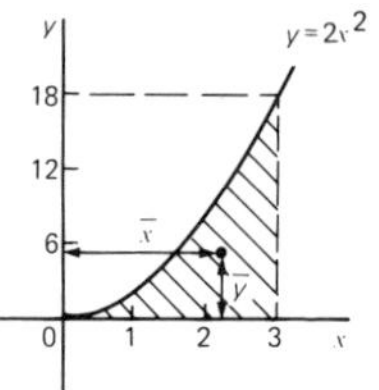

Fig 9

(b) (i) When the shaded area of *Fig 9* is revolved 360° about the x-axis,
The volume generated

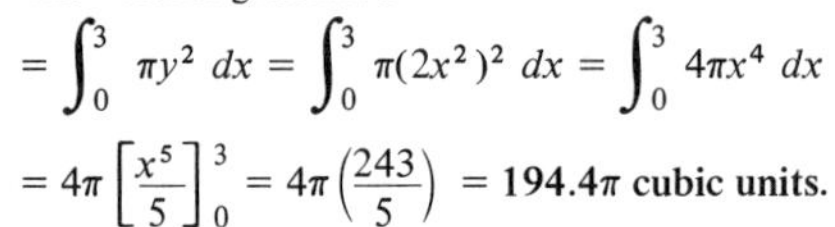

$$= \int_0^3 \pi y^2\,dx = \int_0^3 \pi(2x^2)^2\,dx = \int_0^3 4\pi x^4\,dx$$

$$= 4\pi\left[\frac{x^5}{5}\right]_0^3 = 4\pi\left(\frac{243}{5}\right) = \mathbf{194.4\pi\ cubic\ units.}$$

(ii) When the shaded area of *Fig 9* is revolved 360° about the y-axis, the volume generated = (volume generated by $x = 3$)–(volume generated by $y = 2x^2$)

$$= \int_0^{18} \pi(3)^2\,dy - \int_0^{18} \pi\left(\frac{y}{2}\right)dy$$

$$= \pi\int_0^{18}\left(9 - \frac{y}{2}\right)dy = \pi\left[9y - \frac{y^2}{4}\right]_0^{18} = \mathbf{81\pi\ cubic\ units}$$

(c) If the co-ordinates of the centroid of the shaded area in *Fig 9* are $(\bar{x}, \bar{y})$ then:

(i) by integration,
$$\bar{x} = \frac{\int_0^3 xy\,dx}{\int_0^3 y\,dx} = \frac{\int_0^3 x(2x^2)dx}{18} = \frac{\int_0^3 2x^3\,dx}{18}$$

$$= \frac{\left[\frac{2x^4}{4}\right]_0^3}{18} = \frac{81}{36} = 2.25$$

$$\bar{y} = \frac{\frac{1}{2}\int_0^3 y^2\,dx}{\int_0^3 y\,dx} = \frac{\frac{1}{2}\int_0^3 (2x^2)^2\,dx}{18} = \frac{\frac{1}{2}\int_0^3 4x^4\,dx}{18}$$

$$= \frac{\frac{1}{2}\left[\frac{4x^5}{5}\right]_0^3}{18} = 5.4$$

(ii) Using the theorem of Pappus:
Volume generated when shaded area is revolved about OY = (area)$(2\pi\bar{x})$

i.e. $81\pi = (18)(2\pi\bar{x})$, from which, $\bar{x} = \dfrac{81\pi}{36\pi} = 2.25$

Volume generated when shaded area is revolved about OX = (area)$(2\pi\bar{y})$

i.e., $194.4\pi = (18)(2\pi\bar{y})$, from which, $\bar{y} = \dfrac{194.4\pi}{36\pi} = 5.4$

Hence the centroid of the shaded area in *Fig 9* is at (2.25, 5.4).

Problem 8 Locate the position of the centroid enclosed by the curves $y = x^2$ and $y^2 = 8x$.

Fig 10 shows the two curves intersecting at (0, 0) and (2, 4). These are the same curves as used in *Problem 12,* chapter 18, where the shaded area was calculated as $2\frac{2}{3}$ square units. Let the co-ordinates of centroid C be $\bar{x}$ and $\bar{y}$.

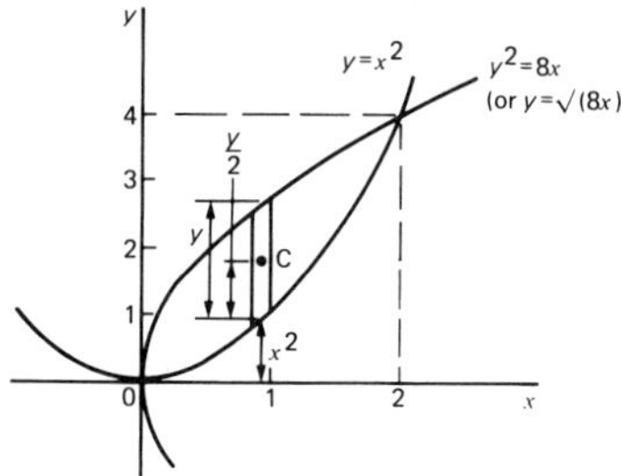

Fig 10

By integration, $\bar{x} = \dfrac{\int_0^2 xy\,dx}{\int_0^2 y\,dx}$.

The value of y is given by the height of the typical strip shown in *Fig 10,* i.e., $y = [\sqrt{(8x)} - x^2]$

$$\text{Hence, } \bar{x} = \frac{\int_0^2 x\,[\sqrt{(8x)} - x^2]\,dx}{2\frac{2}{3}} = \frac{\int_0^2 (\sqrt{(8)}x^{3/2} - x^3)dx}{2\frac{2}{3}}$$

$$= \frac{\left[\sqrt{(8)}\dfrac{x^{5/2}}{\frac{5}{2}} - \dfrac{x^4}{4}\right]_0^2}{2\frac{2}{3}} = \frac{\left(\sqrt{(8)}\dfrac{\sqrt{2^5}}{\frac{5}{2}} - 4\right)}{2\frac{2}{3}} = \frac{2\frac{2}{5}}{2\frac{2}{3}} = 0.9$$

Care needs to be taken when finding $\bar{y}$ in such examples as this.

From *Fig 10,* $y = \sqrt{(8x)} - x^2$ and $\frac{y}{2} = \frac{1}{2}[\sqrt{(8x)} - x^2]$

The perpendicular distance from centroid C of the strip to OX is $\frac{1}{2}[\sqrt{(8x)} - x^2] + x^2$

Taking moments about OX gives:

$$(\text{total area})(\bar{y}) = \sum_{x=0}^{x=2} (\text{area of strip})(\text{perpendicular distance of centroid of strip to OX})$$

$$\text{Hence (area)}(\bar{y}) = \int_0^2 [\sqrt{(8x)} - x^2]\left[\frac{1}{2}\{\sqrt{(8x)} - x^2\} + x^2\right] dx$$

$$(2\tfrac{2}{3})(\bar{y}) = \int_0^2 [\sqrt{(8x)} - x^2]\left(\frac{\sqrt{(8x)}}{2} + \frac{x^2}{2}\right) dx = \int_0^2 \left(\frac{8x}{2} - \frac{x^4}{2}\right)dx$$

$$= \left[\frac{8x^2}{4} - \frac{x^5}{10}\right]_0^2 = (8 - 3\tfrac{1}{5}) - (0) = 4\tfrac{4}{5}$$

$$\text{Hence } \bar{y} = \frac{4\frac{4}{5}}{2\frac{2}{3}} = 1.8$$

Thus the position of the centroid of the shaded area in *Fig 10* is at (0.9, 1.8).

Problem 9 Use calculus to find the position of the centroid of the sector of a circle of radius r with angle subtended at the centre 2α radians.

The required sector is shown in *Fig 11* with its axis of symmetry on the x-axis. The area of a sector of a circle of radius r, and angle subtended at the centre θ radians, is given by $\frac{1}{2}r^2\theta$. Considering the shaded element:

Area of element $= \frac{1}{2}r^2\ \delta\theta$ (i.e., area of a sector of radius r and subtended angle $\delta\theta$ radians)

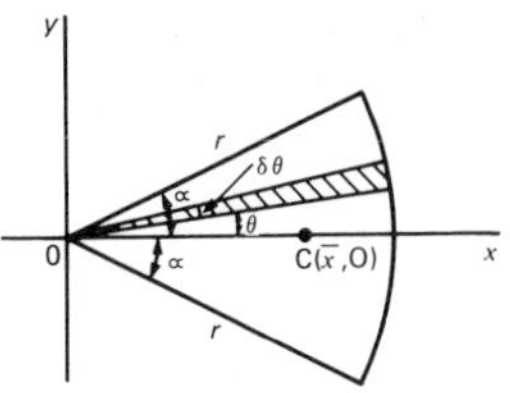

Fig 11

First moment of area of element about OY = (area)(perpendicular distance from centroid of element to OY)

Assuming the small shaded sector to approximate to a triangle, its centroid is $\frac{r}{3}$ from the base of the triangle, i.e. $\frac{2}{3}r$ from 0. Hence the perpendicular distance from the centroid to OY (i.e., length a in *Fig 12*), is $\frac{2}{3}r\cos\theta$.

Hence, first moment of area of element about OY $= (\frac{1}{2}r^2\,\delta\theta)(\frac{2}{3}r\cos\theta)$.

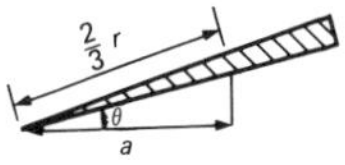

Fig 12

$$\text{Then } (\bar{x})(\text{total area}) = \underset{\delta\theta\to 0}{\text{limit}} \sum_{\theta=-\alpha}^{\theta=\alpha} (\frac{1}{2}r^2\,\delta\theta)(\frac{2}{3}r\cos\theta)$$

$$= \int_{-\alpha}^{\alpha} \frac{1}{3}r^3\cos\theta\, d\theta$$

$$\text{i.e. } \bar{x} = \frac{\int_{-\alpha}^{\alpha} \frac{1}{3}r^3\cos\theta\, d\theta}{\text{total area}} = \frac{\frac{1}{3}r^3[\sin\theta]_{-\alpha}^{\alpha}}{\frac{1}{2}r^2(2\alpha)} = \frac{\frac{1}{3}r^3[\sin\alpha - \sin(-\alpha)]}{r^2\alpha}$$

$$= \frac{\frac{2}{3}r^3\sin\alpha}{r^2\alpha} = \frac{2r\sin\alpha}{3\alpha}$$

Thus the centroid lies on the axis of symmetry, distance $\frac{2r\sin\alpha}{3\alpha}$ from the centre.

C FURTHER PROBLEMS ON CENTROIDS OF SIMPLE SHAPES BY INTEGRATION

In *Problems 1 to 5*, find the position of the centroids of the areas bounded by the given curves, the x-axis and the given ordinates.

1 $y = 2x;\quad x = 0, x = 3$ [(2, 2)]

2 $y = 3x+2;\quad x = 0, x = 4$ $\left[\left(2\frac{1}{2}; 4\frac{3}{4}\right)\right]$

3 $y = 5x^2;\quad x = 1, x = 4$ [(3.036; 24.36)]

4 $y = 2x^3;\quad x = 0, x = 2$ $\left[\left(1\frac{3}{5}; 4\frac{4}{7}\right)\right]$

5 $y = x(3x+1)$; $x = -1, x = 0$ $\left[\left(-\frac{5}{6}, \frac{19}{30}\right)\right]$

6 Determine the position of the centroid of a sheet of metal formed by the curve $y = 4x-x^2$ which lies above the x-axis. [(2, 1.6)]

7 Find the co-ordinates of the centroid of the area which lies between the curve $y/x = x-2$ and the x-axis. [(1, 0.4)]

8 Determine the co-ordinates of the centroid of the area formed between the curve $y = 9-x^2$ and the x-axis. [(0, 3.6)]

9 Determine the centroid of the area lying between $y = 4x^2$, the y-axis and the ordinates $y = 0$ and $y = 4$. $\left[\left(\frac{3}{8}, 2\frac{2}{5}\right)\right]$

10 Find the position of the centroid of the area enclosed by the curve $y = \sqrt{(5x)}$, the x-axis and the ordinate $x = 5$. $\left[\left(3, 1\frac{7}{8}\right)\right]$

11 Sketch the curve $y^2 = 9x$ between the limits $x = 0$ and $x = 4$. Determine the position of the centroid of this area. [(2.4, 0)]

12 Using (a) the theorem of Pappus, and (b) integration, determine the position of the centroid of a metal template in the form of a quadrant of a circle of radius 4 cm. (The equation of a circle, centre 0, radius r is $x^2+y^2 = r^2$.)

[On the centre line, distance 2.40 cm from the centre, i.e. at co-ordinates (1.70, 1.70)]

13 Calculate the points of intersection of the curves $x^2 = 4y$ and $\frac{y^2}{4} = x$, and determine the position of the centroid of the area enclosed by them.

[(0, 0) and (4, 4); (1.8, 1.8)]

14 (a) Determine the area bounded by the curve $y = 5x^2$, the x-axis and the ordinates $x = 0$ and $x = 3$.

(b) If this area is revolved 360° about (i) the x-axis, and (ii) the y-axis, find the volumes of the solids of revolution produced in each case.

(c) Determine the co-ordinates of the centroid of the area using (i) integral calculus, and (ii) the theorem of Pappus.

[(a) 45 square units; (b) (i) 1215π cubic units (ii) 202.5π cubic units; (c) (2.25, 13.5)]

15 Determine the position of the centroid of the sector of a circle of radius 3 cm whose angle subtended at the centre is 40°.

[On the centre line, 1.96 cm from the centre]

16 Sketch the curves $y = 2x^2+5$ and $y-8 = x(x+2)$ on the same axes and determine their points of intersection. Calculate the co-ordinates of the centroid of the area enclosed by the two curves. [(−1, 7) and (3, 23); (1, 10.20)]

22 Calculus (9) – Second moments of areas of regular sections

A MAIN POINTS CONCERNED WITH SECOND MOMENTS OF AREAS OF REGULAR SECTIONS

1 (i) The **first moment of area** about a fixed axis of a lamina of area A, perpendicular distance y from the centroid of the lamina is defined in chapter 21 as Ay cubic units.
(ii) The **second moment of area** of the same lamina as in (i) is given by Ay^2, i.e. the perpendicular distance from the centroid of the area to the fixed axis is squared.
(iii) Second moments of areas are usually denoted by I and have units of mm^4, cm^4, and so on.

2 Several areas, $a_1, a_2, a_3, \ldots\ldots\ldots$ at distances $y_1, y_2, y_3, \ldots\ldots\ldots$ from a fixed axis, may be replaced by a single area A, where $A = a_1 + a_2 + a_3 + \ldots\ldots\ldots$ at distance k from the axis, such that $Ak^2 = \Sigma ay^2$. k is called the radius of gyration of Area A about the given axis. Since $Ak^2 = \Sigma ay^2 = I$ then the **radius of gyration, $k = \sqrt{I/A}$.**

3 The second moment of area is a quantity much used in the theory of bending of beams, in the torsion of shafts, and in calculations involving water planes and centres of pressure.

4 The procedure to determine the second moment of area of regular sections about a given axis is (i) to find the second moment of area of a typical element and (ii) to sum all such second moments of area by integrating between appropriate limits.
For example, the **second moment of area of the rectangle** shown in *Fig 1* about axis PP is found by initially considering an elemental strip of width δx and parallel to and distance x from axis PP. Area of shaded strip = $b\delta x$. Second moment of area of the shaded strip about PP = $(x^2)(b\delta x)$. The second moment of area of the whole rectangle about PP is obtained by summing all such strips between $x = 0$ and $x = l$, i.e. $\sum_{x=0}^{x=l} x^2\, b\delta x$.

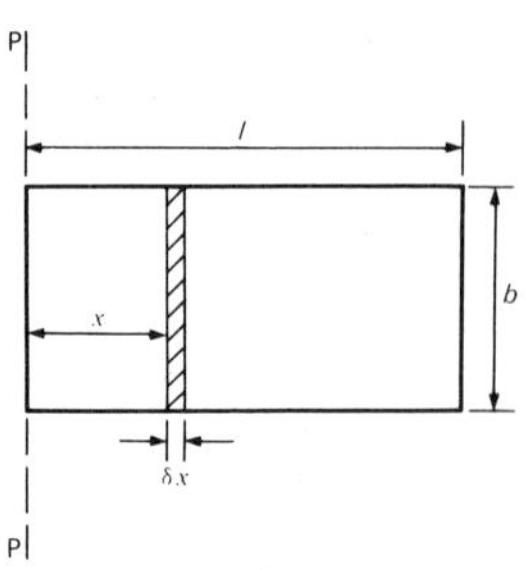

Fig 1

It is a fundamental theorem of integration that $\displaystyle \lim_{\delta x \to 0} \sum_{x=0}^{x=l} x^2 \, b\delta x = \int_0^l x^2 \, b \, dx.$

Thus the second moment of area of the rectangle about PP $= b\displaystyle\int_0^l x^2 \, dx$

$$= b\left[\frac{x^3}{3}\right]_0^l = \frac{bl^3}{3}$$

Since the total area of the rectangle, $A = lb$, then $I_{pp} = (lb)\left(\frac{l^2}{3}\right) = \frac{Al^2}{3}$

$I_{pp} = Ak_{pp}^2$ thus $k_{pp}^2 = \frac{l^2}{3}$

i.e. the radius of gyration about axis PP, $k_{pp} = \sqrt{\frac{l^2}{3}} = \frac{l}{\sqrt{3}}$.

5 **Parallel axis theorem**
In *Fig 2*, axis GG passes through the centroid C of area A. Axes DD and GG are in the same plane, are parallel to each other and distance d apart. The parallel axis theorem states:

$$\boldsymbol{I_{DD} = I_{GG} + Ad^2}$$

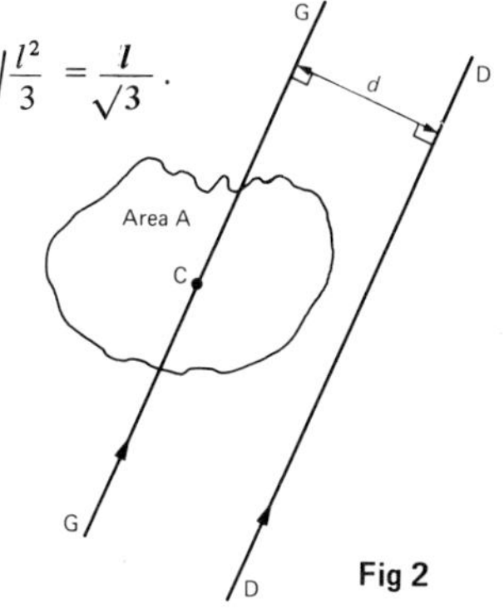

Fig 2

Using the parallel axis theorem the second moment of area of a rectangle about an axis through the centroid may be determined. In the rectangle shown in *Fig 3*,

$I_{pp} = \frac{bl^3}{3}$ (from para. 4).

From the parallel axis theorem

$I_{pp} = I_{GG} + (bl)\left(\frac{l}{2}\right)^2$

i.e. $\frac{bl^3}{3} = I_{GG} + \frac{bl^3}{4}$

from which $I_{GG} = \frac{bl^3}{3} - \frac{bl^3}{4} = \boldsymbol{\frac{bl^3}{12}}$

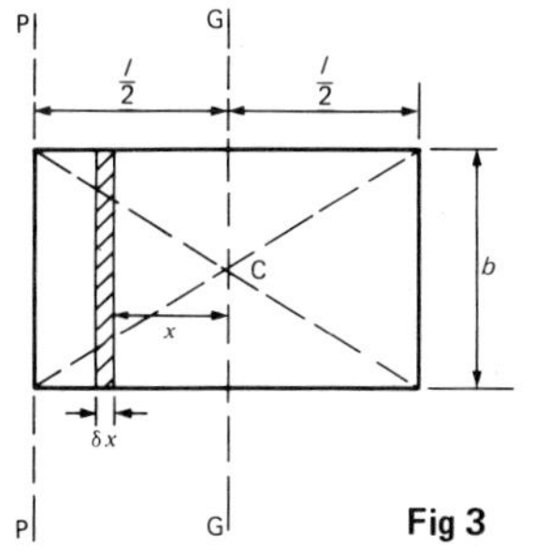

Fig 3

6 **Perpendicular axis theorem**
In *Fig 4*, axes OX, OY and OZ are mutually perpendicular. If OX and OY lie in the plane of area A then the perpendicular axis theorem states:

$$\boldsymbol{I_{OZ} = I_{OX} + I_{OY}}$$

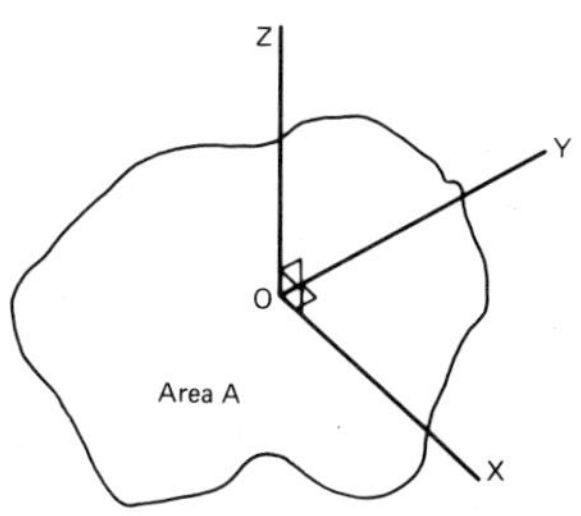

Fig 4

7 Further derivations of the second moment of area and radius of gyration of regular sections may be found in the worked problems in section B.
A summary of derived standard results are listed in *Table 1*.

TABLE 1 Summary of standard results of the second moments of areas of regular sections

Shape	*Position of axis*	*Second moment of area, I*	*Radius of gyration, k*
Rectangle length l	(1) Coinciding with b	$\frac{bl^3}{3}$	$\frac{l}{\sqrt{3}}$
breadth b	(2) Coinciding with l	$\frac{lb^3}{3}$	$\frac{b}{\sqrt{3}}$
	(3) Through centroid, parallel to b	$\frac{bl^3}{12}$	$\frac{l}{\sqrt{12}}$
	(4) Through centroid, parallel to l	$\frac{lb^3}{12}$	$\frac{b}{\sqrt{12}}$
Triangle Perpendicular height h	(1) Coinciding with b	$\frac{bh^3}{12}$	$\frac{h}{\sqrt{6}}$
base b	(2) Through centroid, parallel to base	$\frac{bh^3}{36}$	$\frac{h}{\sqrt{18}}$
	(3) Througn vertex, parallel to base	$\frac{bh^3}{4}$	$\frac{h}{\sqrt{2}}$
Circle radius r	(1) Through centre, perpendicular to plane (i.e. polar axis)	$\frac{\pi r^4}{2}$	$\frac{r}{\sqrt{2}}$
	(2) Coinciding with diameter	$\frac{\pi r^4}{4}$	$\frac{r}{2}$
	(3) About a tangent	$\frac{5\pi}{4} r^4$	$\frac{\sqrt{5}}{2} r$
Semicircle radius r	Coinciding with diameter	$\frac{\pi r^4}{8}$	$\frac{r}{2}$

B WORKED PROBLEMS ON SECOND MOMENTS OF AREAS OF REGULAR SECTIONS

Problem 1 Determine the second moment of area and the radius of gyration about axes AA, BB and CC for the rectangle shown in *Fig 5.*

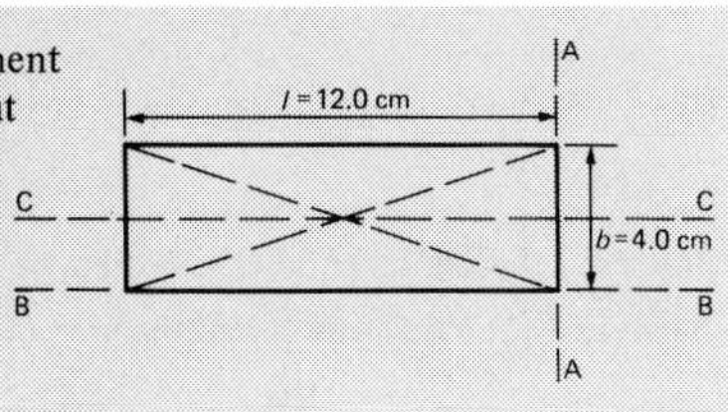

Fig 5

From *Table 1,* the second moment of area about axis AA,

$$I_{AA} = \frac{bl^3}{3} = \frac{(4.0)(12.0)^3}{3} = \mathbf{2304\ cm^4}$$

Radius of gyration, $k_{AA} = \frac{l}{\sqrt{3}} = \frac{12.0}{\sqrt{3}} = \mathbf{6.93\ cm.}$

Similarly, $I_{BB} = \frac{lb^3}{3} = \frac{(12.0)(4.0)^3}{3} = \mathbf{256\ cm^4}$

and $k_{BB} = \frac{b}{\sqrt{3}} = \frac{4.0}{\sqrt{3}} = \mathbf{2.31\ cm.}$

The second moment of area about the centroid of a rectangle is $\frac{bl^3}{12}$ when the axis through the centroid is parallel with the breadth b. In this case, the axis CC is parallel with the length l.

Hence $I_{CC} = \frac{lb^3}{12} = \frac{(12.0)(4.0)^3}{12} = \mathbf{64\ cm^4}$

and $k_{CC} = \frac{b}{\sqrt{12}} = \frac{4.0}{\sqrt{12}} = \mathbf{1.15\ cm.}$

Problem 2 Determine the second moment of area of the rectangle of length l and breadth b shown in *Fig 6*, about axis DD.

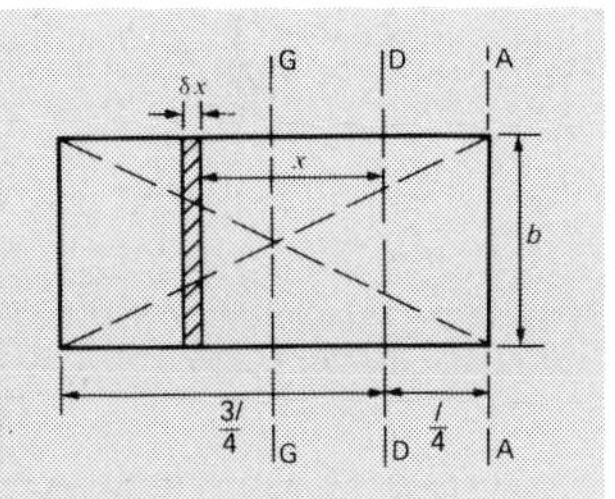

Fig 6

Method 1 Consider an elemental strip, width δx, parallel to and at a distance x from axis DD as shown in *Fig 6.* Area of strip $= b\delta x$.
Second moment of area of strip about YY $= x^2(b\delta x)$.
Hence the second moment of area of the rectangle about DD,

$$I_{DD} = \lim_{\delta x \to 0} \sum_{x=-\frac{l}{4}}^{x=\frac{3l}{4}} x^2\, b\delta x,$$

i.e., from para. 4, $I_{DD} = \int_{-\frac{l}{4}}^{\frac{3l}{4}} x^2 b\,dx = b\left[\frac{x^3}{3}\right]_{-\frac{l}{4}}^{\frac{3l}{4}} = \frac{b}{3}\left[x^3\right]_{-\frac{l}{4}}^{\frac{3l}{4}} = \frac{b}{3}\left[\left(\frac{3l}{4}\right)^3 - \left(-\frac{l}{4}\right)^3\right]$

$$= \frac{b}{3}\left[\frac{27}{64}l^3 + \frac{l^3}{64}\right] = \frac{b}{3}\left(\frac{28}{64}l^3\right) = \frac{7}{48}\boldsymbol{bl^3}$$

Method 2 Using the parallel axis theorem, $I_{DD} = I_{GG} + (bl)(\frac{l}{4})^2$, since the distance between I_{DD} and I_{GG} is $\frac{l}{4}$ and the area of the rectangle is bl.

From *Table 1*, $I_{GG} = \frac{bl^3}{12}$

Hence $I_{DD} = \frac{bl^3}{12} + \frac{bl^3}{16} = \frac{7}{48}\boldsymbol{bl^3}$, as above.

Problem 3 Find the second moment of area and the radius of gyration about axis PP for the rectangle shown in *Fig 7.*

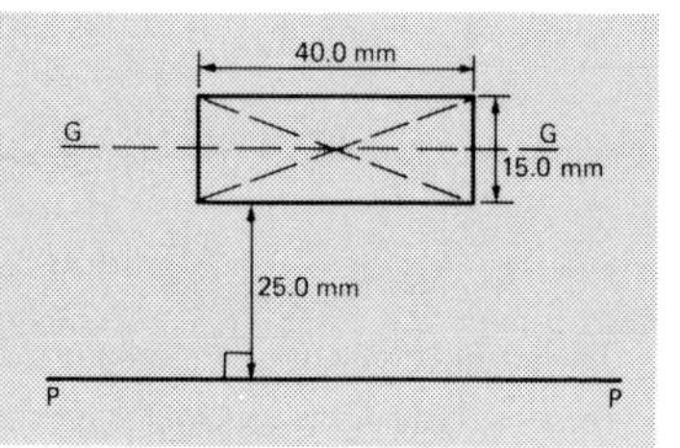

Fig 7

$I_{GG} = \frac{lb^3}{12}$, when $l = 40.0$ mm,
and $b = 15.0$ mm.

Hence $I_{GG} = \frac{(40.0)(15.0)^3}{12} = 11\ 250\ \text{mm}^4$

From the parallel axis theorem, $I_{PP} = I_{GG} + Ad^2$, where $A = 40.0 \times 15.0 = 600\ \text{mm}^2$ and $d = 25.0 + 7.5 = 32.5$ mm, the perpendicular distance between GG and PP.
Hence $I_{PP} = 11\ 250 + (600)(32.5)^2 = \mathbf{645\ 000\ mm^4}$.

$I_{PP} = Ak_{PP}^2$, from which, $k_{PP} = \sqrt{\frac{I_{PP}}{\text{area}}} = \sqrt{\left(\frac{645\ 000}{600}\right)} = \mathbf{32.79\ mm.}$

Problem 4 Show that the second moment of area about the base of a triangle of perpendicular height h is given by $A\frac{h^2}{6}$, where A is the area of the triangle.

Fig 8 shows triangle PQR having base b and perpendicular height h. Consider an elemental strip of length y, width δx, parallel to and distance x from QR. Area of strip $\triangleq y\delta x$. Second moment of area of strip about QR $\triangleq x^2\ (y\delta x)$
Hence the second moment of area of triangle PQR about QR,

$$I_{QR} = \lim_{\delta x \to 0} \sum_{x=0}^{x=h} x^2\, y\delta x = \int_0^h x^2\, y\,dx.$$

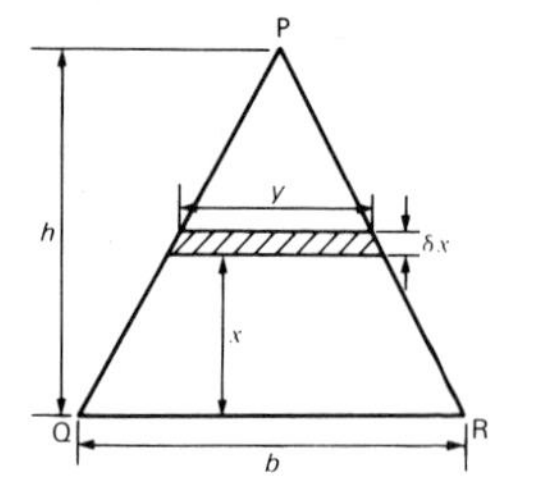

Fig 8

By similar triangles, $\frac{h-x}{y} = \frac{h}{b}$, from which, $y = \frac{b}{h}(h-x)$

$$\text{Thus } I_{QR} = \int_0^h x^2\, y dx = \int_0^h x^2\, \frac{b}{h}(h-x) dx = \frac{b}{h}\int_0^h (x^2 h - x^3) dx$$

$$= \frac{b}{h}\left[\frac{x^3}{3}h - \frac{x^4}{4}\right]_0^h = \frac{b}{h}\left[\frac{h^3}{3}(h) - \frac{h^4}{4}\right]$$

$$= \frac{bh^3}{3} - \frac{bh^3}{4} = \frac{bh^3}{12}$$

Area of triangle PQR, $A = \frac{1}{2}bh$. Hence $I_{QR} = (\frac{1}{2}bh)(\frac{h^2}{6}) = \boldsymbol{A\frac{h^2}{6}}$.

Problem 5 For the triangle shown in *Fig 8* determine the second moment of area and the radius of gyration (a) about an axis through its centroid, parallel to base QR, and (b) about an axis through vertex P and parallel with base QR.

(a) The centroid of a triangle lies at $h/3$ from the base. Let the second moment of area about the centroid be I_{GG}.
Applying the parallel axis theorem to the triangle shown in *Fig 8* gives:

$$I_{QR} = I_{GG} + (\frac{1}{2}bh)(\frac{h}{3})^2$$

i.e. $$\frac{bh^3}{12} = I_{GG} + \frac{bh^3}{18}$$

from which, $$I_{GG} = \frac{bh^3}{12} - \frac{bh^3}{18} = \boldsymbol{\frac{bh^3}{36}}$$

Radius of gyration, $$k_{GG} = \sqrt{\frac{I_{GG}}{\text{area}}} = \sqrt{\left(\frac{\frac{bh^3}{36}}{\frac{1}{2}bh}\right)} = \sqrt{\left(\frac{h^2}{18}\right)} = \boldsymbol{\frac{h}{\sqrt{18}}}$$

(b) Let the second moment of area of triangle PQR about an axis through P be I_P. The distance between P and the centroid of PQR is $\frac{2}{3}h$.
Applying the parallel axis theorem to the triangle shown in *Fig 8* gives:

$$I_P = I_{GG} + (\frac{1}{2}bh)(\frac{2h}{3})^2 = \frac{bh^3}{36} + \frac{2}{9}bh^3 = \boldsymbol{\frac{1}{4}bh^3}.$$

Radius of gyration, $$k_P = \sqrt{\frac{I_P}{\text{area}}} = \sqrt{\left(\frac{\frac{1}{4}bh^3}{\frac{1}{2}bh}\right)} = \sqrt{\left(\frac{h^2}{2}\right)} = \boldsymbol{\frac{h}{\sqrt{2}}}$$

Problem 6 Determine the second moment of area and radius of gyration about axis QQ of the triangle BCD shown in *Fig 9*.

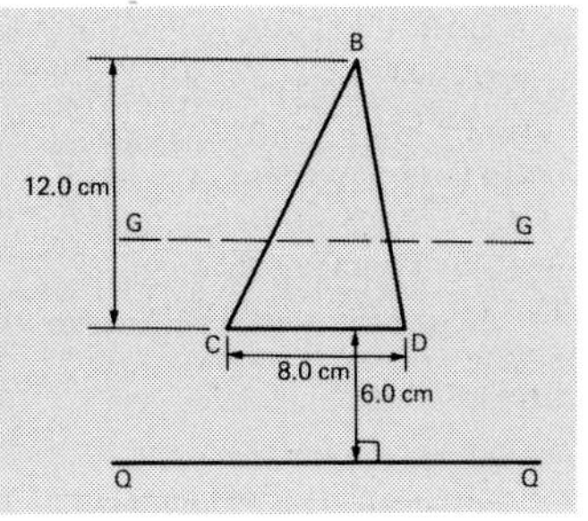

Fig 9

Using the parallel axis theorem: $I_{QQ} = I_{GG} + Ad^2$,
where I_{GG} is the second moment of area about the centroid of the triangle, i.e.,
$\dfrac{bh^3}{36} = \dfrac{(8.0)(12.0)^3}{36} = 384 \text{ cm}^4$,

A is the area of the triangle $= \dfrac{1}{2}bh = \dfrac{1}{2}(8.0)(12.0) = 48 \text{ cm}^2$

and d is the distance between axes GG and QQ $= 6.0 + \dfrac{1}{3}(12.0) = 10$ cm.
Hence the second moment of area about axis QQ, $I_{QQ} = 384 + (48)(10)^2 = \mathbf{5184\ cm^4}$

Radius of gyration, $k_{QQ} = \sqrt{\dfrac{I_{QQ}}{\text{area}}} = \sqrt{\left(\dfrac{5184}{48}\right)} = \mathbf{10.4\ cm.}$

Problem 7 Determine the second moment of area and the radius of gyration of a circle of radius r about its polar axis.

A polar axis is an axis through the centre of a circle, perpendicular to the plane of the circle as shown by axis ZZ in *Fig 10.* Consider the elemental annulus of width δx shown shaded in *Fig 10*, radius x from the centre, 0. If δx is very small, then the area of the elemental annulus is approximately given by (circumference × width), i.e. area of annulus $\triangleq (2\pi x)(\delta x)$.

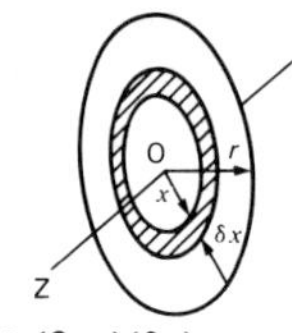

Fig 10

Second moment of area of annulus about the centre of the circle $\triangleq x^2(2\pi x \delta x)$.

Total second moment of area of the circle about ZZ, $I_{ZZ} = \lim_{\delta x \to 0} \sum_{x=0}^{x=r} x^2(2\pi x \delta x)$

i.e., $I_{ZZ} = \int_0^r 2\pi x^3 dx = 2\pi \left[\dfrac{x^4}{4}\right]_0^r = \dfrac{\pi r^4}{2}$

Radius of gyration, $k_{ZZ} = \sqrt{\dfrac{I_{ZZ}}{\text{area}}} = \sqrt{\left(\dfrac{\frac{\pi r^4}{2}}{\pi r^2}\right)} = \dfrac{\mathbf{r}}{\sqrt{\mathbf{2}}}$

Problem 8 Find the second moment of area and radius of gyration of a circle of radius r (a) about a diameter and (b) about a tangent. (c) Deduce the second moment of area and radius of gyration about a diameter for a semicircle of radius r.

A circle of radius r, having three mutually perpendicular axes OX, OY and ZZ is shown in *Fig 11.*

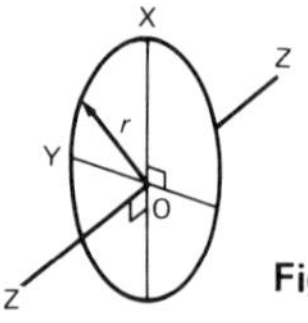

Fig 11

(a) From *Problem 7,* $I_{ZZ} = \dfrac{\pi r^4}{2}$

By symmetry, $I_{OX} = I_{OY}$
Using the perpendicular axis theorem: $I_{ZZ} = I_{OX} + I_{OY} = 2I_{OX}$

i.e., $\dfrac{\pi r^4}{2} = 2I_{OX}$, from which, $I_{OX} = \dfrac{\pi r^4}{4}$

Hence the second moment of area of a circle about a diameter is $\dfrac{\pi r^4}{4}$.

Radius of gyration, $k_{OX} = \sqrt{\frac{I_{OX}}{\text{area}}} = \sqrt{\left(\frac{\frac{\pi r^4}{4}}{\pi r^2}\right)} = \frac{r}{2}$

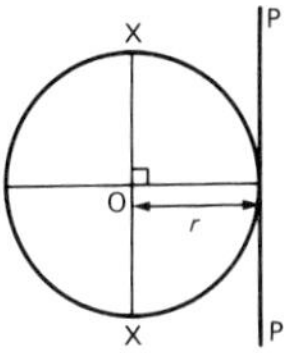

(b) *Fig 12* shows a circle of radius r having a tangent PP parallel to diameter XX.
By the parallel axis theorem: $I_{PP} = I_{XX} + Ar^2$

From (a), $I_{XX} = \frac{\pi r^4}{4}$. Thus $I_{PP} = \frac{\pi r^4}{4} + (\pi r^2)r^2 = \frac{5}{4}\pi r^4$.

Fig 12

Thus the second moment of area of a circle about a tangent is $\frac{5\pi}{4}r^4$.

Radius of gyration, $k_{PP} = \sqrt{\frac{I_{PP}}{\text{area}}} = \sqrt{\left(\frac{\frac{5}{4}\pi r^4}{\pi r^2}\right)} = \frac{\sqrt{5}}{2}r.$

(c) An alternative method to (a) of finding the second moment of area of a circle about a diameter is to take an elemental strip of width δx, distance x from axis YY as shown in *Fig 13.*

Second moment of area of the strip about YY $\triangleq x^2\ (y\delta x)$

Thus the second moment of area of the circle about YY,

$$I_{YY} = \lim_{\delta x \to 0} \sum_{x=0}^{x=r} x^2 y \delta x = \int_{-r}^{r} x^2\ y dx$$

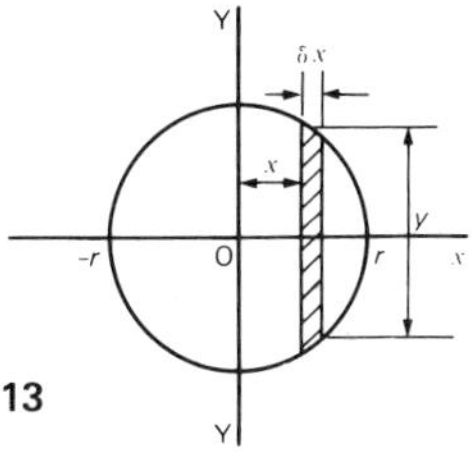

Fig 13

The evaluation of this integral is beyond the scope of this test. However, if the same procedure is adopted for a semicircle, the second moment of area of the semicircle about YY in *Fig 13* is $\int_0^r x^2\ ydx.$

Since $\int_{-r}^{r} x^2\ ydx = 2\int_0^r x^2\ ydx$ it follows that the second moment of area of a semicircle about its diameter is one half of that for a circle about its diameter, i.e., $\frac{1}{2}\left(\frac{\pi r^4}{4}\right) = \frac{\pi r^4}{8}$.

The radius of gyration of a semicircle about its diameter.

$$k = \sqrt{\left(\frac{\frac{\pi r^4}{8}}{\frac{\pi r^2}{2}}\right)} = \frac{r}{2},\ \text{i.e., the same value as for a circle.}$$

Problem 9 Determine the second moment of area and radius of gyration of the circle shown in *Fig 14* about axis YY.

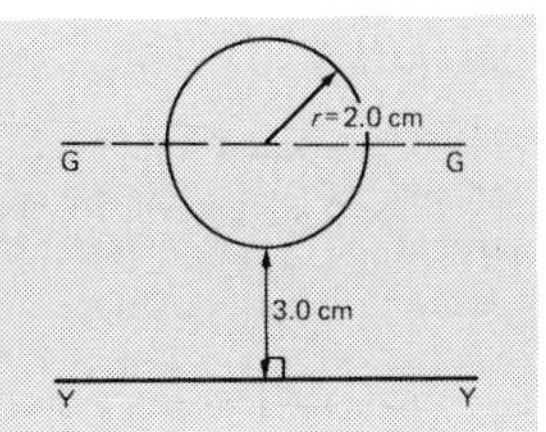

Fig 14

In *Fig 14*, $I_{GG} = \frac{\pi r^4}{4} = \frac{\pi}{4}(2.0)^4 = 4\pi\ \text{cm}^4$.

Using the parallel axis theorem, $I_{YY} = I_{GG} + Ad^2$, where $d = 3.0+2.0 = 5.0$ cm.
Hence $I_{YY} = 4\pi+[\pi(2.0)^2](5.0)^2 = 4\pi+100\pi$
$= 104\pi = \mathbf{327\ cm^4}$.

Radius of gyration, $k_{YY} = \sqrt{\frac{I_{YY}}{\text{area}}} = \sqrt{\left(\frac{104\pi}{\pi(2.0)^2}\right)} = \sqrt{(26)} = \mathbf{5.10\ cm}$.

Problem 10 Determine the second moment of area and radius of gyration for the semi-circle shown in *Fig 15* about axis XX.

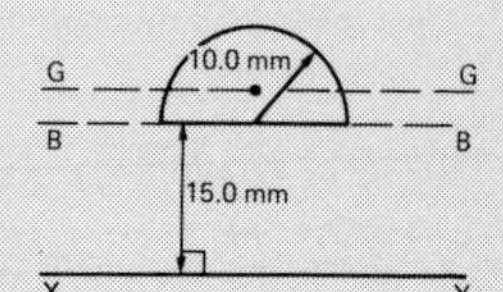

Fig 15

The centroid of a semicircle lies at $\frac{4r}{3\pi}$ from its diameter (see chapter 21). Using the parallel axis theorem: $I_{BB} = I_{GG} + Ad^2$,

where $I_{BB} = \frac{\pi r^4}{8}$ (from *Problem 8*) $= \frac{\pi(10.0)^4}{8} = 3927\ \text{mm}^4$,

$A = \frac{\pi r^2}{2} = \frac{\pi(10.0)^2}{2} = 157.1\ \text{mm}^2$

and $d = \frac{4r}{3\pi} = \frac{4(10.0)}{3\pi} = 4.244$ mm.

Hence $3927 = I_{GG} + (157.1)(4.244)^2$
i.e. $3927 = I_{GG} + 2830$, from which, $I_{GG} = 3927-2830 = 1097\ \text{mm}^4$.
Using the parallel axis theorem again: $I_{XX} = I_{GG} + A(15.0+4.244)^2$
i.e. $I_{XX} = 1097+(157.1)(19.244)^2 = 1097+58179 = 59276\ \text{mm}^4$
or **59280 mm⁴**, correct to 4 significant figures.

Radius of gyration, $k_{XX} = \sqrt{\frac{I_{XX}}{\text{area}}} = \sqrt{\left(\frac{59276}{157.1}\right)} = \mathbf{19.42\ mm}$.

Problem 11 Determine the polar second moment of area of the propeller shaft cross-section shown in *Fig 16*.

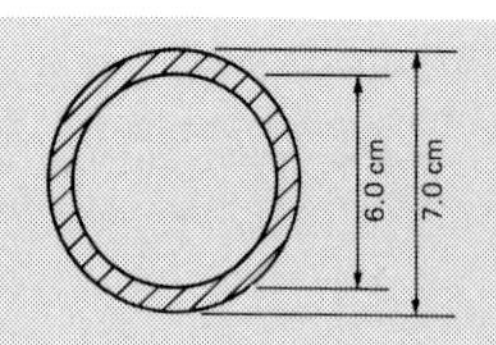

Fig 16

The polar second moment of area of a circle $= \frac{\pi r^4}{2}$.

The polar second moment of area of the shaded area is given by the polar second moment of area of the 7.0 cm diameter circle minus the polar second moment of area of the 6.0 cm diameter circle.

Hence the polar second moment of area of the cross-section shown

$$= \frac{\pi}{2}\left(\frac{7.0}{2}\right)^4 - \frac{\pi}{2}\left(\frac{6.0}{2}\right)^4 = 235.7-127.2 = \mathbf{108.5\ cm^4}$$

Problem 12 Determine the second moment of area and radius of gyration of a rectangular lamina of length 40 mm and width 15 mm about an axis through one corner, perpendicular to the plane of the lamina.

The lamina is shown in *Fig 17*.
From the perpendicular axis theorem:
$I_{ZZ} = I_{XX} + I_{YY}$

$$I_{XX} = \frac{lb^3}{3} = \frac{(40)(15)^3}{3} = 45\ 000\ \text{mm}^4;$$

$$I_{YY} = \frac{bl^3}{3} = \frac{(15)(40)^3}{3} = 320\ 000\ \text{mm}^4$$

Hence $I_{ZZ} = 45\ 000 + 320\ 000$

$= \mathbf{365\ 000\ mm^4}$ **or** $\mathbf{36.5\ cm^4}$.

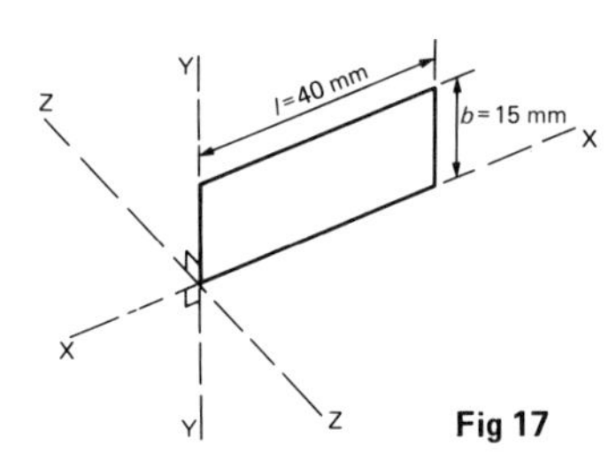

Fig 17

Radius of gyration, $k_{ZZ} = \sqrt{\frac{I_{ZZ}}{\text{area}}} = \sqrt{\left(\frac{365\ 000}{40 \times 15}\right)} = \mathbf{24.7\ mm}$ **or** $\mathbf{2.47\ cm}$.

Problem 13 Determine the polar second moment of area of the shaded cross-sectional area shown in *Fig 18*.

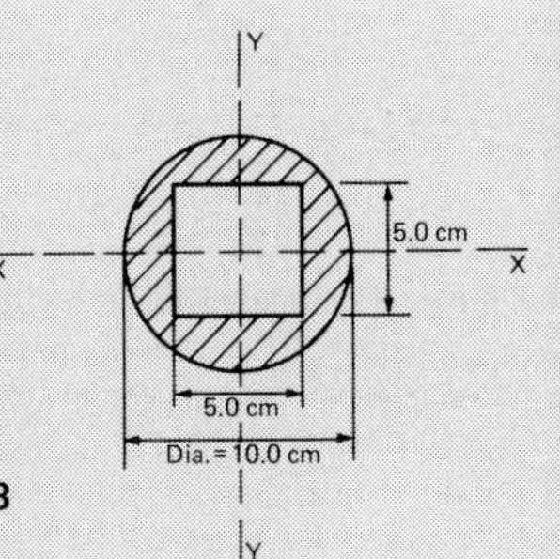

Fig 18

$$I_{XX} = \frac{\pi r^4}{4} - \frac{bl^3}{3} = \frac{\pi}{4}\left(\frac{10.0}{2}\right)^4 - \frac{(5.0)(5.0)^3}{3}$$

$$= 490.9 - 208.3 = 282.6\ \text{cm}^4.$$

Since the cross-sectional area is symmetrical $I_{XX} = I_{YY}$.
From the perpendicular axis theorem:
$I_{\text{polar}} = I_{XX} + I_{YY} = 282.6 + 282.6 = \mathbf{565.2\ cm^4}$

Problem 14 In *Fig 19*, the shaded circular area of radius 6.0 cm is removed from the larger circular lamina. If length OB = 7.5 cm, determine the second moment of area and radius of gyration of the remaining section about axis PP, each correct to 4 significant figures.

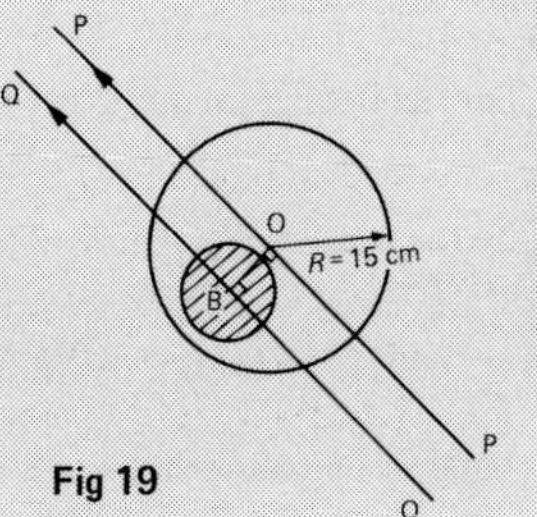

Fig 19

For the 15 cm circle $I_{PP} = \frac{\pi R^4}{4} = \frac{\pi(15)^4}{4} = 39\,760 \text{ cm}^4$.

For the 6.0 cm circle, $I_{QQ} = \frac{\pi r^4}{4} = \frac{\pi(6.0)^4}{4} = 1018 \text{ cm}^4$

Using the parallel axis theorem, the second moment of area of the 6.0 cm radius circle about PP is given by: $I_{PP} = I_{QQ} + Ad^2$

$= 1018+[\pi(6.0)^2](7.5)^2$, since OB = 7.5 cm

$= 7380 \text{ cm}^4$.

Hence the second moment of area of the lamina remaining about PP = 39 760–7380 = **32 380 cm⁴**.

(Note that the second moment of area of the smaller circle is subtracted since this area has been removed.)

Radius of gyration, $k_{XX} = \sqrt{\left(\frac{32\,380}{\text{area of lamina}}\right)} = \sqrt{\left(\frac{32\,380}{\pi 15^2 - \pi 6.0^2}\right)}$

$=$ **7.385 cm**, correct to 4 significant figures.

Problem 15 Determine, correct to 3 significant figures, the second moment of area about axis XX for the composite area shown in *Fig 20.*

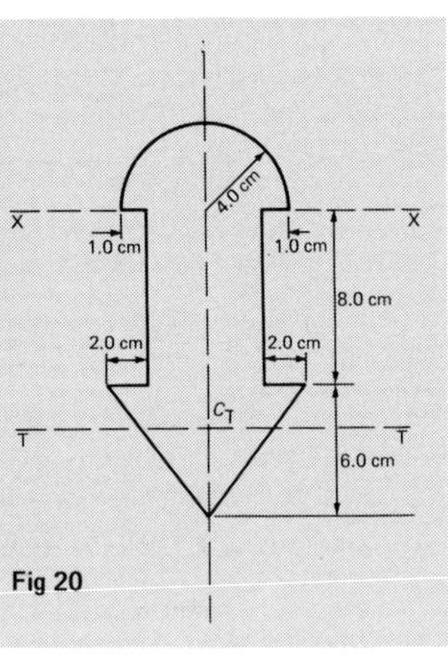

Fig 20

For the semicircle, $I_{XX} = \frac{\pi r^4}{8} = \frac{\pi(4.0)^4}{8} = 100.5 \text{ cm}^4$.

For the rectangle, $I_{XX} = \frac{bl^3}{3} = \frac{(6.0)(8.0)^3}{3} = 1024 \text{ cm}^4$.

For the triangle, about axis TT through centroid C_T, $I_{TT} = \frac{bh^3}{36} = \frac{(10.0)(6.0)^3}{36}$

$= 60 \text{ cm}^4$.

By the parallel axis theorem, the second moment of area of the triangle about axis XX,

$$= 60+\left[\frac{1}{2}(10.0)(6.0)\right]\left[8.0+\frac{1}{3}(6.0)\right]^2 = 360 \text{ cm}^4.$$

Total second moment of area about XX = 100.5+1024+360

= 1484.5 = **1480 cm⁴**, correct to 3 significant figures.

Problem 16 (a) Determine the second moment of area and the radius of gyration about axis XX for the I-section shown in *Fig 21*; (b) Determine the position of the centroid of the I-section; (c) Calculate the second moment of area and radius of gyration about an axis CC through the centroid of the section, parallel to axis XX.

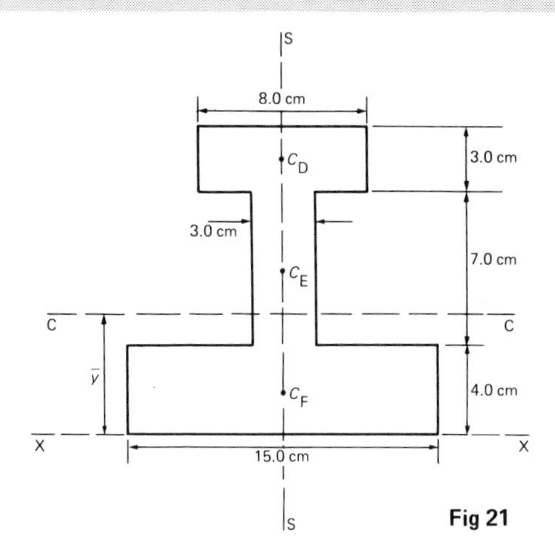

Fig 21

The I-section is divided into three rectangles, D, E and F and their centroids denoted by C_D, C_E and C_F respectively.

(a) *For rectangle D:*

The second moment of area about C_D (an axis through C_D parallel to XX) $= \dfrac{bl^3}{12}$

$$= \frac{(8.0)(3.0)^3}{12} = 18 \text{ cm}^4.$$

Using the parallel axis theorem: $I_{XX} = 18 + Ad^2$,
where $A = 8.0 \times 3.0 = 24 \text{ cm}^2$ and $d = 12.5$ cm.
Hence $I_{XX} = 18 + 24(12.5)^2 = 3768 \text{ cm}^4$

For rectangle E:

The second moment of area about C_E (an axis through C_E parallel to XX)

$$= \frac{bl^3}{12} = \frac{(3.0)(7.0)^3}{12} = 85.75 \text{ cm}^4.$$

Using the parallel axis theorem: $I_{XX} = 85.75 + (7.0 \times 3.0)(7.5)^2 = 1267 \text{ cm}^4$.

For rectangle F:

$$I_{XX} = \frac{lb^3}{3} = \frac{(15.0)(4.0)^3}{3} = 320 \text{ cm}^4.$$

Total second moment of area for the I-section about axis XX,
$I_{XX} = 3768 + 1267 + 320 = \mathbf{5355 \text{ cm}^4}$.

Total area of I-section $= (8.0 \times 3.0) + (3.0 \times 7.0) + (15.0 \times 4.0) = 105 \text{ cm}^2$.

Radius of gyration, $k_{XX} = \sqrt{\dfrac{I_{XX}}{\text{area}}} = \sqrt{\left(\dfrac{5355}{105}\right)} = \mathbf{7.14 \text{ cm}}$.

(b) The centroid of the I-section will lie on the axis of symmetry, shown as axis SS in *Fig 21*. (Centroids of composite shapes are discussed in *Mathematics 2 Checkbook*, chapter 7.)

Part	*Area* (a cm²)	*Distance of centroid from XX* (i.e. y cm)	*Moment about XX* (i.e. ay cm³)
D	24	12.5	300
E	21	7.5	157.5
F	60	2.0	120
$\Sigma a = A = 105$			$\Sigma ay = 577.5$

$A\bar{y} = \Sigma ay$, from which, $\bar{y} = \dfrac{\Sigma ay}{A} = \dfrac{577.5}{105} = 5.5$ cm.

Thus the centroid is positioned on the axis of symmetry 5.5 cm from axis XX.

(c) From the parallel axis theorem: $I_{XX} = I_{CC} + Ad^2$

i.e., $5355 = I_{CC} + (105)(5.5)^2$

$= I_{CC} + 3176$

Hence $I_{CC} = 5355 - 3176 =$ **2179 cm⁴**.

Radius of gyration, $k_{CC} = \sqrt{\dfrac{I_{CC}}{\text{area}}} = \sqrt{\left(\dfrac{2179}{105}\right)} =$ **4.56 cm.**

C FURTHER PROBLEMS ON SECOND MOMENTS OF AREAS OF REGULAR SECTIONS

1 Determine the second moment of area and radius of gyration for the rectangle shown in *Fig 22*, about (a) axis AA; (b) axis BB; and (c) axis CC.

[(a) 72 cm⁴, 1.73 cm; (b) 128 cm⁴, 2.31 cm; (c) 512 cm⁴, 4.62 cm;]

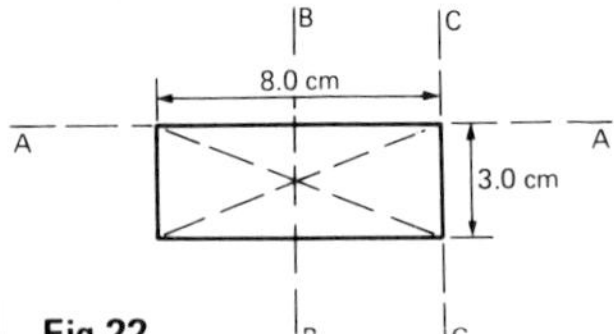

Fig 22

2 Determine the second moment of area and radius of gyration for the triangle shown in *Fig 23* about (a) axis DD; (b) axis EE; and (c) an axis through the centroid of the triangle parallel to axis DD.

[(a) 729 mm⁴, 3.67 mm; (b) 2187 mm⁴, 6.36 mm; (c) 243 mm⁴, 2.12 mm.]

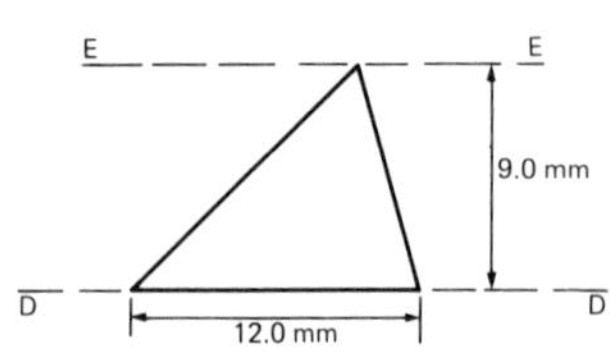

Fig 23

3 For the circle shown in *Fig 24*, find the second moment of area and radius of gyration about (a) axis FF and (b) axis HH.

[(a) 201 cm⁴, 2.0 cm; (b) 1005 cm⁴, 4.47 cm]

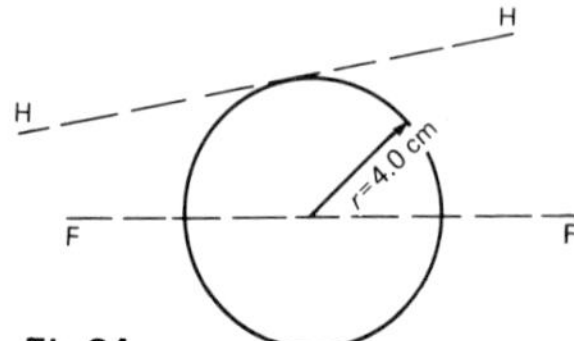

Fig 24

4 For the semicircle shown in *Fig 25*, find the second moment of area and radius of gyration about axis JJ.

[3927 mm⁴, 5.0 mm]

5 Show that the second moment of area of a rectangular lamina having dimensions l and b about an axis parallel to b at a distance $\frac{l}{3}$ from one end is $\frac{bl^3}{9}$.

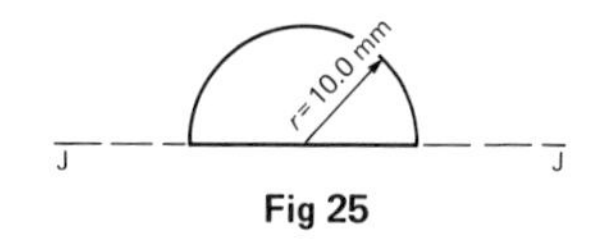

Fig 25

6 *State the parallel axis theorem*. For each of the areas shown in *Fig 26* determine the second moment of area and radius of gyration about axis LL, by using the parallel axis theorem. [(a) 335 cm^4, 4.73 cm; (b) 22 030 cm^4, 14.3 cm; (c) 628 cm^4, 7.07 cm]

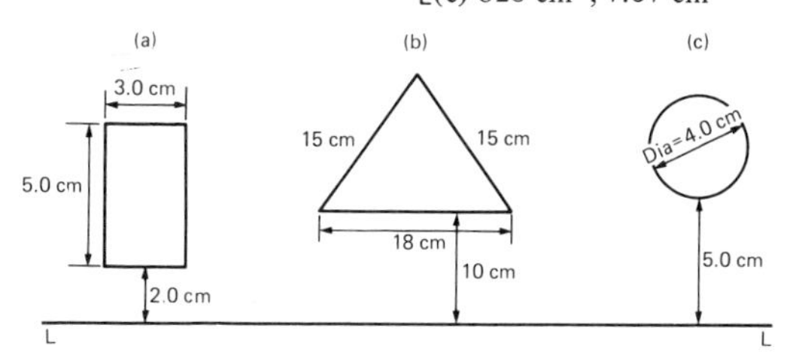

Fig 26

7 *State the perpendicular axis theorem.* Calculate the second moment of area and radius of gyration of a rectangular cover of length 20.0 mm and width 12.0 mm about an axis through one corner perpendicular to the plane of the rectangle. [43 520 mm^4; 13.47 mm]

8 Calculate the radius of gyration of a rectangular door 2.0 m high by 1.5 m wide about a vertical axis through its hinge. [0.866 m]

9 A circular door of a boiler is hinged so that it turns about a tangent. If its diameter is 1.0 m, determine its second moment of area and radius of gyration about the hinge. [0.245 m^4; 0.559 m]

10 Determine the second moment of area and the radius of gyration of a circular wooden cover of diameter 5.0 cm about an axis perpendicular to the plane of the cover and passing through the centre. [61.4 cm^4; 1.77 cm]

11 Find the second moment of area and radius of gyration about axes AA and BB for the composite shape shown in *Fig 27*.
[I_{AA} = 3133 cm^4; k_{AA} = 5.13 cm; I_{BB} = 1069 cm^4; k_{BB} = 3.00 cm]

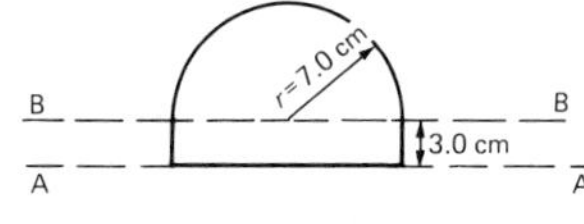

Fig 27

12 A circular cover, centre 0, has a radius of 12.0 cm. A hole of radius 4.0 cm and centre X, where OX = 6.0 cm, is cut in the cover. Determine the second moment of area and the radius of gyration of the remainder about a diameter through 0 perpendicular to OX. [14 280 cm^4; 5.96 cm]

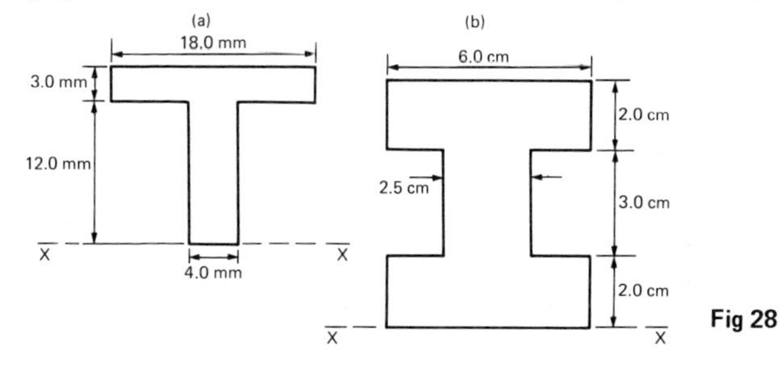

Fig 28

13 For the sections shown in *Fig 28*, find the second moment of area and radius of gyration about axis XX.

[(a) 12 190 mm^4; 10.9 mm; (b) 549.5 cm^4; 4.18 cm]

14 Determine the second moment of area and the radius of gyration about axis XX for the composite shape shown in *Fig 29*. [38 100 cm^4; 12.8 cm]

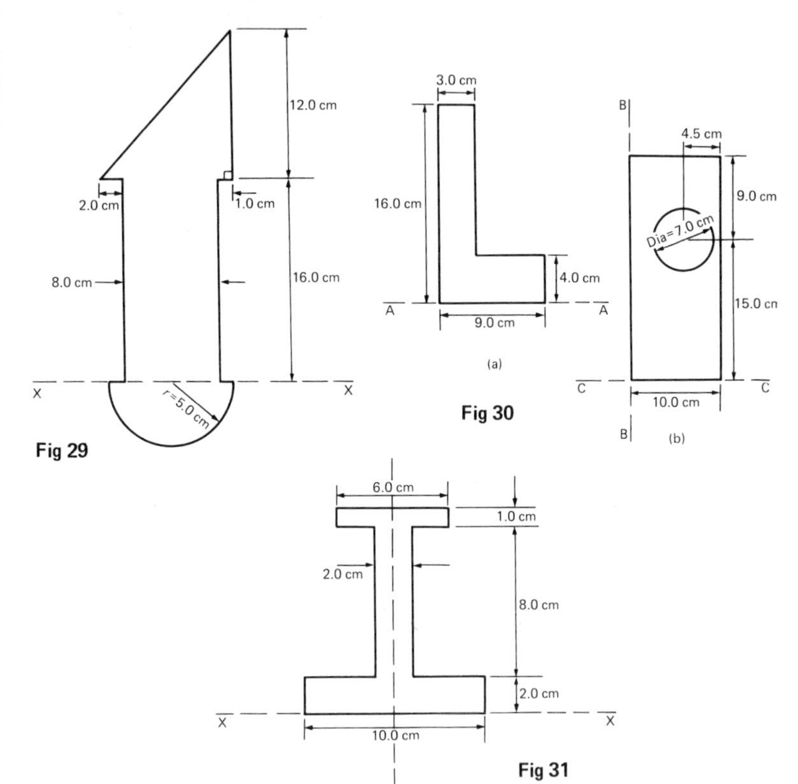

Fig 29

Fig 30

Fig 31

15 Show that the radius of gyration of a square plate of side k about a diagonal is given by $\dfrac{k}{2\sqrt{3}}$.

16 Determine the second moments of areas about the given axes for the shapes shown in *Fig 30*. (In *Fig 30(b)*, the circular area is removed.)

[$I_{AA} = 4224$ cm^4; $I_{BB} = 6718$ cm^4; $I_{CC} = 37\,300$ cm^4]

17 (a) Find the second moment of area and radius of gyration about the axis XX for the beam section shown in *Fig 31*.
(b) Determine the position of the centroid of the section.
(c) Calculate the second moment of area and radius of gyration about an axis through the centroid parallel to axis XX.

[(a) 1351 cm^4; 5.67 cm; (b) 4.26 cm from axis XX on the axis of symmetry; (c) 588.8 cm^4; 3.74 cm]

23 Calculus (10) – Differential equations

A MAIN POINTS CONCERNED WITH DIFFERENTIAL EQUATIONS

1 (i) Integrating both sides of the derivative $\frac{dy}{dx} = 3$ with respect to x gives $y = \int 3dx$ i.e., $y = 3x+c$, where c is an arbitrary constant. $y = 3x+c$ represents a **family of curves**, each of the curves in the family depending on the value of c. Examples include $y = 3x+8$, $y = 3x+3$, $y = 3x$ and $y = 3x-10$ and these are shown in *Fig 1*. Each are straight lines of gradient 3.

(ii) A particular curve of a family may be determined when a point on the curve is specified. Thus, if $y = 3x+c$ passes through the point (1, 2) then $2 = 3(1)+c$, from which, $c = -1$. The equation of the curve passing through (1, 2) is therefore $y = 3x-1$. (See *Problem 1*.)

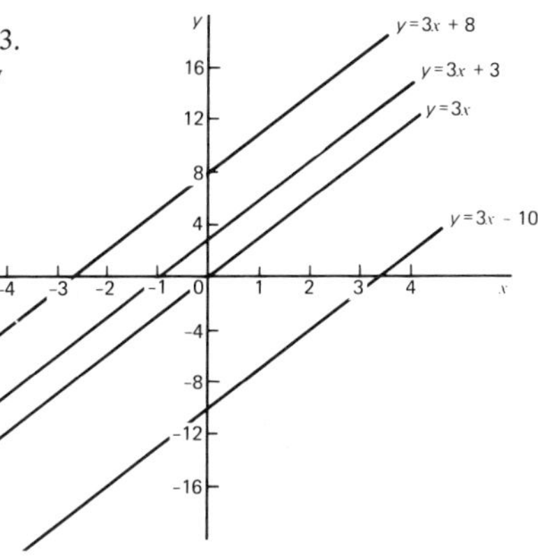

Fig 1

2 A **differential equation** is one that contains differential coefficients.

Examples include (i) $\frac{dy}{dx} = 7x$ and (ii) $\frac{d^2y}{dx^2} + 5\frac{dy}{dx} + 2y = 0$.

3 Differential equations are classified according to the highest derivative which occurs in them. Thus example (i) above is a **first order differential equation**, and example (ii) is a **second order differential equation**.

4 Starting with a differential equation it is possible, by integration and by being given sufficient data to determine unknown constants, to obtain the original function. This process is called '**solving the differential equation**'.

5 (i) A solution to a differential equation which contains one or more arbitrary constants of integration is called the **general solution** of the differential equation.

(ii) When additional information is given so that constants may be calculated the **particular solution** of the differential equation is obtained. The additional information is called **boundary conditions.**

(iii) From para. 1, $y = 3x+c$ is the general solution of the differential equation $\frac{dy}{dx} = 3$. Given the boundary conditions $x = 1$ when $y = 2$, produces the particular solution of $y = 3x-1$.

6 Differential equations are widely used in engineering and science. There are several different types of differential equation and each requires their own method of solution. The solution of two first order differential equations are dealt with in this chapter.

(i) **Differential equations of the form $\frac{dy}{dx} = f(x)$.**

An equation of the form $\frac{dy}{dx} = f(x)$ may be solved by integration.

The solution is $y = \int f(x)dx$. (See *Problems 2 to 5.*)

(ii) **Differential equations of the form $\frac{dQ}{dt} = kQ$.**

The general solution of an equation of the form $\frac{dQ}{dt} = kQ$ is $Q = Ae^{kt}$, where A is a constant.

This solution may be checked: Differentiating $Q = Ae^{kt}$ with respect to t gives $\frac{dQ}{dt} = k(Ae^{kt}) = kQ$. (See *Problems 6 to 11.*)

7 Examples of the natural laws of the form $\frac{dQ}{dt} = kQ$ include:

(i) Newtons law of cooling: $\frac{d\theta}{dt} = -k\theta$. The law is $\theta = \theta_0 e^{-kt}$.

(ii) Decay of current in an inductive circuit: $\frac{di}{dt} = ki$. The law is $i = Ae^{kt}$.

(iii) Linear expansion: $\frac{dl}{d\theta} = kl$. The law is $l = l_0 e^{k\theta}$.

B WORKED PROBLEMS ON DIFFERENTIAL EQUATIONS

Problem 1 Sketch the family of curves given by the equation $dy/dx = 4x$ and determine the equation of one of these curves which passes through the point (2, 3).

Integrating both sides of $\frac{dy}{dx} = 4x$ with respect to x gives: $\int \frac{dy}{dx}\, dx = \int 4x\, dx$

i.e. $y = 2x^2+c$.

Some members of the family of curves having an equation $y = 2x^2+c$ include $y = 2x^2+15$, $y = 2x^2+8$, $y = 2x^2$ and $y = 2x^2-6$, and these are shown in *Fig 2*. To determine the equation of the curve passing through the point (2, 3), $x = 2$ and $y = 3$ are substituted into the equation $y = 2x^2+c$. Thus $3 = 2(2)^2+c$, from which, $c = 3-8 = -5$.

Hence the equation of the curve passing through the point (2, 3) is $y = 2x^2-5$.

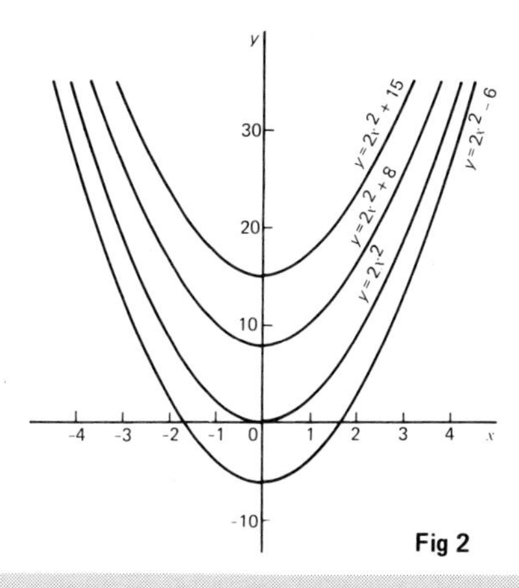

Fig 2

Problem 2 Determine the general solution of the equation $\frac{dy}{dx} = 2x^2 - x + 1$. Given the boundary conditions $y = 2$ when $x = 3$, find the particular solution.

Integrating both sides of $\frac{dy}{dx} = 2x^2 - x + 1$ with respect to x gives:

$$y = \int (2x^2 - x + 1)\,dx$$

i.e. $\boldsymbol{y = \frac{2x^3}{3} - \frac{x^2}{2} + x + c}$, **which is the general solution.**

When $y = 2$ and $x = 3$ then $2 = \frac{2(3)^3}{3} - \frac{(3)^2}{2} + 3 + c$,

from which, $c = 2 - 18 + 4\frac{1}{2} - 3 = -14\frac{1}{2}$.

Hence the particular solution is $\boldsymbol{y = \frac{2x^3}{3} - \frac{x^2}{2} + x - 14\frac{1}{2}}$.

Problem 3 Determine the general solutions of the following equations:
(a) $3\frac{dy}{dx} + \frac{4}{x} = 2x$; (b) $x\frac{dy}{dx} = 3 - 4x^2$; (c) $\frac{1}{4}\cos 2\theta + \frac{dM}{d\theta} = \frac{1}{2}\sin 3\theta$.

(a) Rearranging $3\frac{dy}{dx} + \frac{4}{x} = 2x$ gives $3\frac{dy}{dx} = 2x - \frac{4}{x}$ and $\frac{dy}{dx} = \frac{2}{3}x - \frac{4}{3x}$

Integrating with respect to x gives: $y = \int \left(\frac{2}{3}x - \frac{4}{3x}\right) dx$

i.e. $\boldsymbol{y = \frac{1}{3}x^2 - \frac{4}{3}\ln x + c}$ which is the general solution.

(b) Rearranging $x\frac{dy}{dx} = 3 - 4x^2$ gives $\frac{dy}{dx} = \frac{3}{x} - 4x$

Integrating with respect to x gives: $y = \int \left(\frac{3}{x} - 4x\right) dx$

i.e. $\boldsymbol{y = 3\ln x - 2x^2 + c}$, which is the general solution.

(c) Rearranging $\frac{1}{4}\cos 2\theta + \frac{dM}{d\theta} = \frac{1}{2}\sin 3\theta$ gives $\frac{dM}{d\theta} = \frac{1}{2}\sin 3\theta - \frac{1}{4}\cos 2\theta$.

Integrating with respect to θ gives: $M = \int\left(\frac{1}{2}\sin 3\theta - \frac{1}{4}\cos 2\theta\right)d\theta$

i.e., $M = -\frac{1}{6}\cos 3\theta - \frac{1}{8}\sin 2\theta + c$, **which is the general solution.**

Problem 4 Find the particular solutions of the following differential equations:

(a) $4\frac{dr}{d\theta} + 2\cos\theta = 0$ given $r = 6\frac{1}{2}$ when $\theta = \frac{\pi}{2}$

(b) $4\sin 2x + \frac{dy}{dx} = 3e^x$ given $y = 4$ when $x = \frac{\pi}{4}$.

(a) Rearranging $4\frac{dr}{d\theta} + 2\cos\theta = 0$ gives $4\frac{dr}{d\theta} = -2\cos\theta$

and $\frac{dr}{d\theta} = -\frac{1}{2}\cos\theta$

Integrating with respect to θ gives: $r = \int(-\frac{1}{2}\cos\theta)d\theta$

i.e. $r = -\frac{1}{2}\sin\theta + c$, which is the general solution

When $r = 6\frac{1}{2}$, $\theta = \frac{\pi}{2}$. Thus $6\frac{1}{2} = -\frac{1}{2}\sin\frac{\pi}{2} + c$, from which, $c = 7$.

Hence the particular solution is $r = -\frac{1}{2}\sin\theta + 7$.

(b) Rearranging $4\sin 2x + \frac{dy}{dx} = 3e^x$ gives $\frac{dy}{dx} = 3e^x - 4\sin 2x$.

Integrating with respect to x gives: $y = \int(3e^x - 4\sin 2x)dx$

i.e. $y = 3e^x + 2\cos 2x + c$

When $y = 4$, $x = \frac{\pi}{4}$. Thus $4 = 3e^{\frac{\pi}{4}} + 2\cos 2(\frac{\pi}{4}) + c$

from which, $c = 4 - 3e^{\frac{\pi}{4}}$.

Hence the particular solution is $y = 3e^x + 2\cos 2x + 4 - 3e^{\frac{\pi}{4}}$.

Problem 5 The bending moment M of the beam is given by $\frac{dM}{dx} = -w(l-x)$, where w and x are constants. Determine M in terms of x given: $M = \frac{1}{2}wl^2$ when $x = 0$.

$\frac{dM}{dx} = -w(l-x) = -wl + wx$

Integrating with respect to x gives: $M = -wlx + \frac{wx^2}{2} + c$, which is the general solution.

When $M = \frac{1}{2}wl^2$, $x = 0$. Thus $\frac{1}{2}wl^2 = -wl(0) + \frac{w(0)^2}{2} + c$

from which, $c = \frac{1}{2}wl^2$.

Hence the particular solution is $M = -wlx + \frac{wx^2}{2} + \frac{1}{2}wl^2$

i.e. $M = \frac{1}{2}w(l^2 - 2lx + x^2)$ or $\mathbf{M = \frac{1}{2}w(l - x)^2}$.

Problem 6 Find the particular solution of the differential equation $dQ/dt = 2Q$ given that $Q = 4$ when $t = 0.5$.

The general solution of $\frac{dQ}{dt} = 2Q$ is $Q = Ae^{2t}$, where A is a constant.
Given the boundary conditions $Q = 4$ when $t = 0.5$ enables constant A to be determined. Thus $4 = Ae^{2(0.5)}$, i.e. $A = \frac{4}{e^1}$ or $4e^{-1}$.

Hence the particular solution is $Q = (4e^{-1})e^{2t}$, i.e., $\mathbf{Q = 4e^{2t-1}}$.

Problem 7 Solve the equation $dy/dx = 5y$ given that $y = 2$ when $x = 1$.

$\frac{dy}{dx} = 5y$, is of the form $\frac{dQ}{dt} = kQ$ (where $y = Q$, $x = t$ and $k = 5$).

The solution of $\frac{dQ}{dt} = kQ$ is $Q = Ae^{kt}$

Hence the general solution of $\frac{dy}{dx} = 5y$ is $y = Ae^{5x}$.
Given the boundary conditions $y = 2$ when $x = 1$ enables constant A to be determined. Thus $2 = Ae^{5(1)}$, i.e., $A = \frac{2}{e^5}$ or $2e^{-5}$.

Hence the particular solution is $y = (2e^{-5})e^{5x}$, i.e., $\mathbf{y = 2e^{5(x-1)}}$.

Problem 8 Determine the particular solution of the differential equation $\frac{1}{5}\frac{dw}{dx} + \frac{1}{4}w = 0$, given $w = 3.52$ when $x = 1.16$.

Rearranging $\frac{1}{5}\frac{dw}{dx} + \frac{1}{4}w = 0$ gives $\frac{1}{5}\frac{dw}{dx} = -\frac{1}{4}w$, i.e. $\frac{dw}{dx} = -\frac{5}{4}w$.

The general solution is $w = Ae^{-\frac{5}{4}x}$.

When $w = 3.52$, $x = 1.16$. Thus $3.52 = Ae^{-\frac{5}{4}(1.16)}$

from which, $A = \frac{3.52}{e^{-\frac{5}{4}(1.16)}} = 15.0$ correct to 3 significant figures

Hence the particular solution is $w = 15.0e^{-\frac{5}{4}x}$.

Problem 9 The rate of cooling of a body is proportional to the excess of its temperature above that of its surroundings, θ° C, and the equation is $d\theta/dt = k\theta$, where k is a constant. A body cools from 85°C to 65°C in 4.0 minutes at a surrounding temperature of 15°C. Determine how long, to the nearest second, the body will take to cool to 55°C.

The general solution of $d\theta/dt = k\theta$ is $\theta = Ae^{kt}$.
Let the temperature of 85°C correspond to time $t = 0$. The excess of body temperature above the surroundings at $t = 0$ is (85–15), i.e., 70°C.
Hence $70 = Ae^{k(0)}$, from which, $A = 70$.
4.0 minutes later the general solution becomes $(65-15) = 70e^{(k)(4.0)}$

$$\text{i.e.} \quad e^{4.0k} = \frac{50}{70}$$

Taking Naperian logarithms gives $4.0k = \ln\left(\frac{50}{70}\right)$,

from which, $$k = \frac{1}{4.0}\ln\left(\frac{50}{70}\right) = -0.08412$$

At 55°C, $(55-15) = 70e^{-0.08412t}$, i.e., $\frac{40}{70} = e^{-0.08412t}$ or $\frac{70}{40} = e^{0.08412t}$

Taking Naperian logarithms gives: $\ln\frac{70}{40} = 0.08412\,t$,

from which, $t = \frac{1}{0.08412}\ln\left(\frac{70}{40}\right) = 6.6526$ minutes.

Thus the time for the body to cool to 55°C is 6 minutes 39 seconds, correct to the nearest second.

Problem 10 The decay of current in an electrical circuit containing resistance R ohms and inductance L henrys is given by: $L — + iR = 0$, where i is the current flowing at time t seconds. Determine the general solution of the equation.

The current falls to 4 A in 0.8 ms in a circuit containing $R = 4 \times 10^3$ ohms and $L = 2$ henrys. How long will it take for the current to fall to 1 A?

Rearranging $L\frac{di}{dt} + iR = 0$ gives $\frac{di}{dt} = \frac{-iR}{L} = \left(\frac{-R}{L}\right)i$, which is of the form $\frac{dQ}{dt} = kQ$.

The general solution is $i = Ae^{\left(-\frac{R}{L}\right)t}$, i.e., $\boldsymbol{i = Ae^{-\frac{Rt}{L}}}$

When $R = 4 \times 10^3$, $L = 2$, $i = 4$ and $t = 0.8 \times 10^{-3}$ then:

$$4 = Ae^{\frac{-(4 \times 10^3)(0.8 \times 10^{-3})}{2}} = Ae^{-1.6} \text{ from which, } A = \frac{4}{e^{-1.6}} = 4e^{1.6} = 19.812$$

When $i = 1$, $1 = 19.812e^{-\frac{(4 \times 10^3)}{2}t}$ or $1 = 19.812e^{-2t}$, where t is in milliseconds.

Rearranging gives: $e^{-2t} = \frac{1}{19.812}$, or $e^{2t} = 19.812$

Taking Naperian logarithms gives: $2t = \ln 19.812$ and $t = \frac{1}{2}\ln 19.812 = 1.493$

Hence the time for the current to fall to 1 A is 1.49 ms, correct to 3 significant figures.

Problem 11 The rate of decay of a radioactive material is given by $dN/dt = -\lambda N$, where λ is a decay constant and λN the number of radioactive atoms disintegrating per second. Determine the half-life, in years, of a nickel isotope, assuming the decay constant is 1.832×10^{-10} atoms per second and a 365 day year. (Half-life means the time for N to become one half of its original value.)

The general solution of $\frac{dN}{dt} = -\lambda N$ is $N = Ae^{-\lambda t}$, where constant A represents the original number of radioactive atoms present (i.e. when $t = 0$).

For half-life, $N = \frac{1}{2}A$. Hence $\frac{1}{2}A = Ae^{-\lambda t}$ at half-life,

from which $\frac{1}{2} = e^{-\lambda t}$ i.e. $2 = e^{\lambda t}$

$\ln 2 = \lambda t$ and $t = \frac{1}{\lambda} \ln 2$.

Hence the half-life of a nickel isotope having $\lambda = 1.832 \times 10^{-10}$ atoms per second

is $\frac{1}{1.832 \times 10^{-10}} \ln 2 = 3.784 \times 10^9$ seconds $= \frac{3.784 \times 10^9}{60 \times 60 \times 24 \times 365}$ years

$= \mathbf{120}$ **years.**

C FURTHER PROBLEMS ON DIFFERENTIAL EQUATIONS

1 Sketch a family of curves represented by each of the following differential equations:

(a) $\frac{dy}{dx} = 6$; (b) $\frac{dy}{dx} = 3x$; (c) $\frac{dy}{dx} = x+2$.

2 Sketch the family of curves given by the equation $\frac{dy}{dx} = 2x+3$ and determine the equation of one of these curves which passes through the point (1, 3). $[y = x^2+3x-1]$

In *Problems 3 to 7*, find the general solution of the differential equations.

3 $\frac{dy}{dx} = 2x - \frac{4}{x^2}$ $\quad [y = x^2 + \frac{4}{x} + c]$

4 $\frac{dy}{dx} - 4x = \sin x$ $\quad [y = 2x^2 - \cos x + c]$

5 $x\frac{dy}{dx} = 5 - x^2$ $\quad [y = 5 \ln x - \frac{x^2}{2} + c]$

6 $2\frac{dv}{dt} + \frac{1}{2}t^2 = 4$ $\quad [v = 2t - \frac{1}{12}t^3 + c]$

7 $3\frac{dy}{dx} - 2x^4 = e^{2x}$ $\quad [y = \frac{e^{2x}}{6} + \frac{2}{15}x^5 + c]$

In *Problems 8 to 12*, find the particular solution of the differential equations.

8 $\frac{dy}{dx} + 2x^3 = 1$; $y = 1$ when $x = 2$ $\quad [y = x - \frac{1}{2}x^4 + 7]$

9 $x\left(x - \frac{dy}{dx}\right) = 2$; $y = 3\frac{1}{2}$ when $x = 1$ $\quad [y = \frac{x^2}{2} - 2 \ln x + 3]$

10 $\frac{ds}{dt} - 3t^2 = 8$; $s = 16$ when $t = 2$ $\quad [s = t^3 + 8t - 8]$

11 $\frac{dy}{d\theta} - \sin 2\theta = 4$; $y = \pi$ when $\theta = \frac{\pi}{4}$ $\quad [y = 4\theta - \frac{1}{2}\cos 2\theta]$

12 $\frac{2}{(x+1)^2} = 3 - \frac{dy}{dx}$; $y = 10$ when $x = 1$. $\quad \left[y = 3x + \frac{2}{(x+1)} + 6\right]$

13 The velocity, v, of a body is equal to its rate of change of distance, $\frac{ds}{dt}$. Determine an equation for s in terms of t given that $v = 3+2t$, and $s = 0$ when $t = 0$. $[s = 3t+t^2]$

14 The angular velocity ω of a flywheel of moment of inertia I is given by $I(d\omega/dt)+k = 0$, where k is a constant. Determine ω in terms of t given that $\omega = \omega_0$ when $t = 0$. $\left[\omega = \omega_0 - \dfrac{kt}{I}\right]$

15 An object is thrown vertically upwards with an initial velocity, u, of 20 m/s. The motion of the object follows the differential equation $ds/dt = u - gt$, where s is the height of the object in metres at time t seconds and $g = 9.8$ m/s^2. Determine the height of the object after 3 seconds if $s = 0$ when $t = 0$. [15.9 m]

16 Find the general solutions of the following differential equations

(a) $\dfrac{dQ}{dt} = 5Q$; (b) $\dfrac{dy}{dx} = 8y$; (c) $\dfrac{1}{4}\dfrac{d\theta}{dt} + \dfrac{2}{3}\theta = 0$.

[(a) $Q = Ae^{5t}$; (b) $y = Ae^{8x}$; (c) $\theta = Ae^{-\frac{8}{3}t}$]

17 Determine the particular solution of the differential equation $dQ/dt = 4Q$, given $Q = 3$ when $t = 2 \times 10^{-3}$. [$Q = 2.976e^{4t}$]

18 Find the particular solution of the equation $\dfrac{1}{5}\dfrac{dl}{dm} - \dfrac{1}{2}l = 0$ given $l = 1.5 \times 10^4$ when $m = 3 \times 10^{-2}$. [$l = 13\,920e^{\frac{5}{2}m}$]

19 An aluminium conductor heats up to 60°C when carrying a current of 100 A. If the temperature coefficient of linear expansion α_0 for aluminium is 23×10^{-6}/°C at 0°C and the equation relating temperature θ with length l is $dl/d\theta = \alpha l$, find the increase in length of the conductor at 60°C, correct to the nearest centimetre, when l is 500 m at 0°C. [69 cm]

20 The rate of decay of a radioactive substance is given by $\dfrac{dN}{dt} = -\lambda N$, where λ is the decay constant and λN the number of radioactive atoms disintegrating per second. Determine the half-life of a mercury isotope, in hours, (i.e., the time for N to become one half of its original value), assuming the decay constant for mercury to be 2.917×10^{-6} atoms per second. [66 hours]

21 The current i amperes in a conductor at a time t seconds is given by the differential equation $(di/dt)+ki = 0$, where k is a constant. If $i = 4$ A when $t = 0$ and $i = 2.5$ A when $t = 1.2$ s, find the value of k, correct to 3 significant figures, and determine the current flowing when $t = 3$ s. [$k = 0.392$; $i = 1.234$ A]

22 A differential equation relating the difference in tensions T, pulley contact angle θ and coefficient of friction μ is $dT/d\theta = \mu T$. When $\theta = 0$, $T = 150$ N, and $\mu = 0.30$ as slipping starts. Determine the tension at the point of slipping when $\theta = 2$ radians. Determine also the value of θ when T is 300 N. [273 N; 2.31 rads]

23 The charge Q coulombs at time t seconds for a capacitor C farads when discharging through a resistance R ohms is given by $R\dfrac{dQ}{dt} + \dfrac{Q}{C} = 0$. A circuit contains a resistance of 200×10^3 ohms and a capacitance of 10×10^{-6} farads, and after 0.1 s the charge falls to 5 coulombs. Determine the initial charge and the charge after 1 s, each correct to 3 significant figures. [5.26 C; 3.19 C]

24 The rate of cooling of a body is given by $d\theta/dt = k\theta$, where k is a constant. If $\theta = 60$°C when $t = 2$ minutes and $\theta = 50$°C when $t = 5$ minutes, determine the time taken for θ to fall to 40°C, correct to the nearest second. [8 minutes 40 seconds]

24 Statistics (1) – Probability

A. MAIN POINTS CONCERNED WITH PROBABILITY

1 The **probability** of something happening is the likelihood or chance of it happening. Values of probability lie between 0 and 1, where 0 represents an absolute impossibility and 1 represents an absolute certainty. The probability of an event happening usually lies somewhere between these two extreme values and is expressed either as a proper or a decimal fraction. Examples of probability are:

that a length of copper wire has zero resistance at 100°C	0
that a fair, six-sided dice will stop with a 3 upwards	$\frac{1}{6}$ or 0.1667
that a fair coin will land with a head upwards	$\frac{1}{2}$ or 0.5
that a length of copper wire has some resistance at 100°C	1.

2 If p is the probability of an event happening and q is the probability of the same event not happening, then the total probability is $p+q$ and is equal to unity, since it is an absolute certainty that the event either does or does not occur, i.e., $p+q = 1$.

3 The **expectation,** E, of an event happening is defined in general terms as the product of the probability p of an event happening and the number of attempts made, n, i.e., $E = pn$. Thus, since the probability of obtaining a 3 upwards when rolling a fair dice is 1/6, the expectation of getting a 3 upwards on four throws of the dice is

$$\frac{1}{6} \times 4, \text{ i.e., } \frac{2}{3}.$$

Thus expectation is the average occurrence of an event.

4 A **dependent event** is one in which the probability of one event happening affects the probability of another event happening. Let 5 transistors be taken at random from a batch of 100 transistors for test purposes, and the probability of there being a defective transistor, p_1, be determined. At some later time, let another 5 transistors be taken at random from the 95 remaining transistors in the batch and the probability of there being a defective transistor, p_2, be determined. The value of p_2 is different from p_1 since the batch size has effectively altered from 100 to 95, i.e., the probability p_2 is dependent on probability p_1. Since 5 transistors are drawn, and then another 5 transistors are drawn without replacing the first 5, the second random selection is said to be **without replacement**.

5 An **independent event** is one in which the probability of an event happening does **not** affect the probability of another event happening. If 5 transistors are taken at random from a batch of transistors and the probability of a defective transistor p_1

is determined and the process is repeated after the original 5 have been replaced in the batch to give p_2, then p_1 is equal to p_2. Since the 5 transistors are replaced between draws, the second selection is said to be **with replacement.**

6 **The addition law of probability**

The addition law of probability is recognised by the word '**or**' joining the probabilities. If p_A is the probability of event A happening and p_B is the probability of event B happening, the probability of event A **or** event B happening is given by $p_A + p_B$. Similarly, the probability of events A **or** B **or** C **or** . . . . N happening is given by

$$p_A + p_B + p_C + \ldots\ldots\ldots + p_N.$$

7 **The multiplication law of probability**

The multiplication law of probability is recognised by the word '**and**' joining the probabilities. If p_A is the probability of event A happening and p_B is the probability of event B happening, the probability of event A **and** event B happening is given by $p_A \times p_B$. Similarly, the probability of events A **and** B **and** C **and** . . . . . N happening is given by $p_A \times p_B \times p_C \times \ldots\ldots \times p_N$.

B. WORKED PROBLEMS ON PROBABILITY

Problem 1 Determine the probabilities of selecting at random (a) a man, and (b) a woman from a crowd containing 20 men and 33 women.

(a) The probability of selecting at random a man, p, is given by the ratio

$$\frac{\text{number of men}}{\text{number in crowd}}, \text{ i.e., } p = \frac{20}{20+33} = \frac{20}{53}.$$

(b) The probability of selecting the winning horse in the first race is $\frac{1}{10}$.

$$\frac{\text{number of women}}{\text{number in crowd}}, \text{ i.e., } q = \frac{33}{20+33} = \frac{33}{53}.$$

[Check: the total probability should be equal to 1. $p = \frac{20}{53}$ and $q = \frac{33}{53}$.

The total probability, $p+q = \frac{20}{53} + \frac{33}{53} = 1$, hence no obvious error has been made.]

Problem 2 Find the expectation of obtaining a 4 upwards with 3 throws of a fair dice.

Expectation is the average occurrence of an event and is defined as the probability times the number of attempts, (see para. 3). The probability, p, of obtaining a 4 upwards for one throw of the dice is 1/6. Also, 3 attempts, n, are made, hence the expectation, E, is pn, i.e. $\frac{1}{6} \times 3 = \frac{1}{2}$.

Problem 3 Calculate the probabilities of selecting at random:
(a) the winning horse in a race in which 10 horses are running,
(b) the winning horses in both the first and second races if there are 10 horses in each race.

(a) Since only one of the ten horses can win, the probability of selecting at random the winning horse is

$$\frac{\text{number of winners}}{\text{number of horses}}, \text{ i.e. } \frac{1}{10}.$$

(b) The probability of selecting the winning horse in the first race is $\frac{1}{10}$.

The probability of selecting the winning horse in the second race is $\frac{1}{10}$.

The probability of selecting the winning horses in the first **and** second race is given by the multiplication law of probability, (see para. 7),

i.e., probability $= \frac{1}{10} \times \frac{1}{10} = \mathbf{\frac{1}{100}}$ **or 0.01**

Problem 4 The probability of a component failing in one year due to excessive temperature is 1/20, due to excessive vibration is 1/25 and due to excessive humidity is 1/50. Determine the probabilities that during a one-year period, a component:
(a) fails due to excessive temperature and excessive vibration,
(b) fails due to excessive vibration or excessive humidity and
(c) will not fail because of both excessive temperature and excessive humidity.

Let p_A be the probability of failure due to excessive temperature, then
$p_A = \frac{1}{20}$ and $\bar{p}_A = \frac{19}{20}$.

Let p_B be the probability of failure due to excessive vibration, then
$p_B = \frac{1}{25}$ and $\bar{p}_B = \frac{24}{25}$.

Let p_C be the probability of failure due to excessive humidity, then
$p_C = \frac{1}{50}$ and $\bar{p}_C = \frac{49}{50}$.

(a) The probability of a component failing due to excessive temperature **and** excessive vibration is given by

$p_A \times p_B$, i.e., $\frac{1}{20} \times \frac{1}{25} = \mathbf{\frac{1}{500}}$ **or 0.002**

(b) The probability of a component failing due to excessive vibration or excessive humidity is

$p_B + p_C$, i.e., $\frac{1}{25} + \frac{1}{50} = \mathbf{\frac{3}{50}}$ **or 0.06**

(c) The probability that a component will not fail due to excessive temperature and will not fail due to excessive humidity is

$\bar{p}_A \times \bar{p}_C$, i.e. $\frac{19}{20} \times \frac{49}{50} = \mathbf{\frac{931}{1000}}$ **or 0.931**

Problem 5 A batch of 100 capacitors contains 73 which are within the required tolerance values, 17 which are below the required tolerance values, and the remainder are above the required tolerance values. Determine the probabilities that when randomly selecting a capacitor and then a second capacitor:
(a) both are within the required tolerance values when selecting with replacement and (b) the first one drawn is below and the second one drawn is above the required tolerance value, when selection is without replacement.

(a) The probability of selecting a capacitor within the required tolerance values is 73/100. The first capacitor drawn is now replaced and a second one is drawn from the batch of 100. The probability of this capacitor being within the required tolerance values is also 73/100. Thus, the probability of selecting a capacitor within the required tolerance values for both the first **and** the second draw is

$\frac{73}{100} \times \frac{73}{100} = \mathbf{\frac{5329}{10\,000}}$ **or 0.5329**

(b) The probability of obtaining a capacitor below the required tolerance values on the first draw is 17/100. There are now only 99 capacitors left in the batch, since the first capacitor is not replaced. The probability of drawing a capacitor above the required tolerance values on the second draw is 10/99, since there are (100–73–17), i.e. 10 capacitors above the required tolerance value. Thus, the probability of randomly selecting a capacitor below the required tolerance values and followed by randomly selecting a capacitor above the tolerance values is

$$\frac{17}{100} \times \frac{10}{99} = \frac{170}{9900} = \frac{17}{990} = 0.0172$$

Problem 6 A batch of 40 components contains 5 which are defective. If a component is drawn at random from the batch and tested and then a second component is drawn, determine the probability that neither of the components is defective.

With replacement:
The probability that the component selected on the first draw is satisfactory is 35/40, i.e. 7/8. The component is now replaced and a second draw is made. The probability that this component is also satisfactory is 7/8. Hence, the probability that both the first component drawn and the second component drawn are satisfactory is

$$\frac{7}{8} \times \frac{7}{8} = \frac{49}{64} \text{ or } 0.7656$$

Without replacement:
The probability that the first component drawn is satisfactory is 7/8. There are now only 34 satisfactory components left in the batch and the batch number is 39. Hence, the probability of drawing a satisfactory component on the second draw is 34/39. Thus the probability that the first component drawn and the second component drawn are satisfactory is

$$\frac{7}{8} \times \frac{34}{39} = \frac{238}{312} \text{ or } 0.7628$$

Problem 7 A batch of 40 components contains 5 which are defective. If a component is drawn at random from the batch and tested and then a second component is drawn at random, calculate the probability of having one defective component, both with and without replacement.

The probability of having one defective component can be achieved in two ways. If p is the probability of drawing a defective component and q is the probability of drawing a satisfactory component, then the probability of having one defective component is given by drawing a satisfactory component and then a defective component or by drawing a defective component and then a satisfactory one, i.e., by $q \times p + p \times q$.

With replacement:

$$p = \frac{5}{40} = \frac{1}{8} \text{ and } q = \frac{35}{40} = \frac{7}{8}.$$

Hence, probability of having a defective component is $\frac{1}{8} \times \frac{7}{8} + \frac{7}{8} \times \frac{1}{8}$,

i.e. $\frac{7}{64} + \frac{7}{64} = \frac{14}{64} = \frac{7}{32}$ or 0.2188

Without replacement:

$p_1 = \frac{1}{8}$ and $q_1 = \frac{7}{8}$ on the first of the two draws. The batch number is now 39 for the second draw, thus $p_2 = \frac{5}{39}$ and $q_2 = \frac{35}{39}$

$$p_1q_2 + q_1p_2 = \frac{1}{8} \times \frac{35}{39} + \frac{7}{8} \times \frac{5}{39} = \frac{35+35}{312} = \mathbf{\frac{70}{312}} \text{ or } \mathbf{0.2244}$$

Problem 8 A box contains seventy-four brass washers, eighty-six steel washers and forty aluminium washers. Three washers are drawn at random from the box without replacement. Determine the probability that all three are steel washers.

Assume, for clarity of explanation, that a washer is drawn at random, then a second, then a third, (although this assumption does not affect the results obtained). The total number of washers if 74+86+40, i.e., 200.

The probability of randomly selecting a steel washer on the first draw is 86/200. There are now 85 steel washers in a batch of 199. The probability of randomly selecting a steel washer on the second draw is 85/199. There are now 84 steel washers in a batch of 198. The probability of randomly selecting a steel washer on the third draw is 84/198. Hence the probability of selecting a steel washer on the first draw and the second draw and the third draw is

$$\frac{86}{200} \times \frac{85}{199} \times \frac{84}{198} = \frac{307\,020}{3\,940\,200} = \mathbf{\frac{5\,117}{65\,670}} \text{ or } \mathbf{0.078}$$

Problem 9 For the box of washers given in *Problem 8* above, determine the probability that there are no aluminium washers drawn, when three washers are drawn at random from the box without replacement.

The probability of not drawing an aluminium washer on the first draw is 1–(40/200), i.e. 160/200. There are now 199 washers in the batch of which 159 are not aluminium washers. Hence, the probability of not drawing an aluminium washer on the second draw is 159/199. Similarly, the probability of not drawing an aluminium washer on the third draw is 158/198. Hence the probability of not drawing an aluminium washer on the first and second and third draws is

$$\frac{160}{200} \times \frac{159}{199} \times \frac{158}{198} = \frac{100\,488}{197\,010} = \mathbf{\frac{50\,244}{98\,505}} \text{ or } \mathbf{0.510}$$

Problem 10 For the box of washers in *Problem 8* above, find the probability that there are two brass washers and either a steel or an aluminium washer when three are drawn at random, without replacement.

Two brass washers (A) and one steel washer (B) can be obtained in any of the following ways:

1st draw	*2nd draw*	*3rd draw*
A	A	B
A	B	A
B	A	A

Two brass washers and one aluminium washer (C) can also be obtained in any of the following ways:

1st draw	*2nd draw*	*3rd draw*
A	A	C
A	C	A
C	A	A

Thus there are six possible ways of achieving the combinations specified. If A represents a brass washer, B a steel washer and C an aluminium washer, then the combinations and their probabilities are as shown:

DRAW			*PROBABILITY*
First	Second	Third	
A	A	B	$\frac{74}{200} \times \frac{73}{199} \times \frac{86}{198} = 0.0590$
A	B	A	$\frac{74}{200} \times \frac{86}{199} \times \frac{73}{198} = 0.0590$
B	A	A	$\frac{86}{200} \times \frac{74}{199} \times \frac{73}{198} = 0.0590$
A	A	C	$\frac{74}{200} \times \frac{73}{199} \times \frac{40}{198} = 0.0274$
A	C	A	$\frac{74}{200} \times \frac{40}{199} \times \frac{73}{198} = 0.0274$
C	A	A	$\frac{40}{200} \times \frac{74}{199} \times \frac{73}{198} = 0.0274$

The probability of having the first combination **or** the second, **or** the third and so on is given by the sum of the probabilities, i.e., by $3 \times 0.0590 + 3 \times 0.0274$, that is, **0.2592**

C. FURTHER PROBLEMS ON PROBABILITY

1 In a batch of 45 lamps there are 10 faulty lamps. If one lamp is drawn at random, find the probability of it being (a) faulty and (b) satisfactory.

$$\left[\text{(a) } \frac{2}{9};\ \text{(b) } \frac{7}{9}\right]$$

2 A box of fuses are all of the same shape and size and comprises 23 2-A fuses, 47 5-A fuses and 69 13-A fuses. Determine the probability of selecting at random (a) a 2-A fuse, (b) a 5-A fuse and (c) a 13-A fuse.

$$\left[\text{(a) } \frac{23}{139} \text{ or } 0.1655;\ \text{(b) } \frac{47}{139} \text{ or } 0.3381;\ \text{(c) } \frac{69}{139} \text{ or } 0.4964\right]$$

3 (a) Find the probability of having a 2 upwards when throwing a fair 6-sided dice.
(b) Find the probability of having a 5 upwards when throwing a fair 6-sided dice.
(c) Determine the probability of having a 2 and then a 5 on two throws of a fair 6-sided dice.

$$\left[\text{(a) } \frac{1}{6};\ \text{(b) } \frac{1}{6};\ \text{(c) } \frac{1}{36}\right]$$

4 The probability of event A happening is 3/5 and the probability of event B happening is 2/3. Calculate the probabilities of (a) both A and B happening, (b) only event A happening, i.e. event A happening and event B not happening, (c) only event B happening and (d) either A, or B, or A and B happening.

$$\left[\text{(a) } \frac{2}{5};\ \text{(b) } \frac{1}{5};\ \text{(c) } \frac{4}{15};\ \text{(d) } \frac{13}{15}\right]$$

5 When testing 1000 soldered joints, 4 failed during a vibration test and 5 failed due to having a high resistance. Determine the probability of a joint failing due to (a) vibration, (b) high resistance, (c) vibration or high resistance and (d) vibration and high resistance.

$$\left[(a)\ \frac{1}{250};\ (b)\ \frac{1}{200};\ (c)\ \frac{9}{1000};\ (d)\ \frac{1}{50\,000}\right]$$

6 The probability that component A will operate satisfactorily for 5 years is 4/5 and that B will operate satisfactorily over the same period of time is 3/4. Find the probabilities that in a 5 year period:
(a) both components operate satisfactorily,
(b) only component A will operate satisfactorily, and
(c) only component B will operate satisfactorily.

$$\left[(a)\ \frac{3}{5};\ (b)\ \frac{1}{5};\ (c)\ \frac{3}{20}\right]$$

7 In a particular street, 80% of the houses have telephones. If two houses selected at random are visited, calculate the probabilities that (a) they both have a telephone and (b) one has a telephone but the other does not have a telephone.

$$\left[(a)\ \frac{16}{25};\ (b)\ \frac{8}{25}\right]$$

8 Veroboard pins are packed in packets of 20 by a machine. In a thousand packets, 40 have less than 20 pins. Find the probability that if 2 packets are chosen at random, one will contain less than 20 pins and the other will contain 20 pins or more.

$$\left[\frac{48}{625}\right]$$

9 A batch of 1-kW fire elements contain 16 which are within a power tolerance and 4 which are not. If 3 elements are selected at random from the batch, calculate the probabilities that (a) all three are within the power tolerance and (b) two are within but one is not within the power tolerance.

$$\left[(a)\ \frac{28}{57};\ (b)\ \frac{8}{19}\right]$$

10 The statistics on numbers of boys and girls of a group of families each having 3 children are analysed. Assuming equal probability for the birth of a boy or a girl, determine the percentage of the group likely to have (a) two boys and a girl, (b) at least one boy, (c) no girls and (d) at most two girls.
[(a) 12.5%; (b) 87.5%; (c) 12.5%; (d) 87.5%]

11 An amplifier is made up of three transistors, A, B and C. The probabilities of A, B or C being defective are 1/20, 1/25 and 1/50 respectively. Calculate the percentage of amplifiers produced (a) which work satisfactorily and (b) which have just one defective transistor. [(a) 89.38%; (b) 10.25%]

12 A box contains 14 40-W lamps, 28 60-W lamps and 58 25-W lamps, all the lamps being of the same shape and size. Three lamps are drawn at random from the box, first one, then a second, then a third. Determine the probabilities of: (a) getting one 25-W, one 40-W and one 60-W lamp, with replacement, (b) getting one 25-W, one 40-W and one 60-W lamp without replacement, (c) getting either one 25-W and two 40-W or one 60-W and two 40-W lamps with replacement and (d) getting either one 25-W and two 40-W or one 60-W and two 40-W lamps without replacement.

$$\left[\begin{array}{ll}(a)\ \dfrac{1421}{62\,500}\text{ or }0.0227; & (b)\ \dfrac{2842}{121\,275}\text{ or }0.0234;\\ (c)\ \dfrac{1421}{125\,000}\text{ or }0.0114; & (d)\ \dfrac{11\,739}{242\,550}\text{ or }0.0484\end{array}\right]$$

25 Statistics (2) – The Binomial and Poisson distributions

A MAIN POINTS CONCERNING THE BINOMIAL AND POISSON PROBABILITY DISTRIBUTIONS

The Binomial Distribution

1 The binomial distribution deals with two numbers only, these being the probability that an event will happen, p, and the probability that an event will not happen, q. Thus, when a coin is tossed, if p is the probability of the coin landing with a head upwards, q is the probability of the coin landing with a tail upwards. $p+q$ must always be equal to unity. A binomial distribution can be used for finding, say, the probability of getting three heads in seven tosses of the coin, or in industry for determining defect rates as a result of sampling.

2 One way of defining a binomial distribution is as follows:
'if p is the probability that an event will happen and q is the probability that the event will not happen, then the probabilities that the event will happen 0, 1, 2, 3, ... , n times in n trials are given by the successive terms of the expansion of $(q+p)^n$, *taken from left to right'.*
The binomial expansion introduced in chapter 4 is used to obtain the terms of $(q+p)^n$. This concept of a binomial distribution is used in *Problems 1 and 2.*

3 In industrial inspection, p is often taken as the probability that a component is defective and q is the probability that the component is satisfactory. In this case, a binomial distribution may be defined as:
'the probabilities that 0, 1, 2, 3, ..., n components are defective in a sample of n components, drawn at random from a large batch of components, are given by the successive terms of the expansion of $(q+p)^n$, *taken from left to right.'*
This definition is used in *Problems 3 and 4.*

4 The terms of a binomial distribution may be represented pictorially by drawing a histogram, (see *Problem 5*).

The Poisson Distribution

5 (i) When the number of trials, n, in a binomial distribution becomes large, (usually taken as larger than 10), the calculations associated with determining the values of the terms becomes laborious. If n is large, p is small and the product np is less than 5, a very good approximation to a binomial distribution is given by the corresponding Poisson distribution, in which calculations are usually simpler.

(ii) The Poisson approximation to a binomial distribution may be defined as follows:
'the probabilities that an event will happen 0, 1, 2, 3, ..., n times in n

trials are given by the successive terms of the expression $e^{-\lambda}(1+\lambda+\frac{\lambda^2}{2!}+\frac{\lambda^3}{3!}+\ldots\ldots)$ *taken from left to right'.*

The symbol λ is the expectation of an event happening and is equal to np, (see *Problem 6*).

6 The principal use of a Poisson distribution is to determine the theoretical probabilities when p, the probability of an event happening, is known, but q, the probability of the event not happening is not known. For example, the average number of goals scored per match by a football team can be calculated, but it is not possible to quantify the number of goals which were not scored. In this type of problem, a Poisson distribution may be defined as follows:
'the probabilities of an event occurring 0, 1, 2, 3, ... times are given by the successive terms of the expression $e^{-\lambda}(1+\lambda+\frac{\lambda^2}{2!}+\frac{\lambda^3}{3!}+\ldots)$, *taken from left to right.'*
The symbol λ is the value of the average occurrence of the event, (see *Problem 7*).

7 The terms of a Poisson distribution may be represented pictorially by drawing a histogram, (see *Problem 8*).

B WORKED PROBLEMS ON BINOMIAL AND POISSON DISTRIBUTIONS

Problem 1 Determine the probabilities of having, (a) at least 1 girl and (b) at least 1 girl and 1 boy, in a family of 4 children, assuming equal probability of male and female birth.

The probability of a girl being born, p, is 0.5 and the probability of a girl not being born, (male birth), q, is also 0.5. The number in the family, n, is 4. From para. 2, the probabilities of 0, 1, 2, 3, 4 girls in a family of 4 are given by the successive terms of the expansion of $(q+p)^4$ taken from left to right.

From chapter 4, $(q+p)^4 = q^4+4q^3p+6q^2p^2+4qp^3+p^4$

Hence the probability of no girls is q^4, i.e., $0.5^4 = 0.0625$
the probability of 1 girl is $4q^3p$, i.e., $4 \times 0.5^3 \times 0.5 = 0.2500$
the probability of 2 girls is $6q^2p^2$, i.e., $6 \times 0.5^2 \times 0.5^2 = 0.3750$
the probability of 3 girls is $4qp^3$, i.e., $4 \times 0.5 \times 0.5^3 = 0.2500$
the probability of 4 girls is p^4, i.e., $0.5^4 = 0.0625$

Total probability, $(q+p) = 1.0000$

(a) The probability of having at least one girl is the sum of the probabilities of having 1, 2, 3 and 4 girls, i.e.,
0.2500+0.3750+0.2500+0.0625 = **0.9375**
[Alternatively, the probability of having at least 1 girl is:
1–(the probability of having no girls), i.e., 1–0.0625, giving 0.9375, as obtained previously.]

(b) The probability of having at least 1 girl and 1 boy is given by the sum of the probabilities of having: 1 girl and 3 boys, 2 girls and 2 boys and 3 girls and 2 boy, i.e.,
0.2500+0.3750+0.2500 = **0.8750**
[Alternatively, this is also the probability of having 1–(probability of having no girls + probability of having no boys), i.e., $1-2 \times 0.0625 = 0.8750$, as obtained previously.]

Problem 2 A dice is rolled 9 times. Find the probabilities of having a 4 upwards, (a) 3 times and (b) less than 4 times.

Let p be the probability of having a 4 upwards. Then $p = 1/6$, since dice have six sides.
Let q be the probability of not having a 4 upwards. Then $q = 5/6$.
From para. 2, the probabilities of having a 4 upwards 0, 1, 2, ... , n times are given by the successive terms of the expansion of $(q+p)^n$, taken from left to right.
From chapter 4, $(q+p)^9 = q^9+9q^8p+36q^7p^2+84q^6p^3+\ldots\ldots$
The probability of having a 4 upwards no times is $q^9 = 0.1938$
The probability of having a 4 upwards 1 time is $9q^8p = 0.3489$
The probability of having a 4 upwards twice is $36q^7p^2 = 0.2791$
The probability of having a 4 upwards 3 times is $84q^6p^3 = 0.1302$

(a) The probability of having a 4 upwards 3 times is **0.1302**
(b) The probability of having a 4 upwards less than 4 times is the sum of the probabilities of having a 4 upwards 0, 1, 2, and 3 times, i.e., 0.1938+0.3489+0.2791+0.1302 = **0.9518**

Problem 3 A machine is producing a large number of bolts automatically. In a box of these bolts, 95% are within the allowable tolerance values with respect to diameter, the remainder being outside of the diameter tolerance values. Seven bolts are drawn at random from the box. Determine the probabilities that (a) two and (b) more than two of the seven bolts are outside of the diameter tolerance values.

Let p be the probability that a bolt is outside of the allowable tolerance values, i.e., is defective, and let q be the probability that a bolt is within the tolerance values, i.e., is satisfactory. Then $p = 5\%$, i.e., 0.05 per unit and $q = 95\%$, i.e., 0.95 per unit. The sample number is 7.

From para. 3, the probabilities of drawing 0, 1, 2, ... , n defective bolts are given by the successive terms of the expansion of $(q+p)^n$, taken from left to right. In this problem
$(q+p)^n = (0.95+0.05)^7 = 0.95^7+7\times0.95^6\times0.05+21\times0.95^5\times0.05^2+\ldots\ldots$
(see chapter 4).
Thus the probability of no defective bolts is $0.95^7 = 0.6983$
the probability of 1 defective bolt is $7\times0.95^6\times0.05 = 0.2573$
the probability of 2 defective bolts is $21\times0.95^5\times0.05^2 = 0.0406$,
and so on.

(a) The probability that two bolts are outside of the diameter tolerance values is **0.0406.**
(b) To determine the probability that more than two bolts are defective, the sum of the probabilities of 3 bolts, 4 bolts, 5 bolts, 6 bolts and 7 bolts being defective can be determined. An easier way to find this sum is to find 1−(the sum of 0 bolts, 1 bolt and 2 bolts being defective), since the sum of all the terms is unity. Thus, the probability of there being more than two bolts outside of the tolerance values is:
1−(0.6983+0.2573+0.0406), i.e., **0.0038**

Problem 4 A package contains 50 similar components and inspection shows that four have been damaged during transit. If six components are drawn at random from the contents of the package, determine the probabilities that in this sample (a) one and (b) less than three are damaged.

The probability of a component being damaged, p, is 4 in 50, i.e., 0.08 per unit. Thus, the probability of a component not being damaged, q, is 1–0.08, i.e., 0.92. From para. 3, the probability of there being 0, 1, 2, ... , 6 damaged components is given by the successive terms of $(q+p)^6$, taken from left to right.

$(q+p)^6 = q^6+6q^5p+15q^4p^2+20q^3p^3+\ldots\ldots$, (see chapter 4).

(a) The probability of one damaged component is $6q^5p = 6 \times 0.92^5 \times 0.08$, i.e., **0.3164**

(b) The probability of less than three damaged components is given by the sum of the probabilities of 0, 1 and 2 damaged components.

$$q^6+6q^5p+15q^4p^2 = 0.92^6+6 \times 0.92^5 \times 0.08+15 \times 0.92^4 \times 0.08^2 = 0.6064+0.3164+0.0688 = \mathbf{0.9916}$$

Problem 5 The probability of a student successfully completing a course of study in three years is 0.45. Draw a histogram showing the probabilities of 0, 1, 2, ... , 10 students successfully completing the course in three years.

Let p be the probability of a student successfully completing a course of study in three years and q be the probability of not doing so. Then $p = 0.45$ and $q = 0.55$. The number of students, n, is 10.

From para. 2, the probabilities of 0, 1, 2, ... , 10 students successfully completing the course are given by the successive terms of the expansion of $(q+p)^{10}$, taken from left to right.

$$(q+p)^{10} = q^{10}+10q^9p+45q^8p^2+120q^7p^3+210q^6p^4+252q^5p^5+210q^4p^6 +120q^3p^7+45q^2p^8+10qp^9+p^{10}.$$

Substituting $q = 0.55$ and $p = 0.45$ in this expansion gives the values of the successive terms as:

0.0025, 0.0207, 0.0763, 0.1665, 0.2384, 0.2340, 0.1597, 0.0746, 0.0229, 0.0042 and 0.0003. The histogram depicting these probabilities is shown in *Fig 1*.

Problem 6 If 3% of the gearwheels produced by a company are defective, determine the probabilities that in a sample of 80 gearwheels, (a) two and (b) more than two will be defective.

The sample number, n, is large, the probability of a defective gearwheel, p, is small and the product np is 80×0.03, i.e., 2.4, which is less than 5. Hence a Poisson approximation to a binomial distribution may be used. The expectation of a defective gearwheel, $\lambda = np = 2.4$.

From para. 5(ii), the probabilities of 0, 1, 2, ... defective gearwheels are given by the successive terms of the expression

$e^{-\lambda}\left(1+\lambda+\frac{\lambda^2}{2!}+\ldots\ldots\right)$, taken from left to right, i.e., by $e^{-\lambda}$, $\lambda e^{-\lambda}$, $\frac{\lambda^2 e^{-\lambda}}{2!}$, ... Thus:

probability of no defective gearwheels is $e^{-\lambda} = e^{-2.4} = 0.0907$

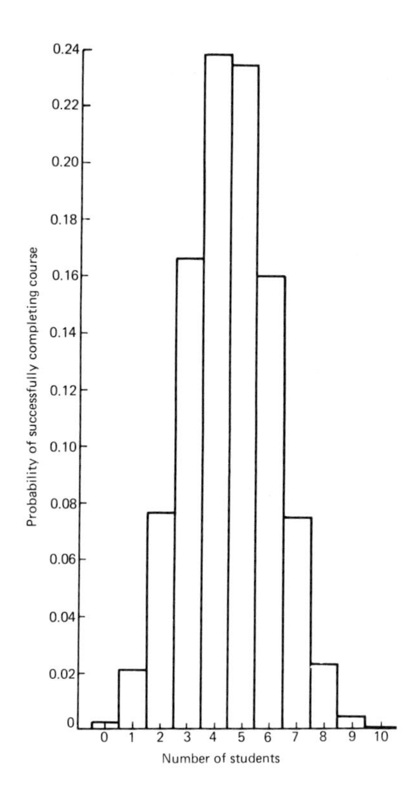

Fig 1

probability of 1 defective gearwheel is $\lambda e^{-\lambda} = 2.4e^{-2.4} \quad = 0.2177$

probability of 2 defective gearwheels is $\dfrac{\lambda^2 e^{-\lambda}}{2!} = \dfrac{2.4^2 e^{-2.4}}{2 \times 1} \quad = 0.2613$

(a) The probability of having 2 defective gearwheels is **0.2613**
(b) The probability of having more than 2 defective gearwheels is
1–(the sum of the probabilities of having 0, 1 and 2 defective gearwheels),
i.e., 1–(0.0907+0.2177+0.2613), that is, **0.4303**

Problem 7 A production department has 35 similar milling machines. The number of breakdowns on each machine averages 0.06 per week. Determine the probabilities of having (a) one and (b) less than three machines breaking down in any week.

Since the average occurrence of a breakdown is known but the number of times when a machine did not break down is not known, a Poisson distribution must be used.

The expectation of a breakdown for 35 machines is 35×0.06, i.e., 2.1 breakdowns per week. From para. 6, the probabilities of a breakdown occurring 0, 1, 2, ... times are given by the successive terms of the expression

$e^{-\lambda}(1+\lambda+\frac{\lambda^2}{2!}+\ldots\ldots)$, taken from left to right. Hence:

probability of no breakdowns is $e^{-\lambda} = e^{-2.1} = 0.1225$
probability of 1 breakdown is $\lambda e^{-\lambda} = 2.1e^{-2.1} = 0.2572$

probability of 2 breakdowns is $\frac{\lambda^2 e^{-\lambda}}{2!} = \frac{2.1^2 e^{-2.1}}{2 \times 1} = 0.2700$

(a) The probability of 1 breakdown per week is **0.2572**
(b) The probability of less than 3 breakdowns per week is the sum of the probabilities of 0, 1 and 2 breakdowns per week,
i.e., 0.1225+0.2572+0.2700, i.e., **0.6497**

Problem 8 The probability of a person having an accident in a certain period of time is 0.0003. For a population of 7500 people, draw a histogram showing the probabilities of 0, 1, 2, 3, 4, 5 and 6 people having an accident in this period.

From para. 6, the probabilities of 0, 1, 2, ... people having an accident are given by the successive terms of the expression

$e^{-\lambda}(1+\lambda+\frac{\lambda^2}{2!}+\ldots\ldots)$, taken from left to right.

The average occurrence of the event, λ, is 7500×0.0003, i.e., 2.25.
The probability of no people having an accident is $e^{-\lambda} = e^{-2.25} = 0.1054$
The probability of 1 person having an accident is $\lambda e^{-\lambda} = 2.25e^{-2.25} = 0.2371$

The probability of 2 people having an accident is $\frac{\lambda^2 e^{-\lambda}}{2!} = \frac{2.25^2 e^{-2.25}}{2!} = 0.2668$

and so on, giving probabilities of 0.2001, 0.1126, 0.0506 and 0.0190 for 3, 4, 5 and 6 respectively having an accident. The histogram for these probabilities is shown in *Fig 2.*

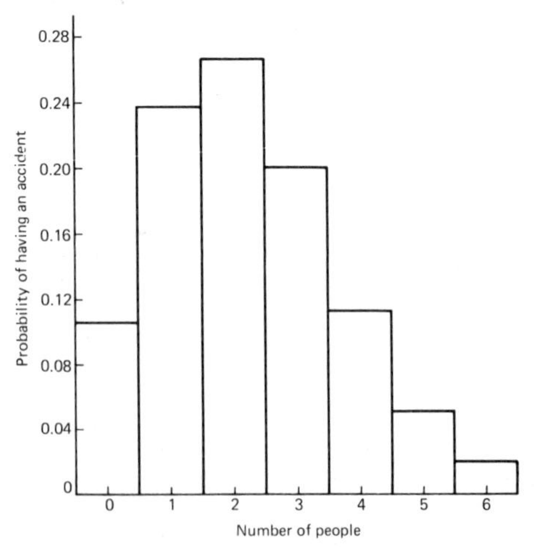

Fig 2

C FURTHER PROBLEMS ON THE BINOMIAL AND POISSON DISTRIBUTIONS

1 Concrete blocks are tested and it is found that, on average, 7% fail to meet the required specification. For a batch of 9 blocks, determine the probabilities that (a) three blocks and (b) less than four blocks will fail to meet the specification. [(a) 0.0186; (b) 0.9976]

2 If the failure rate of the blocks in *Problem 1* rises to 15%, find the probabilities that (a) no blocks and (b) more than two blocks will fail to meet the specification in a batch of 9 blocks. [(a) 0.2316; (b) 0.1408]

3 In a public examination, the failure rate is 25%. Find the probability that three candidates from a random group of eight will fail the examination. [0.2076]

4 The average number of employees absent from a firm each day is 4%. An office within the firm has seven employees. Determine the probabilities that (a) no employee and (b) three employees will be absent on a particular day. [(a) 0.7514; (b) 0.0019]

5 A manufacturer estimates that 3% of his output of a small item is defective. Find the probabilities that in a sample of 10 items, (a) less than two, and (b) more than two items will be defective. [(a) 0.9655; (b) 0.0028]

6 Five coins are tossed simultaneously. Determine the probabilities of having 0, 1, 2, 3, 4 and 5 heads upwards, and draw a histogram depicting the results.
[Vertical adjacent rectangles, whose heights are proportional to 0.0313, 0.1563, 0.3125, 0.3125, 0.1563 and 0.0313]

7 If the probability of rain falling during a particular period is 2/5, find the probabilities of having 0, 1, 2, 3, 4, 5, 6 and 7 wet days in a week. Show these results on a histogram.
[Vertical adjacent rectangles, whose heights are proportional to 0.0280, 0.1306, 0.2613, 0.2903, 0.1935, 0.0774, 0.0172 and 0.0016]

8 An automatic machine produces, on average, 10% of its components outside of the tolerance required. In a sample of 10 components from this machine, determine the probabilities of having three components outside of the tolerance required by assuming (a) a binomial distribution and (b) a Poisson distribution. [(a) 0.0574; (b) 0.0613]

9 The probability that an employee will go to hospital in a certain period of time is 0.0015. Use a Poisson distribution to determine the probability of more than two employees going to hospital during this period of time if there are 2000 employees on the pay-roll. [0.5768]

10 When packaging a product, a manufacturer finds that one packet in twenty is underweight. Determine the probabilities that in a box of 72 packets (a) two and (b) less than four will be underweight. [(a) 0.1771; (b) 0.5153]

11 Enquiries for train times are made, on average, at a rate of 1.3 per minute during a normal working day at an information kiosk. Determine the probabilities of there being (a) four and (b) more than four enquiries in any particular minute. [(a) 0.0324; (b) 0.0107]

12 A manufacturer estimates that 0.25% of his output of a component are defective. The components are marketed in packets of 200. Determine the probability of a packet containing less than three defective components. [0.9856]

13 The demand for a particular tool from a store is, on average, five times a day and the demand follows a Poisson distribution. How many of these tools should be kept in the stores so that the probability of there being one available when required is greater than 10%?

[The probabilities of the demand for 0, 1, 2, ... tools are 0.0067, 0.0337, 0.0842, 0.1404, 0.1755, 0.1755, 0.1462, 0.1044, 0.0653, ... This shows that the probability of wanting a tool 8 times a day is 0.0653, i.e., less than 10%. Hence 7 tools should be kept in the store.]

14 Failure of a group of particular machine tools follows a Poisson distribution with a mean value of 0.7. Determine the probabilities of 0, 1, 2, 3, 4 and 5 failures in a week and present these results on a histogram.

[Vertical adjacent rectangles having heights proportional to 0.4966, 0.3476, 0.1217, 0.0284, 0.0050 and 0.0007]

26 Statistics (3) – The normal distribution

A MAIN POINTS CONCERNED WITH THE NORMAL DISTRIBUTION

1 When data is obtained, it can frequently be considered to be a sample, (i.e., a few members), drawn at random from a large population, (i.e., a set having many members). If the sample number is large, it is theoretically possible to choose class intervals which are very small, but which still have a number of members falling within each class. A frequency polygon of this data then has a large number of small line segments and approximates to a continuous curve. Such a curve is called a **frequency or a distribution curve**.

2 An extremely important symmetrical distribution curve is called the **normal curve** and is as shown in *Fig 1*. This curve can be described by a mathematical equation and is the basis of much of the work done in more advanced statistics. Many natural occurrences such as the heights or weights of a group of people, the sizes of components produced by a particular machine and the life length of certain components, approximate to a normal distribution.

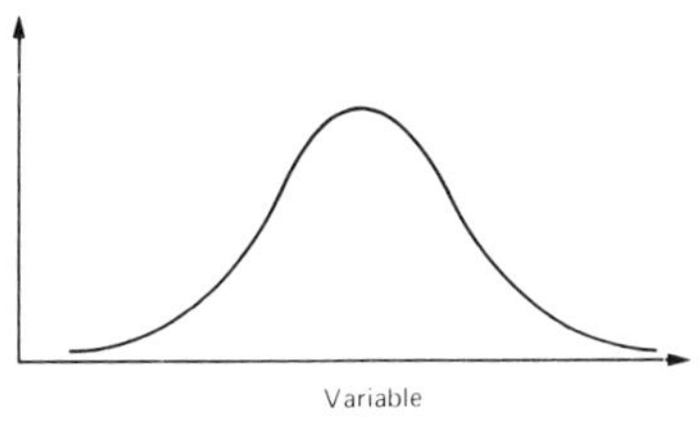

Fig 1

3 Normal distribution curves can differ from one another in the following four ways: (a) by having different mean values; (b) by having different values of standard deviations; (c) the variables having different values and different units; and (d) by having different areas between the curve and the horizontal axis.

4 A normal distribution curve is **standardised** as follows:

(a) The mean value of the unstandardised curve is made the origin, thus making the mean value, $\overline{x}$, zero.

(b) The horizontal axis is scaled in standard deviations. This is done by letting $z = \frac{x-\overline{x}}{\sigma}$, where z is called the **normal standard variate**, x is the value of the variable, $\overline{x}$ is the mean value of the distribution and σ is the standard deviation of the distribution.

(c) The area between the normal curve and the horizontal axis is made equal to unity.

When a normal distribution curve has been standardised, the normal curve is called a **standardised normal curve** or a **normal probability curve**, and any normally distributed data may be represented by the **same** normal probability curve.

TABLE 1 Partial areas under the standardised normal curve

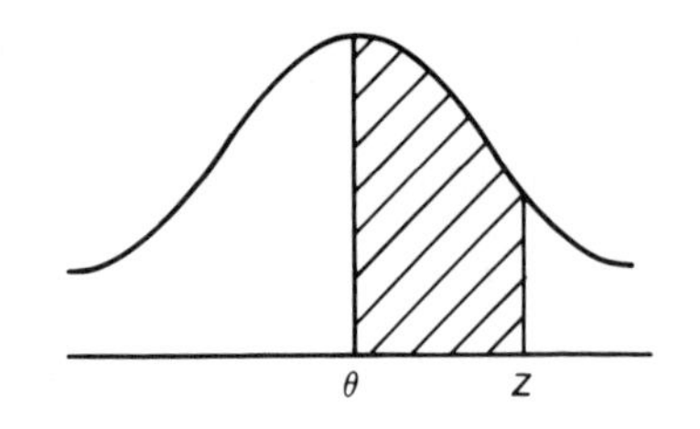

$z = \frac{x-\bar{x}}{\sigma}$	0	1	2	3	4	5	6	7	8	9
0.0	0.0000	0.0040	0.0080	0.0120	0.0159	0.0199	0.0239	0.0279	0.0319	0.0359
0.1	0.0398	0.0438	0.0478	0.0517	0.0557	0.0596	0.0636	0.0678	0.0714	0.0753
0.2	0.0793	0.0832	0.0871	0.0910	0.0948	0.0987	0.1026	0.1064	0.1103	0.1141
0.3	0.1179	0.1217	0.1255	0.1293	0.1331	0.1388	0.1406	0.1443	0.1480	0.1517
0.4	0.1554	0.1891	0.1628	0.1664	0.1700	0.1736	0.1772	0.1808	0.1844	0.1879
0.5	0.1915	0.1950	0.1985	0.2019	0.2054	0.2086	0.2123	0.2157	0.2190	0.2224
0.6	0.2257	0.2291	0.2324	0.2357	0.2389	0.2422	0.2454	0.2486	0.2517	0.2549
0.7	0.2580	0.2611	0.2642	0.2673	0.2704	0.2734	0.2760	0.2794	0.2823	0.2852
0.8	0.2881	0.2910	0.2939	0.2967	0.2995	0.3023	0.3051	0.3078	0.3106	0.3133
0.9	0.3159	0.3186	0.3212	0.3238	0.3264	0.3289	0.3215	0.3340	0.3365	0.3389
1.0	0.3413	0.3438	0.3451	0.3485	0.3508	0.3531	0.3554	0.3577	0.3599	0.3621
1.1	0.3643	0.3665	0.3686	0.3708	0.3729	0.3749	0.3770	0.3790	0.3810	0.3830
1.2	0.3849	0.3869	0.3888	0.3907	0.3925	0.3944	0.3962	0.3980	0.3997	0.4015
1.3	0.4032	0.4049	0.4066	0.4082	0.4099	0.4115	0.4131	0.4147	0.4162	0.4177
1.4	0.4192	0.4207	0.4222	0.4236	0.4251	0.4265	0.4279	0.4292	0.4306	0.4319
1.5	0.4332	0.4345	0.4357	0.4370	0.4382	0.4394	0.4406	0.4418	0.4430	0.4441
1.6	0.4452	0.4463	0.4474	0.4484	0.4495	0.4505	0.4515	0.4525	0.4535	0.4545
1.7	0.4554	0.4564	0.4573	0.4582	0.4591	0.4509	0.4608	0.4616	0.4625	0.4633
1.8	0.4641	0.4649	0.4656	0.4664	0.4671	0.4678	0.4686	0.4693	0.4699	0.4706
1.9	0.4713	0.4719	0.4726	0.4732	0.4738	0.4744	0.4750	0.4756	0.4762	0.4767
2.0	0.4772	0.4778	0.4783	0.4785	0.4793	0.4798	0.4803	0.4808	0.4812	0.4817
2.1	0.4821	0.4826	0.4830	0.4834	0.4838	0.4842	0.4846	0.4850	0.4854	0.4857
2.2	0.4861	0.4864	0.4868	0.4871	0.4875	0.4878	0.4881	0.4884	0.4882	0.4890
2.3	0.4893	0.4896	0.4898	0.4901	0.4904	0.4906	0.4909	0.4911	0.4913	0.4916
2.5	0.4938	0.4940	0.4941	0.4943	0.4945	0.4946	0.4948	0.4949	0.4951	0.4952
2.6	0.4953	0.4955	0.4956	0.4957	0.4959	0.4960	0.4961	0.4962	0.4963	0.4964
2.7	0.4965	0.4966	0.4967	0.4968	0.4969	0.4970	0.4971	0.4972	0.4973	0.4974
2.8	0.4974	0.4975	0.4076	0.4977	0.4977	0.4978	0.4979	0.4980	0.4980	0.4981
2.9	0.4981	0.4982	0.4982	0.4983	0.4984	0.4984	0.4985	0.4985	0.4986	0.4986
3.0	0.4987	0.4987	0.4987	0.4988	0.4988	0.4989	0.4989	0.4989	0.4990	0.4990
3.1	0.4990	0.4991	0.4991	0.4991	0.4992	0.4992	0.4992	0.4992	0.4993	0.4993
3.2	0.4993	0.4993	0.4994	0.4994	0.4994	0.4994	0.4994	0.4995	0.4995	0.4995
3.3	0.4995	0.4995	0.4995	0.4996	0.4996	0.4996	0.4996	0.4996	0.4996	0.4997
3.4	0.4997	0.4997	0.4997	0.4997	0.4997	0.4997	0.4997	0.4997	0.4997	0.4998
3.5	0.4998	0.4998	0.4998	0.4998	0.4998	0.4998	0.4998	0.4998	0.4998	0.4998
3.6	0.4998	0.4998	0.4999	0.4999	0.4999	0.4999	0.4999	0.4999	0.4999	0.4999
3.7	0.4999	0.4999	0.4999	0.4999	0.4999	0.4999	0.4999	0.4999	0.4999	0.4999
3.8	0.4999	0.4999	0.4999	0.4999	0.4999	0.4999	0.4999	0.4999	0.4999	0.4999
3.9	0.5000	0.5000	0.5000	0.5000	0.5000	0.5000	0.5000	0.5000	0.5000	0.5000

5 The area under part of a normal probability curve is directly proportional to probability and the value of the shaded area shown in *Fig 2* can be determined by evaluating:

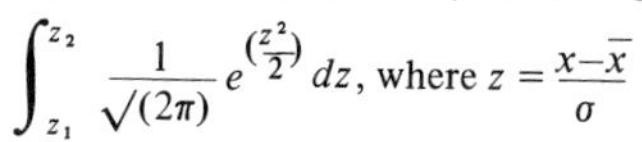

$$\int_{z_1}^{z_2} \frac{1}{\sqrt{(2\pi)}} e^{\left(\frac{z^2}{2}\right)} dz, \text{ where } z = \frac{x-\bar{x}}{\sigma}$$

(see para. 4).
To save repeatedly determining the values of this function, tables of partial areas under the standardised normal curve are available in many mathematical formulae books, and such a table is shown in *Table 1.*

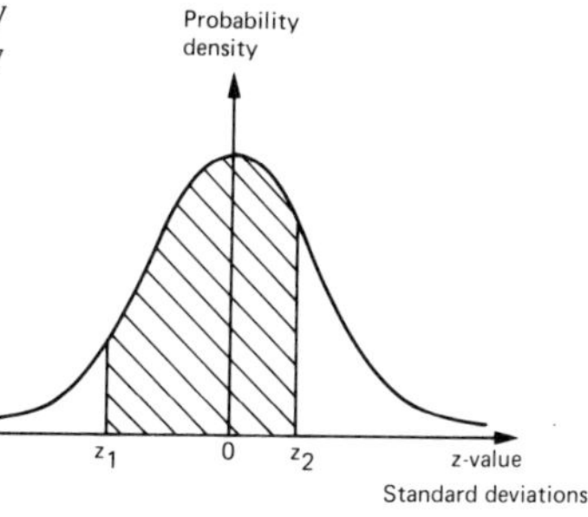

Fig 2

6 It should never be assumed that because data is continuous it automatically follows that it is normally distributed. One way of checking that data is normally distributed is by using **normal probability paper**, often just called **probability paper.** This is special graph paper which has linear markings on one axis and percentage probability values from 0.01 to 99.99 on the other axis, (see *Figs 6 and 7*). The divisions on the probability axis are such that a straight line graph results for normally distributed data when percentage cumulative frequency values are plotted against upper class boundary values. If the points do not lie in a reasonably straight line, then the data is not normally distributed. The method used to test the normality of a distribution is shown in *Problems 5 and 6.*

7 The mean value and standard deviation of normally distributed data may be determined using normal probability paper. For normally distributed data, the area beneath the standardised normal curve and a z-value of unity (i.e. one standard deviation) may be obtained from *Table 1.* For one standard deviation, this area is 0.3413, i.e., 34.13%. An area of ±1 standard deviation is symmetrically placed on either side of the $z = 0$ value, i.e., is symmetrically placed on either side of the 50 per cent cumulative frequency value. Thus an area corresponding to ±1 standard deviation extends from percentage cumulative frequency values of (50+34.13)% to (50−34.13)%, i.e., from 84.13% to 15.87%. For most purposes, these values are taken as 16% and 84%. Thus, when using normal probability paper, the standard deviation of the distribution is given by

$$\frac{(\text{variable value for 84\% cumulative frequency})-(\text{variable value for 16\% cumulative frequency})}{2}$$

(see *Problem 5*).

B WORKED PROBLEMS ON THE NORMAL DISTRIBUTION

Problem 1 The mean height of 500 people is 170 cm and the standard deviation is 9 cm. Assuming the heights are normally distributed, determine the number of people likely to have heights between 150 cm and 195 cm.

The mean value, $\bar{x}$, is 170 cm and corresponds to a normal standard variate value, z, of zero on the standardised normal curve, (see para. 4(a)). A height of 150 cm has a z-value given by $z = \frac{x-\bar{x}}{\sigma}$ standard deviations, from para. 4(b), i.e. $\frac{150-170}{9}$ or -2.22 standard deviations. Using a table of partial areas beneath the standardised normal curve (see *Table 1*), a z-value of -2.22 corresponds to an area of 0.4868 between the mean value and the ordinate $z = -2.22$. The negative z-value shows that it lies to the left of the $z = 0$ ordinate. This area is shown shaded in *Fig 3(a)*. Similarly, 195 cm has a z-value of $\frac{195-170}{9}$, that is 2.78 standard deviations. From *Table 1*, this value of z corresponds to an area of 0.4973, the positive value of z showing that it lies to the right of the $z = 0$ ordinate. This area is shown shaded in *Fig 3(b)*. The total area shaded in *Figs 3(a) and 3(b)* is shown in *Fig 3(c)* and is 0.4868+0.4973, i.e., 0.9841 of the total area beneath the curve.

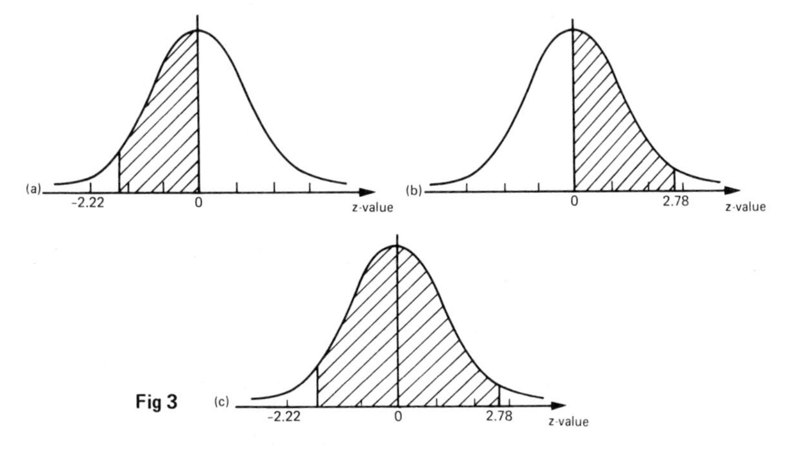

Fig 3

However, from para. 5, the area is directly proportional to probability. Thus, the probability that a person will have a height of between 150 and 195 cm is 0.9841. For a group of 500 people, 500×0.9841, i.e. **492 people** are likely to have heights in this range. The value of 500×0.9841 is 492.05, but since answers based on a normal probability distribution can only be approximate, results are usually given correct to the nearest whole number.

Problem 2 For the group of people given in *Problem 1*, find the numbers of people likely to have heights of less than 165 cm.

From para. 4(b), a height of 165 cm corresponds to $\frac{165-170}{9}$, i.e., -0.56 standard deviations. The area between $z = 0$ and $z = -0.56$ (from *Table 1*) is 0.2123, shown shaded in *Fig 4(a)*. The total area under the standardised normal curve is unity and since the curve is symmetrical, it follows that the total area to the left of the $z = 0$ ordinate is 0.5000. Thus the area to the left of the $z = -0.56$ ordinate, ('left' means 'less than', 'right' means 'more than'), is 0.5000–0.2123, i.e., 0.2877 of the total area, which is shown shaded in *Fig 4(b)*.

From para. 5, the area is directly proportional to probability and since the total area beneath the standardised normal curve is unity, the probability of a person's height being less than 165 cm is 0.2877. For a group of 500 people, 500×0.2877, i.e., **144 people** are likely to have heights of less than 165 cm.

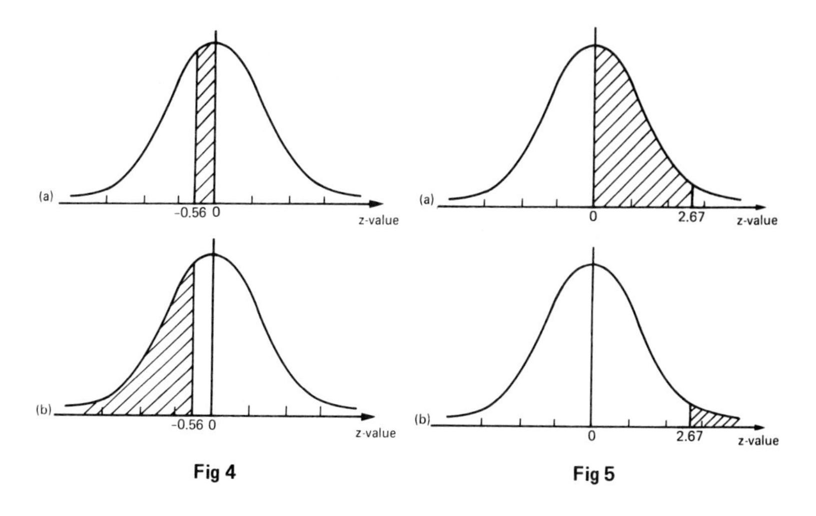

Fig 4

Fig 5

Problem 3 For the group of people given in *Problem 1*, find how many people are likely to have heights of more than 194 cm.

From para. 4, 194 cm corresponds to a z-value of $\dfrac{194-170}{9}$, that is, 2.67 standard deviations. From *Table 1*, the area between $z = 0$, $z = 2.67$ and the standardised normal curve is 0.4962, shown shaded in *Fig 5(a).* Since the standardised normal curve is symmetrical, the total area to the right of the $z = 0$ ordinate is 0.5000, hence the shaded area shown in *Fig 5(b)* is 0.5000–0.4962, i.e., 0.0038. From para. 5, this area represents the probability of a person having a height of more than 194 cm, and for 500 people, the number of people likely to have a height of more than 194 cm is 0.0038×500, i.e., **2 people.**

Problem 4 A batch of 1500 lemonade bottles have an average contents of 753 ml and the standard deviation of the contents if 1.8 ml. If the volumes of the contents are normally distributed, find
(a) the number of bottles likely to contain less than 750 ml,
(b) the number of bottles likely to contain between 751 and 754 ml,
(c) the number of bottles likely to contain more than 757 ml, and
(d) the number of bottles likely to contain between 750 and 751 ml.

(a) From para. 4, the z-value corresponding to 750 ml is given by $\dfrac{x-\bar{x}}{\sigma}$, i.e., $\dfrac{750-753}{1.8} = -1.67$ standard deviations. From *Table 1*, the area between $z = 0$ and $z = -1.67$ is 0.4525. Thus the area to the left of the $z = -1.67$ ordinate is 0.5000–0.4525, (see *Problem 2*), i.e., 0.0475. This is the probability of a bottle containing less than 750 ml. Thus, for a batch of 1500 bottles, it is likely that 1500×0.0475, i.e., **71 bottles** will contain less than 750 ml.

(b) The z-value corresponding to 751 and 754 ml are $\dfrac{751-753}{1.8}$ and $\dfrac{754-753}{1.8}$ i.e., -1.11 and 0.56 respectively. From *Table 1*, the areas corresponding to these values are 0.3665 and 0.2123 respectively. Thus the probability of a bottle containing between 751 and 754 ml is 0.3665+0.2123, (see *Problem 1*), i.e., 0.5788. For 1500 bottles, it is likely that 1500 X 0.5788, i.e., **868 bottles** will contain between 751 and 754 ml.

(c) The z-value corresponding to 757 ml is $\frac{757-753}{1.8}$, i.e., 2.22 standard deviations. From *Table 1*, the area corresponding to a z-value of 2.22 is 0.4868. The area to the right of the $z = 2.22$ ordinate is 0.5000–0.4868, (see *Problem 3*), i.e., 0.0132. Thus, for 1500 bottles, it is likely that 1500 X 0.0132, i.e., **20 bottles** will have contents of more than 757 ml.

(d) The z-value corresponding to 750 ml is -1.67, (see part (a)), and the z-value corresponding to 751 ml is -1.11, (see part (b)). The areas corresponding to these z-values are 0.4525 and 0.3665 respectively, and both these areas lie on the left of the $z = 0$ ordinate. The area between $z = -1.67$ and $z = -1.11$ is 0.4525–0.3665, i.e., 0.0860 and this is the probability of a bottle having contents between 750 and 751 ml. For 1500 bottles, it is likely that 1500 X 0.0860, i.e., **129 bottles** will be in this range.

Problem 5 Use normal probability paper to determine whether the data given below, which refers to the masses of 50 copper ingots, is approximately normally distributed. If the data is normally distributed, determine the mean and standard deviation of the data from the graph drawn.

Class mid-point value (kg)	29.5	30.5	31.5	32.5	33.5	34.5	35.5	36.5	37.5	38.5
Frequency	2	4	6	8	9	8	6	4	2	1

To test the normality of a distribution, the upper class boundary/percentage cumulative frequency values are plotted on normal probability paper. The upper class boundary values are: 30, 31, 32, ... , 38, 39. The corresponding cumulative frequency values, (for 'less than' the upper class boundary values), are: 2, (4+2) = 6, (6+4+2) = 12, 20, 29, 37, 43, 47, 49 and 50. The corresponding percentage cumulative frequency values are $\frac{2}{50} \times 100 = 4$, $\frac{6}{50} \times 100 = 12$, 24, 40, 58, 74, 86, 94, 98 and 100%.

The co-ordinates of upper class boundary/percentage cumulative frequency values are plotted as shown in *Fig 6*. When plotting these values, it will always be found that the co-ordinate for the 100% cumulative frequency value cannot be plotted, since the maximum value on the probability scale is 99.99. Since the points plotted in *Fig 6* lie very nearly in a straight line, the data is approximately normally distributed.

The mean value and standard deviation can be determined from *Fig 6*. Since a normal curve is symmetrical, the mean value is the value of the variable corresponding to a 50% cumulative frequency value, shown as point P on the graph. This shows that **the mean value is 33.6 kg**. The standard deviation is determined using the 84% and 16% cumulative frequency values, (see para. 7), shown as Q and R in *Fig 6*. The variable values for Q and R are 35.7 and 31.4 respectively; thus two standard deviations correspond to 35.7–31.4, i.e., 4.3, showing that the standard deviation of the distribution is approximately $\frac{4.3}{2}$, i.e. **2.15 standard deviations.**

The mean value and standard deviation of the distribution can be calculated using mean, $\bar{x} = \frac{\Sigma fx}{\Sigma f}$ and standard deviation, $\sigma = \sqrt{\left\{\frac{\Sigma[f(x-\bar{x})^2]}{\Sigma f}\right\}}$, where f is

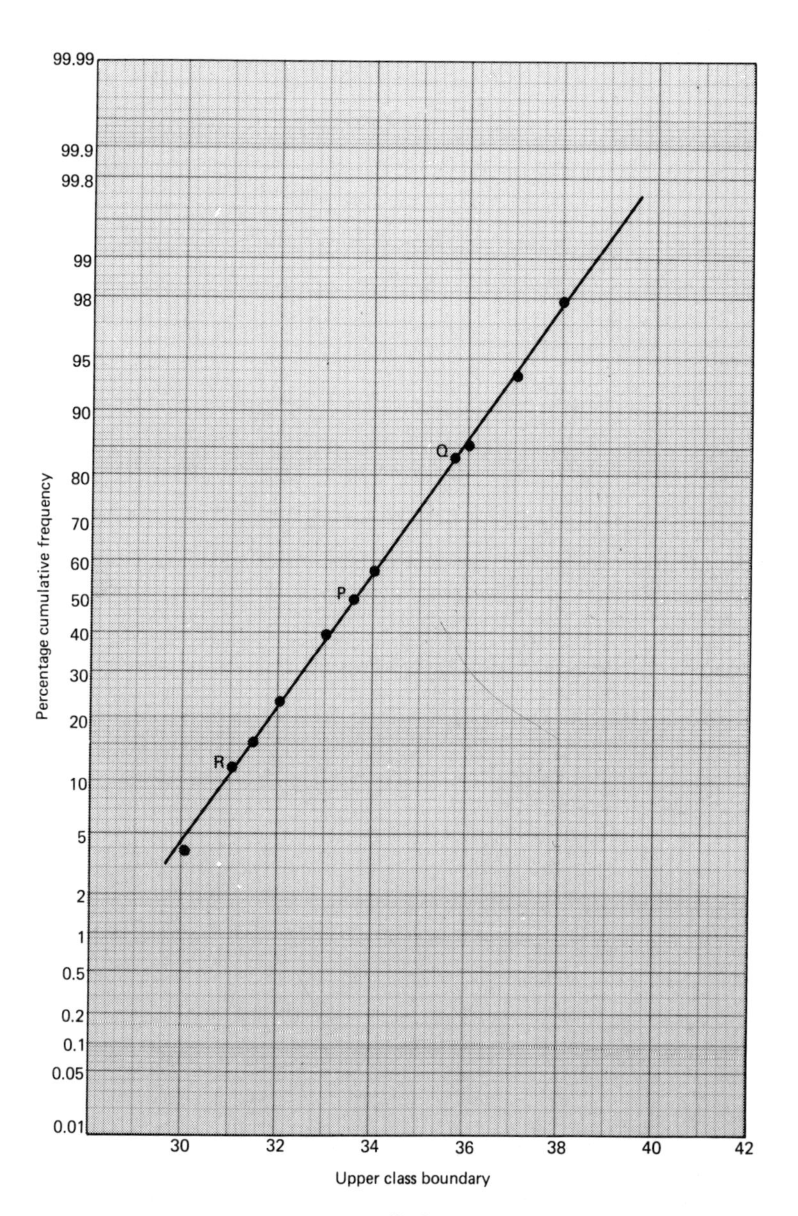
99.99
99.9
99.8
99
98
95
90
80
70
60
50
40
30
20
10
5
2
1
0.5
0.2
0.1
0.05
0.01
Percentage cumulative frequency
Q
P
R
30
32
34
36
38
40
42
Upper class boundary

Fig 6

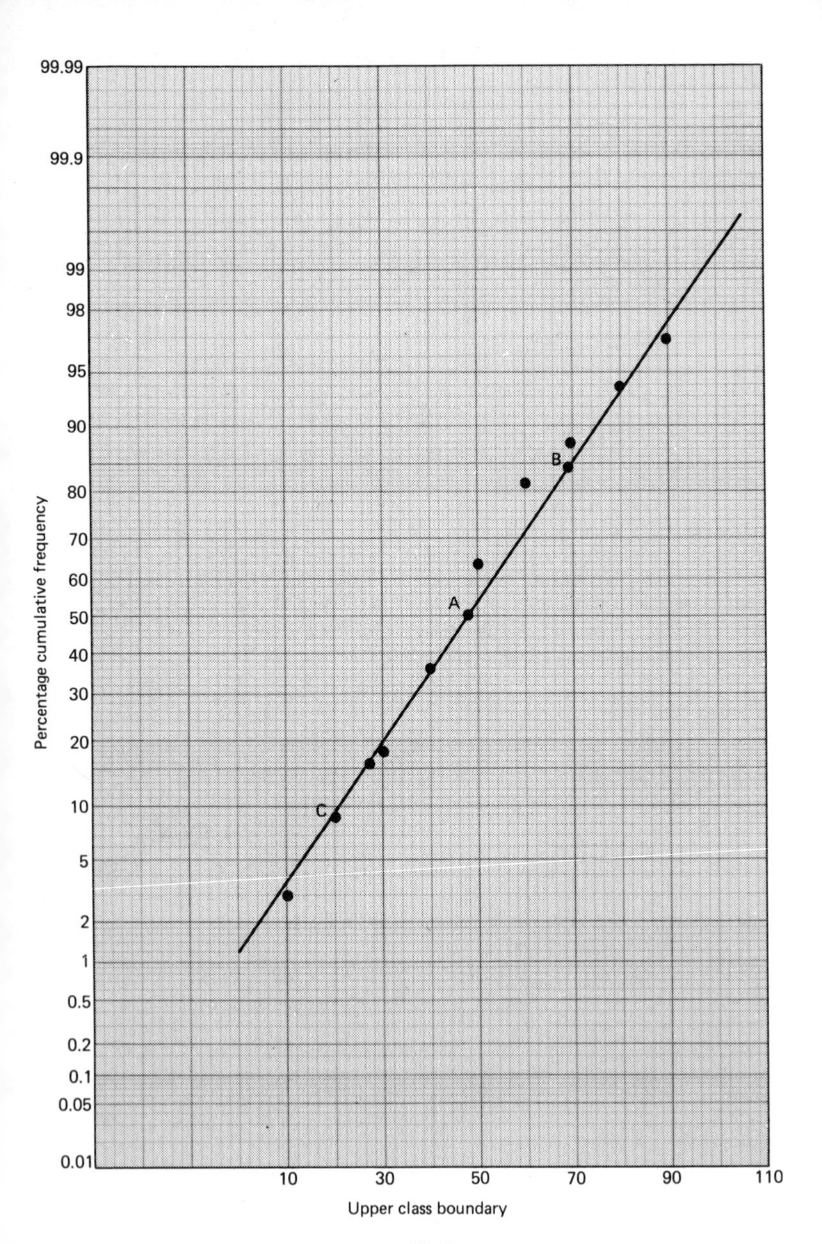
99.99
99.9
99
98
95
90
80
70
60
50
40
30
20
10
5
2
1
0.5
0.2
0.1
0.05
0.01
Percentage cumulative frequency
A
B
C
10
30
50
70
90
110
Upper class boundary

Fig 7

the frequency of a class and x is the class mid-point value. Using these formulae gives a mean value of the distribution of 33.6, (as obtained graphically) and a standard deviation of 2.12, showing that the graphical method of determining the mean and standard deviation give quite realistic results.]

Problem 6 Use normal probability paper to determine whether the data given below is normally distributed. Use the graph and assume a normal distribution whether this is so or not, to find approximate values of the mean and standard deviation of the distribution.

Class mid-point values	5	15	25	35	45	55	65	75	85	95
Frequency	1	2	3	6	9	6	2	2	1	1

To test the normality of a distribution, the upper class boundary/percentage cumulative frequency values are plotted on normal probability paper. The upper class boundary values are: 10, 20, 30, ..., 90 and 100. The corresponding cumulative frequency values are 1, $1+2=3$, $1+2+3=6$, 12, 21, 27, 29, 31, 32 and 33. The percentage cumulative frequency values are $\frac{1}{33}\times 100 = 3$, $\frac{3}{33}\times 100 = 9$, 18, 36, 64, 82, 88, 94, 97 and 100.

The co-ordinates of upper class boundary values/percentage cumulative frequency values are plotted as shown in *Fig 7*. Although six of the points lie approximately in a straight line, three points corresponding to upper class boundary values of 50, 60 and 70 are not close to the line and indicate that **the distribution is not normally distributed.** However, if a normal distribution is assumed, the mean value corresponds to the variable value at a cumulative frequency of 50% and from *Fig 7*, point A is 48. The value of the standard deviation of the distribution can be obtained from the variable values corresponding to the 84% and 16% cumulative frequency values, shown as B and C in *Fig 7* and give:
$2\sigma = 69.5-27.5$, that is, the standard deviation is 21.

The calculated values of the mean and standard deviation of the distribution are 45.9 and 19.4 respectively, showing that errors are introduced if the graphical method of determining these values is used for data which is not normally distributed.

C FURTHER PROBLEMS ON THE NORMAL DISTRIBUTION

(Answers on probabilities are given correct to the nearest whole number.)

1 A component is classed as defective if it has a diameter of less than 69 mm. In a batch of 350 components, the mean diameter is 75 mm and the standard deviation is 2.8 mm. Assuming the diameters are normally distributed, determine how many are likely to be classed as defective. [6]

2 The masses of 800 people are normally distributed, having a mean value of 64.7 kg and a standard deviation of 5.4 kg. Find how many people are likely to have masses of less than 54.4 kg. [22]

3 500 tins of paint have a mean content of 1010 ml and the standard deviation of the contents is 8.7 ml. Assuming the volumes of the contents are normally distributed, calculate the number of tins likely to have contents whose volumes are less than (a) 1025 ml; (b) 1000 ml and (c) 995 ml. [(a) 479; (b) 63; (c) 21]

4 For the 350 components in *Problem 1*, if those having a diameter of more than

81.5 mm are rejected, find, correct to the nearest component, the number likely to be rejected due to being oversized. [4]

5 For the 800 people in *Problem 2*, determine how many are likely to have masses of more than (a) 70 kg and (b) 62 kg. [(a) 131; (b) 553]

6 The mean diameter of holes produced by a drilling machine bit is 4.05 mm and the standard deviation of the diameters if 0.0028 mm. For twenty holes drilled using this machine, determine, correct to the nearest whole number, how many are likely to have diameters of between (a) 4.048 and 4.0553 mm and (b) 4.052 and 4.056 mm, assuming the diameters are normally distributed.
[(a) 15; (b) 4]

7 The intelligence quotients of 400 children have a mean value of 100 and a standard deviation of 14. Assuming that I.Q.'s are normally distributed, determine the number of children likely to have I.Q.'s of between (a) 80 and 90; (b) 90 and 110 and (c) 110 and 130. [(a) 65; (b) 209; (c) 89]

8 The mean mass of active material in tablets produced by a manufacturer is 5.00 g and the standard deviation of the masses is 0.036 g. In a bottle containing 100 tablets, find how many tablets are likely to have masses of (a) between 4.88 and 4.92 g; (b) between 4.92 and 5.04 g and (c) more than 5.04 g.
[(a) 1; (b) 85; (c) 13]

9 A frequency distribution of 150 measurements is as shown:

Class mid-point value	26.4	26.6	26.8	27.0	27.2	27.4	27.6
Frequency	5	12	24	36	36	25	12

Use normal probability paper to show that this data approximates to a normal distribution and hence determine the approximate values of the mean and standard deviation of the distribution. Use the formula for mean and standard deviation to verify the results obtained. [Graphically, $\bar{x} = 27.1, \sigma = 0.3$; by calculation, $\bar{x} = 27.079, \sigma = 0.3001$]

10 A frequency distribution of the class mid-point values of the breaking loads for 275 similar fibres is as shown below:

Load (kN)	17	19	21	23	25	27	29	31
Frequency	9	23	55	78	64	28	14	4

Use normal probability paper to show that this distribution is approximately normally distributed and determine the mean and standard deviation of the distribution, (a) from the graph and (b) by calculations.
[(a) $\bar{x} = 23.5$ kN, $\sigma = 2.9$ kN
(b) $\bar{x} = 23.364$ kN, $\sigma = 2.917$ kN]

Index